STUDIES IN CRYPTOGAMIC BOTANY

Volume–II

RED ALGAE : STRUCTURE, ULTRASTRUCTURE AND REPRODUCTION

RED ALGAE : STRUCTURE, ULTRASTRUCTURE AND REPRODUCTION

M. R. VIJAYARAGHAVAN
BELA BHATIA
Department of Botany
University of Delhi
Delhi-110007

1997

APH PUBLISHING CORPORATION

5, ANSARI ROAD, DARYA GANJ, NEW DELHI - 110002

Published by
S.B. Nangia
A P H Publishing Corporation
4435-36/7, Ansari Road, Daryaganj
New Delhi 110002
Ph.: 23274050
E-mail : aphbooks@gmail.com

2023

Rs. 3995/-

Printed at
Balaji Offset
Navin Shahdara, Delhi 110032

STUDIES IN CRYPTOGAMIC BOTANY

There is obvious need for comprehensive treatment of various aspects of cryptogamic botany including basic principles, methods and application. The potential usefulness of such a series is suggested of the practical significance of researchers in this field by the uniqueness and increasing coherence of the field's conceptial directions and by the international scope of the subject. Individual volumes are being prepared under this series as an attempt to fulfil this need. These volumes present and collate the major aspects of the cryptogamic botany and are designed as a modern standard reference for the work in this field.

Series Editor   Professor K.G. Mukerji
Deprtment of Botany
University of Delhi
Delhi–110007

Assistant Editor :   Mr B.P. Chamola
Department of Botany
University of Delhi
Delhi–110007

Volume-1   The Mushrooms of India – T.N. Lakhanpal

Volume-II   Red Algae : Structure, Ultrastructure And Reproduction – M.R. Vijayraghavan & Bela Bhatia

Volume-III   Advances in Integrated Pest & Disease Management – Rajeev K. Upadhyay, K.G. Mukerji B.P. Chamola & O.P. Dubey

Volume-IV   Brown Algae : Structure, Ultrastructure And Reproduction – Mr Vijayraghavan & Inderdeep Kaur

Forthcoming volumes under the series:

Current concepts in Mycorrhizae – K.G. Mukerji & B.P. Chamola
Lichens – K.G. Mukerji, D.K. Upreti & B.P. Chamola
Morels in India – T.N. Lakhanpal & Onkar S. Shad
Himaliayan Myxomycetes – T.N. Lakhanpal & K.G. Mukerji
Ultrastructure of Microorganisms – K.G. Mukerji & I.S. Bakshi
Fungal Biotechnology – K.G. Mukerji, Neelima Mittal & Surinder Kaur

LIST OF CO-AUTHORS*

BANGIOPHYCIDEAE	VANDANA MEHROTRA AND YAMINI ACHARI
FLORIDEOPHYCIDEAE	
NEMALIALES	ADITYA JHA
CRYPTONEMIALES	INDU NASHIER
GIGARTINALES	MANOJ KUMAR
RHODYMENIALES	VIJAY KUMAR AND MONIKA BHALLA
PALMARIALES	VIJAY KUMAR
CERAMIALES	CHAITANYA MURTI

* Department of Botany
University of Delhi
Delhi - 110007

Series Editor's Note

The explosion of new techniques has nessiciated for the need of concise and authoritative text books for students and researchers in Alage. The aim of this series is to help the novice and the advanced scholars to confront fundamental problems of theoretical and applied nature. Professor M.R. Vijayaraghavan has himself contributed a great deal to the structure, ultrastructure and developmental biology of many algae and angiosperms and he and his colleague Dr Bela Bhatia deserve to be complimented for bringing out this compendium which is source of reference to further detailed information in Red Algal Biology. This book covers the syllabii of many Universities of India, and other neighbouring countries. It should be an excellent and welcome text for students at graduate, postgraduate levels and an important aid for researchers and teachers interested in the reproductive biology of the red algae.

K.G. Mukerji

Preface

The pace of research work in Phycology during the last two decades is fast. There is ample information available on various taxa of Red Algae. This knowledge is, unfortunately, available only in various scientific journals and many of them are out of reach of workers interested in this unique group of algae. My aim in writing this book on Red Algae is to compile the availabe literature for the students at graduate, postgraduate and research levels. This compendium introduce the readers to various aspects of Red algae, covering classification, morphology, reproduction and life histories supplemented with ultrastructural and histochemical details. I hope that my love and enthusiasm, backed by young research group, devoted to red algal research is apparent to you and you will come to share this feelings with me. I strongly feel that there is a need of illustrations for clear depiction of concepts, structure and interactions. Many photomicrographs and line drawings are, therefore included. The large number of selected references enhance the value of this self contained account. I have attempted to write a simple and comprehensive book without compromising the important facts. The text covers the basics as well as currently available information and is amply illustrated to make the text more lucid. I feel that my continuous maintenance of active research in this area has helped me to present a balanced approach.

This book should be an excellent and welcome text for students and teachers as well as useful aid for researchers interested in the biology of the red algae. The book covers the syallabii of many Indian universities and that of neighbouring countries.

M. R. Vijayaraghavan

Acknowledgements

We are thankful to Professor A.R. Mahadevan and Professor V.N. Raja Rao, CAS in Botany, University of Madras; Professor V.S. Jaiswal, CAS in Botany, Banaras Hindu University, Varanasi; Professor V. Krishnamurthy, Krishnamurthy Institute of Algology, Madras; Professor G.S. Venkatraman, formerly, Joint Director, IARI, New Delhi; Professor S. Venugopalan, formerly Head, Department of Botany, Vivekananda College, Madras; Dr Ramalakshmi Dutta and Dr. D.K. Dutta, Vivekananda Institute of Biotechnology, West Bengal; Dr. K.L. Jetani, Gujarat Agricultural Univesity, Port Okha for their help in sending zerox materials of research papers.

We also thank our colleagues Dr. Tripat Kapoor, Dr Gunwant Sokhi, Dr Inderdeep Kaur, Dr Dinabandhu Sahoo, Dr Seema Pasricha, Dr Sudesh, Dr Bharti Chaudry, Dr Sunita Kumari, Ms Sangetha Behl, Ms Priayanka Tyagi, Ms Padma for their help and suggestions during the compilation of this book.

We are indebted to Professor B.M. Johri for his constant encouragement during the preparation of this book.

Our greatful thanks are due to all Phycologists and Editors of various Journals for permission to reproduce their work that has been duly acknowledge at appropriate places in the text.

To Mrs Sita Vijayaraghavan we are thankful for her immense patience and understanding during the compilation of this compendium.

We would also like to thank Mr S.B. Nangia, APH Publishing Corporation, for his assistance and patience through the zestation period of this book.

Department of Botany
University of Delhi
Delhi–110007

M.R. Vijayaraghavan
Bela Bhatia

INTRODUCTION

The red algae comprise about 675 genera and 4,100 species (Kraft, 1981) and form an unique, intriguing and intrinsically interesting group of plants that need critical study. They are least understood group in plant kingdom. The single class Rhodophyceae has two subclasses Bangiophycideae and Florideophycideae (Dixon, 1982b). They comprise unicells, aggregations of unicells, filaments and aggregations of filaments to produce pseudoparenchyma or true parenchyma. The chloroplasts contain various mixtures of chlorophylls, carotenoids and biliprotein pigments (phycocyanin, phycoerythrin), so that the thalli, apart from being red appear violet, brown, black or blue. A few taxa lack chloroplasts or pigments and are regarded as being partially or totally parasitic.

Reproduction occurs by various types of spores and gametes, although vegetative propagation occurs extensively. The gametangia may be highly or slightly specialized, with fertilization occurring in various ways. Gamete fusion is well established for the Florideophycideae. It is obscure in the Bangiophycideae. The product of gamete fusion either develops *in situ* or produces a complex of postfertilization tissue—carposporophyte, which is partially or completely dependent on the female gametangial thallus.

The Bangiophycideae is a small group consisting of four orders and lack both the vegetative and the reproductive complexity of the Florideophycideae. Further a large percentage of freshwater representatives are known and the order Compsopogonales are exclusively nonmarine. The four orders include Porphyridiales (including Goniotrichales), Bangiales, Compsopogonales and Rhodochaetales. The members of the economically important family Bangiaceae (Bangiales) have been extensively investigated but are still enigmatic regarding sexual phase. Karyological studies have been conducted to understand life cycles; to solve the systematic problems of disputed taxa and to discuss possible phylogenetic relationships. In all the four orders ultrastructural studies on chloroplasts, phycobilisomes, pit–connections, mitochondria, golgi bodies and microbodies have been conducted. Pit–connections were earlier thought to be lacking in Bangiophycideae. However, ultrastructural studies reveal that in conchocelis phase of *Porphyra* and *Bangia;* gametophytes of *Rhodochaete* and *Compsopogon* pit–connections are well developed. The Florideophycideae include six orders namely Nemaliales, Cryptonemiales, Gigartinales, Rhodymeniales, Palmariales and Ceramiales. The Gelidiales are now included in Nemaliales as the family Gelidiaceae. Guiry and Irvine (1978) removed the Palmariaceae from the Rhodymeniales and established a new order Palmariales. The Nemaliales are the most primitive taxa; reveal a great range of thallus organization ranging from the simple filamentous to uniaxial/multiaxial forms. Life-cycle is triphasic and its members show both isomorphic and heteromorphic types. Different views exist regarding the position of meiosis in the life-cycle establishing relationship between Florideophycideae and Bangiophycideae. Culture studies have shown that various intermediate stages designated earlier as new genera, are an integral part of the life-cycle.

The Cryptonemiales are fairly large order of marine algae not sharply distinguishable from the Gigartinales. The characteristic feature used as means of separating the two orders is the auxiliary cell. In Cryptonemiales, the auxiliary cell is produced in an accessory or specialized branch of the female thallus whereas in Gigartinales this cell is produced in a vegetative filament. In Cryptonemiales the thalli bearing different types of reproductive structures may be similar, slightly dissimilar or of totally different morphology. Spermatangia are formed single or in clusters, as superficial patches, in internal chains or within the conceptacles. Carpogonium arises from the apical cell of a specially formed filament. Fertilization is always followed by the transfer of the zygote nucleus to an auxiliary cell formed in a specialized filament. Carposporophyte varies in size and may be surrounded by sterile tissue. Life histories are diplobiontic and isomorphic, but *in vitro* studies of a few taxa like *Acrosymphyton, Gloiosiphonia* and *Farlowia* reveal heteromorphic pattern.

In *Gloiosiphonia veticillaris,* ultrastructural studies of carposporophyte show the structural and developmental sequences of the female system and early post–fertilization events. Such studies could also provide some indications of their relationship with other red algae. Besides, they also help in understanding the development and nutrition; structure, composition and function of pit-connection in parasitic algae belonging to the family Choreocolacaceae.

The spore germination in *Fosliella, Preophylla, Melobesia* confirmed that calcification started at an early stage in epilithic and at a later stage in epiphytic taxa. The mode of attachment of six epilithic algae showed three different mechanisms. These are affected after the mucilage surrounds the early sporeling. (a) Penetration and moulding of the growing tips of hypothallial filaments into minute crevices in the rock face. (b) Deposition of a thick layer of calcareous material in an organic matrix which fills in microsopic gaps between the lowermost cell walls and the substratum. (c) Precipitation of large cluster of calcium carbonate crystals into spaces beneath the crust—a process which appears to be influenced by organic substances secreted by the plant. The coralline algae are a major component of the rocky intertidal and shallow subtidal zones.

Gigartinales reveal variable morphological features from very simple to complex forms which range from uniaxial to multiaxial, crustose to discoid and nonprocarpic to procarpic taxa. The reproductive structures, particularly the spermatangia, reveal different types among the various members. Three-celled carpogonial branch is ubiquitous feature of many taxa. Tetrasporangia are mainly of zonate and cruciate types. The former is typical of Solieriaceae and while the latter is found in most other families. The carposporophyte types range from the morphologically most simple to the most complex types.

Members of Furcellariaceae, Gigartinaceae, Gracilariaceae and Solieriaceae are economically important. *Gigartina, Eucheuma, Chondrus* and *Gracilaria* for agar-agar and *Gigartina* and *Chondrus* for K-carrageenan are well known candidates. Various histochemical and ultrastructural studies aim at not only in *in situ* localization of major macromolecules and organelles in cytoplasm but such studies also find usefulness in improving the economically important commercial products like agar and carrageenan. A very recent technique using Nuclear Magnetic Resonance Spectroscopy (NMR) helps to analyse agar polymers.

The order Rhodymeniales is characterized by multitaxial thalli; pit-plugs that lack cap-layers; and auxiliary cells produced prior to fertilization as a terminal cell of an auxiliary-cell branch arising from the same intercalary (supporting) cell that bears the carpogonial branch. The three families Champiaceae, Lomentariaceae, and Rhodymeniaceae are distinguished primarily on vegetative features of the cortex and the medulla and also on the type of tetrasporangia formed namely cruciate or tetrahedral.

Palmaria, Halosaccion, Halosacciocolax and *Rhodophysema* were earlier placed under the family Rhodymeniaceae but Guiry (1974) segregated these genera in a new family Palmariaceae.

Guiry and Irvine (1978) created the order Palmariales to accomodate the taxa placed under the family Palmariaceae. Palmariales are characterized by the presence of a generative stalk cell, the presumed absence of carpogonia and carposporophyte, and the formation of spermatangia in pairs on the spermatangial mother cell. The *in vitro* studies on *Palmaria palmata* and *Halosaccion ramentaceum* (Van der Meer and Todd, 1980); *on Devaleria rametacea* (Van der Meer, 1981); and on *Halosaccion americanum* (Mitman and Phinney, 1985) have revealed the presence of terminal, sessile carpogonium on the vegetative filments. The carposporophyte phase is, however, absent, and the tetrasporophyte (or only the tetrasporangial initial) develops directly from the fertilized carpogonium on the female gametophyte (Hawkes and Scagel, 1986b). Ultrastructural studies on pit-plugs reveal that the plug-core has two cap layers with the cap membrane separating the two layers on either side and provide supporting evidence to justify the creation of the order Palmariales (Pueschel and Cole; 1982).

In Ceramiales the thalli are single filament or assemblage of filaments. The level of organization within the multicellular aggregations is high with cell differentiation. Cells usually contain many small chromatophores and may be uninucleate or multinucleate. The sexual phase and asexual phase (tetrasporophyte) are separate and alternate in the the life cycle. The carpogonial branch is borne laterally on supporting cell and always consists of four cells devoid of any laterals. Auxiliary cell is formed after fertilization. The supporting cell of the carpogonial branch gives rise to the auxiliary cell. The supporting cell in addition to auxiliary cell may also bear one or two sterile branches. Carposporophyte, which is a postfertilization product, is totally dependent on the sexual thallus. A single fusion of male and female nuclei is multiplied by mitosis leading to formation of numerous carpospores which germinate to form the alternate plants, the tetrasporophytes. Meiosis occurs within the tetrasporangia to produce tetraspores, which give rise to the sexual thallus.

Our endeavour has been to provide enough description of structure, ultrastructure, life cycle, ecology and metabolic products enabling the students to visualize and compare algal structures and at the same time to provide enough references so that the research worker can read from the original source the literature for precise details. The current book covers a broad spectrum of morphological, cytological, ultrastructural, cytochemical, physiological, ecological and systematic topics towards better understanding of red algae.

CONTENTS

1 BANGIOPHYCIDEAE

CLASSIFICATION

Fritsch (1945) recognized Bangioideae, a subclass of Rhodophyceae, has a single order Bangiales. The general features of the order are :

a) Simple thalli with uninucleate cells, the nucleus often being ecentric in position.

b) Single axile, stellate chromatophore.

c) Zygote divides directly into diploid carpospores.

d) The thallus shows diffuse growth, and the cells are normally devoid of pit-connections.

e) Photosynthetic products accumulate near the chromatophores as solid granules, resembling Floridean starch.

f) Pigmentation of the cells is due to phycoerythrin and phycocyanin.

g) Membranes consist largely of pectic substances. Cellulose has been demonstrated only in *Porphyridium* and *Erythrotrichia.*

The Bangiales is further divided into two families

(i) Bangiaceae

(ii) Porphyridiaceae

Dixon (1982b), however, recognizes the division Rhodophycota with a single class, Rhodophyceae and two subclasses :

a) Bangiophycideae

b) Floridiophycideae

The subclass Bangiophycideae contains about 1% of the genera of the Rhodophyceae. The general characters of the subclass are :

a) Thalli are unicellular or consist of aggregation of unicells that are filamentous or parenchymatous but of minimal complexity.

b) Cells are uninucleate and have single stellate chloroplast, but in some genera each cell contains numerous parietal and discoid chloroplasts. Pyrenoids are present when only a single chloroplast occurs in the cell. Colour of the thallus depends on pigment composition in the chloroplasts and may be rose red, crimson or blue.

c) Cell division is predominantly intercalary.

d) Pit-connections are reported in various genera, but are not of universal occurrence.

e) Gametangia are not well differentiated.

f) Reproduction occurs through monosporangia or spores formed in packets of 4, 8 or 16.

Bangiophycideae are divided into four orders :

a) Porphyridiales

b) Bangiales

c) Compsopogonales

d) Rhodochaetales

PORPHYRIDIALES : This order is characterized by :

a) Single celled thalli which may sometimes form aggregations with very little differentiation.

b) Single stellate chloroplast with a central pyrenoid.

c) Common method of reproduction in unicellular forms is vegetative cell division.

d) Monosporangia or spores, formed in packets of 4 to 32 per mother cell, may also help in reproduction.

e) Gametes involve light coloured/colourless male gametes and slightly differentiated carpogonia.

Porphyridiales has three families :

i) Porphyridiaceae : Six or seven genera reported from both marine and freshwater locations.

ii) Goniotrichaceae : Members have been reported from marine, brackish and freshwater habitats.

iii) Phragmonemataceae : Genera have been reported from diverse habitats.

The Porphyridiales include taxa like *Asterocystis, Goniotrichum, Kylineilla, Porphyridium, Phragmonema,* and *Rhodella.*

BANGIALES : The general characters of the order are :

a) More than one morphological phases in the life history; major morphological phase is filamentous which later becomes parenchymatous. The minor morphological phase includes either filamentous or minute forms that have a single parietal chloroplast.

b) Each mother cell may produce upto 128 colourless spermatia.

c) Female gamete may be an undifferentiated vegetative cell or has a slightly differentiated protuberance.

Bangiales consist of three families :

i) Erythropeltidaceae - 5 or 6 marine genera are included.

ii) Bangiaceae - Many fresh and marine taxa reported.

iii) Boldiaceae - Monotypic family represented by *Boldia erythrosiphon.*

The important genera included in Bangiales are *Bangia, Boldia, Erythrocladia, Erythrotrichia, Porphyropsis,* and *Porphyra.*

COMPSOPOGONALES : The general features of the order are :

a) Thalli are filamentous that later become parenchymatous, with a central core of large colourless cells bereft of chloroplast surrounded by cells with chloroplast.

b) Both large and small spores occur.

c) A filamentous phase of different morphology resulting due to gamete fusion occurs in the life history of an unnamed species of *Compsopogon*.

The order contains one family :

Compsopogonaceae - Single genus *Compsopogon* is reported to have twelve species.

RHODOCHAETALES : General features of the order are :

a) Thallus is branched, filamentous, composed of elongate cells with pit-connections.

b) Post-fertilization structures are produced which were earlier supposed to be monosporangia.

c) Female gamete is an undifferentiated cell of the thallus. The male gamete is small, colourless and formed from a structure that simulates the monosporangium.

This order contains a single family Rhodochaetaceae and monotypic genus *Rhodochaete parvula.*

Bold and Wynne (1985) consider Rhodophyta with a single class Rhodophyceae which is divided into two subclasses :

a) Bangiophycideae

b) Florideophycideae

Bangiophycideae is further divided into four orders :

a) Porphyridiales

b) Rhodochaetales

c) Compsopogonales

d) Bangiales

Bold and Wynne (1985) further divide these orders as follows :

a) Porphyridiales :

(i) Porphyridiaceae

(ii) Goniotrichaceae

b) Rhodochaetales :

(i) Rhodochaetaceae

c) Compsopogonales :

(i) Compsopogonaceae

(ii) Erythropeltidaceae

(iii) Boldiaceae

d) Bangiales :

(i) Bangiaceae

The classification adopted by Bold and Wynne (1985) does not include the family Phragmonemataceae but Dixon (1982b) has included it under Porphyridiales. The families Erythropeltidaceae and Boldiaceae are included in the order Bangiales by Dixon (1982b) whereas these two families are placed under order Compsopogonales by Bold and Wynne (1985).

Fritsch (1945) has grouped the taxa *Compsopogon, Rhodochaete* and *Conchocelis rosea* as uncertain Bangiales. The former two genera have, however, been placed in two separate orders by both Dixon (1982b), and Bold and Wynne (1985). *Conchocelis rosea* is distinguished from true Bangiales by presence of pit-connections, but resemblance to any particular group of Florideae is not known (Fritsch, 1945). Now it is clearly established that *Conchocelis* is the alternate phase in the life-cycle of *Porphyra*.

MORPHOLOGY AND THALLUS ORGANISATION

Bangiophycideae show a relatively simple morphology as compared to Florideophycideae and comprise of taxa with unicellular (*Porphyridium*), filamentous (*Goniotrichum*) and parenchymatous (*Porphyra*) forms.

BANGIALES : The genus *Erythrotrichia* is one of the least specialized members with heterotrichous filaments (Fritsch, 1945). A discoid one layered base bears unbranched erect threads (Fig. 1 F) *Erythrocladia* is, however, an epiphyte or endophyte on larger seaweeds with prostrate habitat, whereas *Porphyropsis* occurs as small one-layered *Porphyra*–like sheets. Interestingly, *Bangia* and *Porphyra* show greater thallus specialization. *Bangia* shows unbranched threads that may remain uniseriate for sometime and divide later. The older plants produce non-septate rhizoids that spread over the substratum.

Porphyra is a thalloid, blade-like or membranous marine alga (Fig. 1 A,B); grows attached through rhizoids to the rocks and other algae. The frond is parenchymatous or monostromatic in some species (Kumar, 1990). The simple filaments of *Porphyra* in later stages show diffuse growth. *P. naiadum* is the only plant where heterotrichous condition is known (Fritsch, 1945). Based on this exomorphic feature and along with manner of blade production, this taxon has been raised to a separate genus *Smithora* (Fig. 1 C; Hollenberg, 1959).

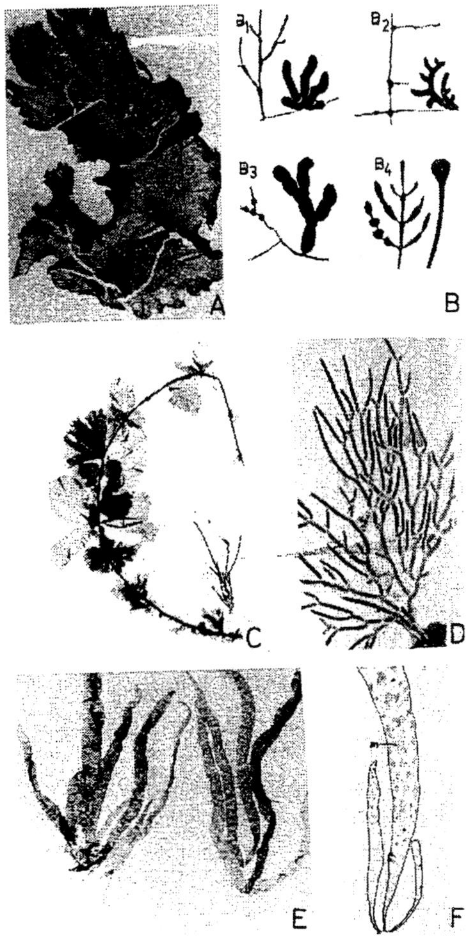

Fig. 1 A-F. Morphology

(m, monosporangial sori)

A *Porphyra,* foliose thallus of a monoecious species. (after Bold and Wynne, 1978).

B Conchocelis phases of *Porphyra abbottae* (B$_1$); *P. torta*(B$_2$); *P. nereocystis*(B$_3$); *P.schizophylla*(B$_4$). The vegetative filaments are towards the left and conchosporangial branch on the right (after Campbell and Cole, 1984).

C *Smithora naiadum* growing as epiphyte on sea-grass. Clusters of blades are produced from persistent, cushion-like bases that regenerate new blades. (after Bold and Wynne, 1978).

D *Goniotrichum alsidii* showing small irregularly branched filamentous thallus. The cells embedded in a well developed mucilaginous matrix. (after Bold and Wynne, 1978).

E *Boldia erythrosiphon* showing saccate and unbranched thallus. (after Bold and Wynne, 1978).

F *Erythrotrichia folliformis* showing a group of simple, parenchymatous blades upon which are distributed monosporangial sori. (after South and Wittick, 1987).

Boldia erythrosiphon shows cylindrical to saccate thalli (Fig. 1 E) which usually grow on snails in streams. The unbranched thalli are usually less than 20 cm in length but rarely reach 30-75 cm. Monospores germinate to produce a prostrate, monostromatic discoid growth or aggregated creeping filaments. The colour of thallus ranges from mauve pink and brownish-red to olive-green.

COMPSOPOGONALES : This order includes only freshwater forms. *Compsopogon* has branched filaments, shows diffuse growth and a uniseriate thallus organisation in apical region. Later, the filaments undergo vertical divisions, become corticated with grey-green to purple or blue hue.

PORPHYRIDIALES : The members are generally unicellular or reveal quasi-filamentous morphology (Kraft, 1981). Both *Porphyridium* and *Rhodella* are unicellular with globose cells that are covered with mucilage envelopes. *Goniotrichum* is a minute epiphyte that shows branched filaments (Fig. 1 D).

RHODOCHAETALES : *Rhodochaete* is monotypic, marine taxon, with apical-cell, branched filaments and occurs as epiphyte on other algae (Fritsch, 1945).

REPRODUCTION

Most Bangiales reproduce asexually through either monospores or neutral spores or gonidia which are nonmotile. The term monospore is found to be more suitable. Sexual reproduction was reported to occur only in *Bangia, Porphyra, Erythrotrichia* and *Erythrocladia* (Fritsch, 1945). The sexuality in *Erythrocladia* was, however, questioned.

ASEXUAL REPRODUCTION

It takes place by non-motile spores. A spore is a specialized cell that can reproduce independently a new plant, and a sporangia is a structure within which the spore arises. According to Guiry (1990) in Bangiophycideae spores were grouped under three types. The classification is not dependent on the origin of spores.

(1) Monospores produced from undifferentiated sporangia.

(2) Monospores produced from undifferentiated vegetative cells.

(3) Mother cells by successive divisions produce packets of 4-128 spores.

Aplanosporangia are the small 'packets' of spores (up to 16) formed as a result of a few divisions of vegetative cells of *Porphyra* (leafy phase) and *Bangia* species. These aplanospores, when released from the thalli show bipolar germination and develop in a manner resembling adult thallus (West and Hommersand, 1981).

According to Guiry (1990) the different types of asexual spores and sporangia encountered in Bangiophycideae are :

Endosporangia : These are uninucleate, mitotic spores and are reported in culture in Phragmonemataceae (Porphyridiales) and *Erythrocladia irregularis.*

Zygotosporangia : The spores arise internally in the carpogonium, by mitosis, after fertilization.

Monosporangia : These form single uninucleate spore on either haploid or diploid plants but are not derived from gonimoblast tissue and are nearly always mitotic. They are generally formed singly but sometimes may be formed in packets. A range of monosporangial types is found in Bangiophycideae. They many be formed by wall formation in an undifferentiated cell (Compsopogonales) or from differentiated sporangia (Erythropeltidaceae).

In *Smithora naiadum* monospores are formed by direct transformation of vegetative cells in terminal position or they may be 'neutral spores' appearing as single sporangia cut off from vegetative cells.

Rhodochaete parvula shows monosporangia on both diploid and haploid plants.

Conchosporangia : These are formed on conchosporangial branches of Conchocelis phase of *Bangia* and *Porphyra*. They are rectangular, have strong pigmentation and form single, uninucleate, spores.

Carposporangia : The Single-spored carposporangia are derived from gonimoblast filaments coming out from zygote and are similar to monosporangia. *Rhodochaete parvula* is the only taxon of Bangiophycideae where carposporangia-like sporangia are known that are formed external to the carpogonium.

SEXUAL REPRODUCTION

Sexual life history has been demonstrated in two orders Bangiales and Rhodochaetales.

Bangiales : The discovery that filamentous alga *Conchocelis rosea,* pro-parte in the life cycle of *Porphyra purpurea,* established that *Porphyra* is biphasic and involves an alternation between macroscopic foliose and a microscopic filamentous phases. Further. in *P. gardneri* the terms a spores and b spores were used to replace carpospores and spermatia respectively. Besides, conclusive evidence of occurrence of sexual fusion was also provided due to the fusion of a small colourless spermatium with a carpogonium, which is a somewhat differentiated thallus cell with an emergent trichogyne providing receptive surface for the speramatia. Fertilization canals or tracts that were observed represent the migration paths of sperm nucleus through the cell wall to the carpogonial protoplast (Hawkes, 1978). The zygote through successive divisions produces many carposporangia each of which forms a carpospore. The carpospores germinate resulting in an entangled mass of uniseriate, branched filaments that usually grow endolithically in shells, This is called the conchocelis phase and represents an alternate, dissimilar phase in the life-history. This stage occasionally regenerates asexually through monospores. Meiosis occurs in the conchosporangial branches (Bold and Wynne, 1985).

Rhodochaetales : Initially the monotypic species *Rhodochaete parvula* was thought to reproduce asexually by monospores but Magne (1960) demonstrated that thalli are sexual and monoecious. Each motile cell produces one spermatium that fertilizes an undifferentiated vegetative cell that functions as the carpogonium which forms a carposporangium and releases one carpospore. The carpospore is presumed to be diploid and its fate is unknown. This is the simplest form of sexual reproduction so far reported in the red algae (West and Hommersand, 1981).

KARYOLOGICAL STUDIES

Karyological studies are necessary for understanding the life histories and to provide data for solving certain systematic problems. This recognition generated the interest of algologists to study the red algal chromosomes. Althrough karyological studies were initiated as early as 1960, very little was known about the subclass Bangiophycideae.

In Rhodophyceae, the haploid number of chromosomes ranges from 2 in Porphyridiaceae (Porphyridiales) and Bangiaceae (Bangiales) to 68-72 in Polyideaceae (Gigartinales). Chromosome numbers as low as (n = 2–10/15) have been reported for taxa belonging to Bangiales, Compsopogonales, Porphyridiales and Rhodochaetales. Many taxa of the Bangiales have the smallest number of chromosome (n = 2-7) in uninucleate cells.

Interestingly *Porphyra* spp. have been studied more extensively cytologically than any other red alga. Ishikawa (1921) made the first attempt to study the nuclear division in both vegetative and reproductive cells of *Porphyra tenera.* Since then considerable amount of work has been carried out and chromosome numbers have been reported for many *Porphyra* species (Krishnamurthy, 1984; Kapraun and Freshwater, 1987).

The most remarkable feature in the karyology of *Porphyra* is the variation in number of chromosomes in different species. At least 43 species of *Porphyra* have been studied for chromosome numbers and it was reported that about twenty five species have the haploid number of 3; in six species the number is 2 and in eight species the number is 4. Yabu (1971) reported the chromosome number in eight species of *Porphyra* as 5; and in one species each, the number as 6 or 7 (see Krishnamurthy, 1984).

In *Porphyra fucicola* the chromosome number is reported to be 3 or 4 (Mumford and Cole, 1977) or as 5 (Krishnamurthy, 1984). Hawkes (1978) reported a diploid chromosome number of 6 in *P. nereocystis* but Krishnamurthy (1984) denied the existence of diploid phase. *P. tenera* may show a chromosome number of 3, 4, or 5.

The nuclear state of carpospores and their formation also reveal variation and a great deal of disagreement exists. A few regard the capospores as diploid structure originating by mitosis from fertilized carpogonia, whereas others consider haploid carpospores to be the products of either meiotic division or by mitosis from a nonfertilized mother cell. Yet another report claimed that meiosis takes place during conchospore formation of the diploid Conchocelis phase (Coll and Oliveira Filho, 1977).

In *Porphyra leucosticta* chromosome counts during the development of 'spermatia' and 'carpospores' showed the absence of meiosis during 'carpospore' formation and that both spermatia and carpospores are haploid. Conchocelis phase was, however, not studied but it was concluded that this phase is also haploid as it arises from haploid 'carpospores'.

The wide range of chromosome numbers in different species of *Porphyra* have led to merging of a few species. Krishnamurthy (1984), in *P. norrisii,* counted 5 chromosomes in speramatia and 10 chromosomes both in carpospores and the Conchocelis phase. This species, however, has been merged with *P. schizophylla* which has 3 chromosomes. Krishnamurthy, however, maintains that *P. norrisii* is distinct from *P. schizophylla.* Karyological studies, have thus, been applied to solve systematic problems and provides additional criteria to distinguish closely related species.

Kapraun and Freshwater (1987) studied the chromosome numbers of Atlantic and Mediterranean *Porphyra* species. *P. leucosticta* and *P. umbilicalis* have trans-Atlantic distribution and *P. carolinensis* and *P. rosengurtii* appear to be endemic to U.S. southeastern coast. Cole and Cox (1977) distinguish the latter two species from other North Atlantic *Porphyra* species.

Porphyra spiralis var. *amplifolia,* a Venezuelan isolate, could not be precisely circumscribed. Kurogi (1972) referred this plant to *P. umbilicalis.* Kapraun and Freshwater (1987), however, did not agree with this new circumscription.

The study of chromosome numbers for five species of *Porphyra* from the warm temperate and tropical North Atlantic and Mediterranean coasts revealed that the basic haploid number of the foliose thallus and spermatia is idential and is 4 in *P. carolinensis, P. leucosticta, P. spiralis* var. *amplifolia, P. umbilicalis* and 3 in *P. rosengurtii* (Kapraun and Freshwater, 1987).

Chromosome size was observed in several stages of development in five *Porphyra* species. Data obtained from.karyological studies were used to correlate ploidy level with the life history phase. In *P. umbilicalis* there is a diploid chromosome number leading to the formation of carpospores. Rest of the four species lack sexual reproduction (Kapraun and Freshwater, 1987).

In the Bangiales, the degree of chromosome contraction and therefore size, varies considerably with stage of nuclear divisions (Mumford and Cole, 1977). The relative size of chromosomes in individual cells, however, should be consistent regardless of particular mitotic stage or fixation procedure. The information for typical karyotypes could, thus, be used to show possible interspecific relationship.

Porphyra carolinensis, P. leucosticta. P. spiralis Var. *amplifolia* and *P. umbilicalis* (Fig. 2 A). The *P. rosengurtii* karyotype (n = 3) seems to differ from the above by deletion of chromosome number 4. The karyological studies have, thus, helped to discuss possible phylogenetic relationships between Atlantic and Pacific species (Freshwater and Kapraun, 1987). The small number of chromosomes rules out the possibility of polyploidy in the genus *Porphyra.*

There is a striking range in chromosome numbers from 2-7 and it becomes difficult to assign the basic chromosome number of the genus. However, it is 3 in large number of species.

The three phenomena that have been considered to play a role in the wide range of chromosome numbers are :

1) Loss or gain of a chromosome due to irregular separation of meiotic chromosomes.
2) Doubling of chromosomes.
3) Hybridization.

Translocations, inversions, fusions, deletions and non-disjunction are responsible for evolution among karyotype, that provides valuable information on the evolutionary processes and trends.

Karyology of *Bangia atropurpurea* from marine and freshwater populations of Mediterranean and Northeastern Atlantic was studied. The gametophytic thalli showed two haploid karyotypes of n = 3 and n = 4. The n = 4 individuals were collected in both marine and inland waters, suggesting that these individuals were aneuploids. Most frequent karyotype n = 3 was observed both in vegetative and spermatangial cells (Fig. 2 B,C). Chromosomes at prophase were positioned to form a circular arrangement or an end to end arrangement (Fig. 2 D,E). Plants of both the populations were morphologically indentical, but n= 4 populations could occur in dry areas. The occurrence of same species in two different habitats has also led to assumptions whether this alga represents a single species or not. The morphological details of the chromosomes observed from the two forms were identical. This similarity in the karyotype of inland and marine individuals supports their conspecificity (Gargiulo *et al.*, 1991).

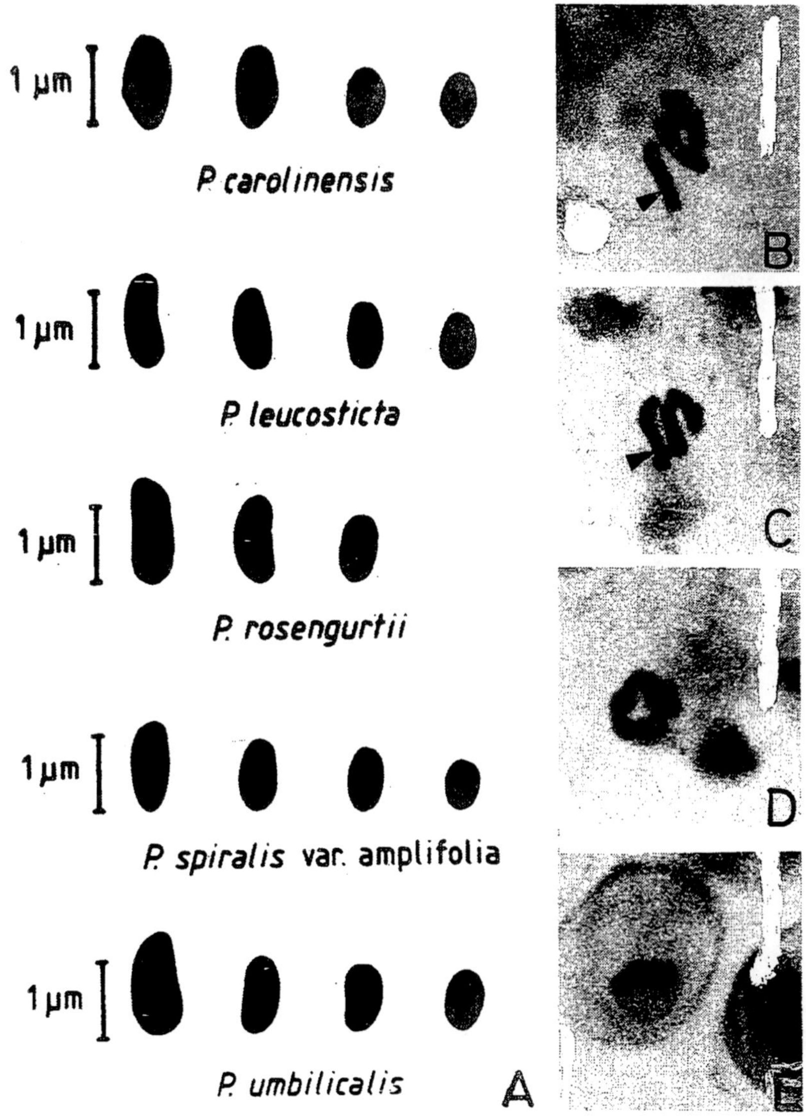

Fig. 2 A-E. Karyological studies of Bangiales.

A Karyotypes of five *Porphyra* species, representing average prophase chromosome preparations (after Kapraun and Freshwater, 1987).

B,C Vegetative cells (B) and spermatangial cell (C) of *Bangia atropurpurea* showing n=3. Arrowheads point to the tentative position of centromere (after Gargiulo *et al.*, 1991).

D,E Spermatangial cells of *B. atropurpurea* with n=3 showing end to end (D) and circular (E) arrangement of chromosomes (after Gargiulo *et al.*, 1991).

Cole *et al.* (1983) studied the chromosome number and morphology of sexual and asexual populations of *Bangia* from British Columbia. Small size and chromosome number (n = 3, 4) was characteristic of Bangiophyceae. The karyograms prepared also showed the tentative position of centromere in *Porphyra papenfusii* (Cole *et. al.,* 1983).

Sexual and asexual populations were distinguished based on chromosome numbers. Sexual populations were of 3 and 4 chromosome numbers whereas the asexual populations were 3 and 6 chromosome numbers. Pairing phenomenon between the two sets of mitotic prophase chromosomes was reported for the first time in *Bangia.* Karyotype studies have been used also for the first time for taxonomic investigations of *Bangia.* It is likely that the karyotypes could be employed to help delimit various *Bangia* species. as all the populations do not have same chromosome number (Cole *et. al.,* 1983).

ULTRASTRUCTURAL AND HISTOCHEMICAL STUDIES

CELL WALL

The Rhodophycean cell has two distinct wall layers. The inner layer, adjoining the cell cavity, is firm and composed of cellulose. The outer layer is made up of pectic substances. Cellulose has, however, been demonstrated in *Porphyridium* and *Erythrotrichia.*

In alternate phases of *Porphyra* the structure of cell wall was analysed and it was found that both the structure and the chemical composition varied depending on the phase of the plant. Mukai et al., (1981) analysed the chemical structure of cell walls of both Conchocelis and thallus phases of *Porphyra tenera* and observed that the Conchocelis cell walls retained their structure even after cell wall purification. The external surface of cell wall appeared structureless and amorphous in contrast to the internal surface which showed randomly arranged microfibrils. Further, the surface replicas of the microfibrils from the conchocelis even showed a small amount of granular material. On the contrary, thallus cell walls retained only a little of its structure after freezing and grinding but showed the presence of microfibrils. The surface replicas of microfibrils had amorphous material.

Cole *et al.,* (1985) studied the cell walls of sexual and asexual phases of *Bangia atropurpurea.* In longitudinal section, the vegetative regions of a multiseriate filament, show adjoining wedge-shaped cells with outer and inner walls. The outer wall is extremely electron-dense and fibrillar materials are embedded in an amorphous substance (Fig. 3 C). The inner wall material surrounds the plasmamembrane in the form of layers. Instead of amorphous matrix, loosely arranged microfibrils were seen in the inner wall (Fig. 3 A,B) and the layers are formed through association of some microfibrils. During growth, a central space develops within the filament. The outer wall, adjacent to this space, is the interior region whereas on the outer most part of the filament is the exterior region. Cuticle which covers the outer wall over the entire length of filament is intermittently sloughed-off.

Histochemical localization of polysaccharides shows clear partitioning of the inner and outer wall. The staining behaviour of the outer wall material showed that it contains acidic, sulphated polysaccharides. PAS technique further demonstrated detectable amounts of neutral polysaccharides in both the outer and the inner walls. The PAS positive material present in the outer coating helped in the attachment of the spermatia and the fertilization channels.

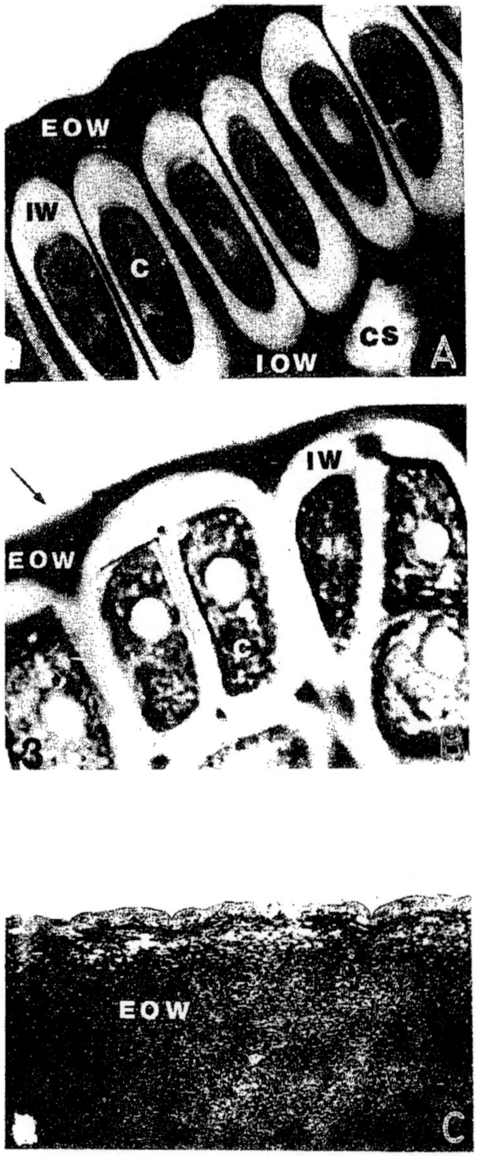

Fig. 3 A-C. Ultrastructure of cell wall in *Bangia atropurpurea* (after Cole *et al.*, 1985).

(C, cell protoplast; CS, central space; EOW, Exterior region of outer wall; IOW, Interior region of outer wall; IW, Inner wall)

A,B Filament consists of wedge shaped cells. Layered inner wall material surrounds immediately the plasmamembrane of each cell.

C The outer wall is electron-dense and fibrillar material embedded in an amorphous substance.

During sexual reproduction there was a change in the nature of acidic wall components shown by decrease in intensity of staining for sulphate and appearance of unidentified anionic groups. Histochemical analysis demonstrated that changes in composition of outer wall occur only in sexually differentiating regions and not in the reproductive areas of asexual filaments and decrease in sulphate reaction is common to both male and female plants (Cole *et al.*, 1985).

PIT-CONNECTIONS

Conspicuous pits on the septa, that connect genetically related cells, forming part of same thread and derived from a common parent-cell by division are called pit-connections.

Absence or presence of pit-connections between cells of kindred origin was a criterion for the separation of two subclasses Bangiophycideae and Florideophycideae of Rhodophyceae. The occurrence of pit-connections in some genera which were otherwise placed in Bangiophycideae led to taxonomic debacle and it was concluded that pit-connections need not be used to distinguish the two subclasses.

Pit-connections were reported in *Bangia* sp., *Compsopogon* sp, *Conchocelis rosea* (now Conchocelis phase), *Rhodochaete* sp. (Dixon, 1963). Pit-connections are ubiquitous in Florideophycideae, but may not occur in Bangiophycideae. According to Fan (1960) pit-connections that occur between juxtaposed cells are more complex than protoplasmic strands. In *Bangia,* on either side of the pit, dark, cushion-like structures, were observed that are identical with those of Florideophycideae (Belcher, 1960).

The pit-connections were detected when the plant material was subjected to certain types of fixatives as revealed by *Compsopogon,* Conchocelis phase and *Rhodochaete.* The ultrastructure of red algal pit-plugs shows a relatively large core bracketed by a pair of thin-caps. The plug-core and plug-caps differ in their cytochemical properties, specially proteins and polysaccharides. Alcian blue, that stains sulphated polysaccharides does not stain the pit-plugs, but the unstained pit-plugs had the appearance of pores in the background cell walls that were stained. The studies, however, demonstrate that protein was the only demonstrable chemical constituent of the plug cores of both Florideophycideae and Bangiophycideae (Pueschel, 1980).

Pueschel and Cole (1982) reported that Conchocelis stage thalli of *Bangia atropurpurea* show pit-core surrounded by a pit-cap consisting of a single thin electron-dense layer (Fig.4A). This structure resembles plugs which have a naked-cap membrane and no plug-caps. This dense line is thicker than a membrane and usually has a granular appearance. This plug-cap layer did not take up PTA-chromic acid stain.

The order Bangiales is unique in that its membranes have pit-connections in the sporophytic or the Conchocelis phase but gametophytic phase lacks them. In *Rhodochaete,* ultrastructural studies reveal that the pit-connections lack pit-plugs. This was, however, proved to be incorrect.

Pit-connections were reported in two species of *Compsopogon.* Cultured plants of *C. coeruleus* revealed 'incompletely deposited walls' that had been misinterpreted as 'true' pit-connections. Scott *et al.*, (1988), through serial sectioning demonstrated equivocally pit-connections in this Bangiophyte.

Light microscopic studies reveal extremely small pit connections which could easily be overlooked in 1µm thick plastic sections. Electron microscopic studies confirmed that pit-connections interconnect juxtaposed cortical cells (Fig. 4B); cortical-axial cells (Fig. 5A) and

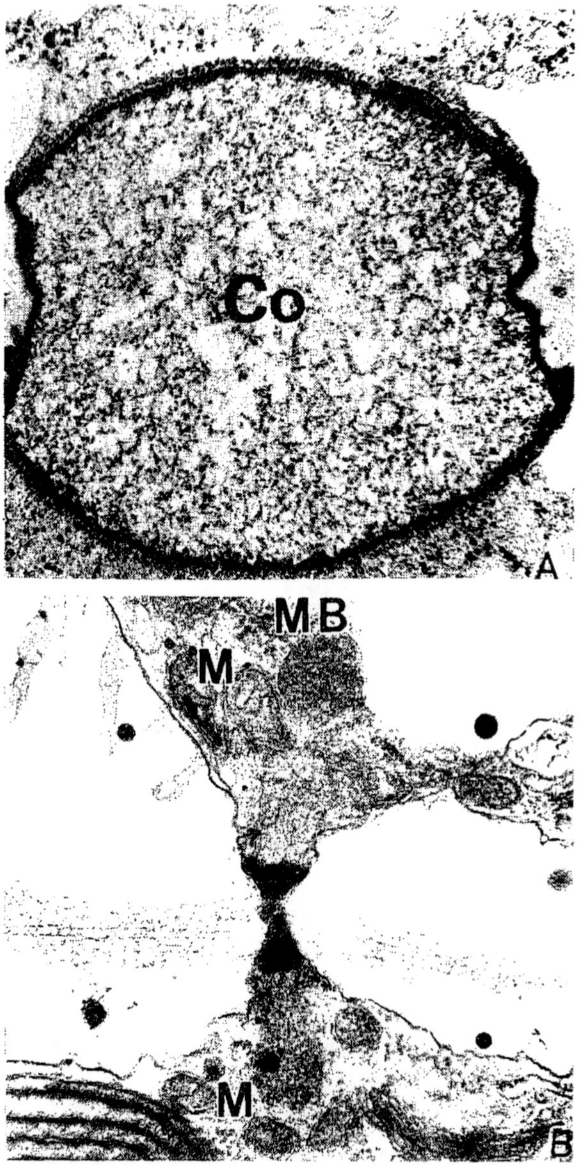

Fig. 4 A,B. Ultrastructure of the pit-connections.

(Co, Plug core; MB, Microbody-like organelle; M, Mitochondria).

A Conchocelis stage of *Bangia atropurpurea* showing pit-core surrounded by a thin layer of pit-cap (after Pueschel and Cole, 1982).

B *Compsopogon coeruleus* showing pit-connections between two cortical cells. Mitochondria and microbody-like organelles are located near the filamentous exclusion zones (after Scott *et al.*, 1988).

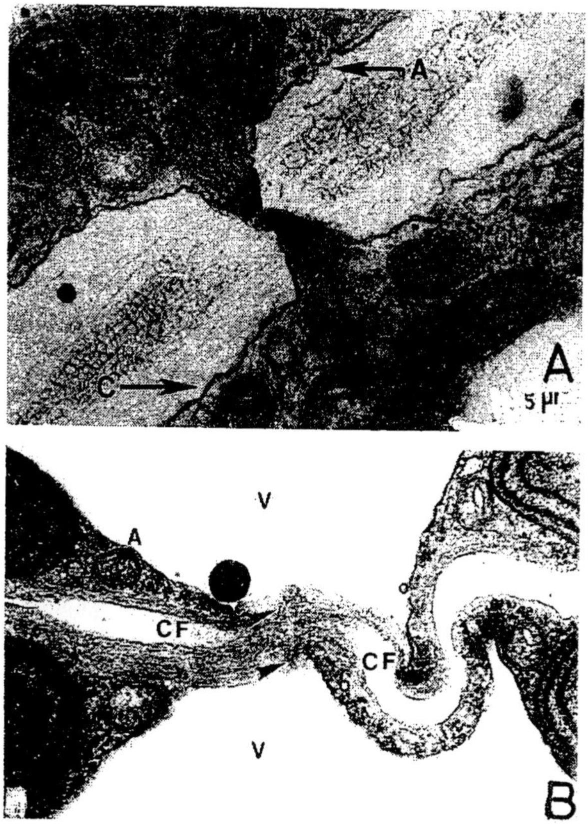

Fig.5 A,B. Ultrastructure of pit-connections in *Compsopogon coeruleus*
(after Scott *et al.*, 1988)

(A, Axial cell; C, cortical cell; CF, cleavage furrow; V, vacuoles).

A Pit-connection between axial and cortical cells. Mitochondria and microbody-like organelles are seen near the exclusion zone.

B A vegetative cell showing pit-plug formation. Cytokinesis occurs in the axial cell. Filamentous edge of the cleavage furrow has cut through the central vacuole of cell. Except for a few ER-like tubules (arrowheads) there is no prominent evidence of pit-connection formation in aperture left behind from incomplete cytokinesis.

axial-axial cells. Secondary pit-connections were, however, not observed. Due to plasmolysis and contraction of cytoplasm from the cell wall very few pit connections were seen intact.

The pit-connections in *Compsopogon coeruleus* and *Rhodochaete* sp. were found to be simple and small (less than 0.2 μm) as compared to the elaborate organization of pit-connections met within higher red algae. The size (0.1 - 0.15 μm in diameter x 0.25 μm in length) and ultrastructure of the pit-connections is the same regardless of their location. These pit-connections bear a little resemblance to the pit-connections reported in Conchocelis phases of both *Porphyra* and *Bangia* (Scott *et al.*, 1988).

Ultrastructural studies in *Compsopogon coeruleus* revealed that pit-connections consist of an extremely electron–dense material, the pit-plug separated from the cell wall by the plasma membrane, which is continuous from cell to cell (Scott *et al.*, 1988). The two opposite cytoplasmic regions of the plug were bordered by a single, distinct, fibrillar–granular, exclusion zone (Figs. 4B, 5A). The presence of cap membrane had not been indicated. Fine and linear structures which appear to extend into the plug proper could be seen in the exclusion zone. Mitochondria are associated with pit–connections and single membrane bound microbody like organelles is occasionally found in same regions (Figs. 4B, 5A).

Stages involved in formation of pit-plug were observed in only two cells of *Compsopogon coeruleus.* Cytokinesis in the axial cells occurs by ingrowth of a very thin cleavage furrow which initially pushes through the peripheral cytoplasmic layer before traversing the large central vacuole and merging in the middle of cell. The advancing edge of the cleavage furrow is somewhat swollen and shows a thin layer of cytoplasm containing filamentous material lying parallel to the swollen portion (Fig. 5B).

Another stage of pit-connection formation showed that the exclusion zones were poorly defined and numerous linear elements that are likely to be equivalent to endoplasmic reticulum cisternae, could be seen in nearly mature plug. In *Compsopogon coeruleus* the electron-dense, plug-materials of pit-connections are produced within or upon these endoplasmic reticula-like components.

CHLOROPLASTS

The distinctive colour of the red algal thallus is due to the water soluble phycobilin pigments which are present in the chromatophores. These phycobiliproteins are arranged in large pigment-protein complexes, named phycobilisomes, which are attached to the outer surface of thylakoid membrane.

Bangiales possess a single axile, massive stellate, chromatophore, and include a conspicuous central pyrenoid (Fritsch, 1945). Earlier studies indicated the association of small granules with the unstacked chloroplast lamellae of red algae, but these granules were absent from the inner and outer chloroplast limiting membrane and also the thylakoids traversing the pyrenoids. These granules have a highly ordered spatial arrangement on the lamellar membranes (Gantt and Conti, 1966)

Mature cells of *Porphyridium cruentum* contain a single, well developed chloroplast that occupies a major portion of the cell (Fig. 6) These chloroplasts are enclosed by a double limiting membrane of about 200 A^0. Certain chloroplast extensions which are in direct continuation of photosynthetic lamellae protrude out from the chloroplasts. Chloroplast lamellae are double membraned and are arranged parallel to each other with a spacing of 500 A^0 or more. Each lamella is 200 A^0 wide and abundant in peripheral area of the chloroplast. Around the pyrenoids, parallel arrangement of lamellae is less and they extend also into and through the pyrenoid. Adjacent chloroplast lamellae in chloroplast extensions show interconnections (Gantt and Conti, 1965).

In *Porphyridium cruentum* both phycobilins and chlorophyll are located in chloroplast lamellae. These granules have been considered as probable sites of phycobilin concentration. Size of these granules varies from less than 320 A^0 to 400 A^0. The size, however, to some extent, depends on the culture conditions and age of the cell. The center to center distance between these granules is 400-500 A^0. These granules are spherical to oblong and have rounded projections which suggest that they may be composed of sub-units having ring-like appearance (Fig. 7. A,B; Gantt and Conti , 1965)

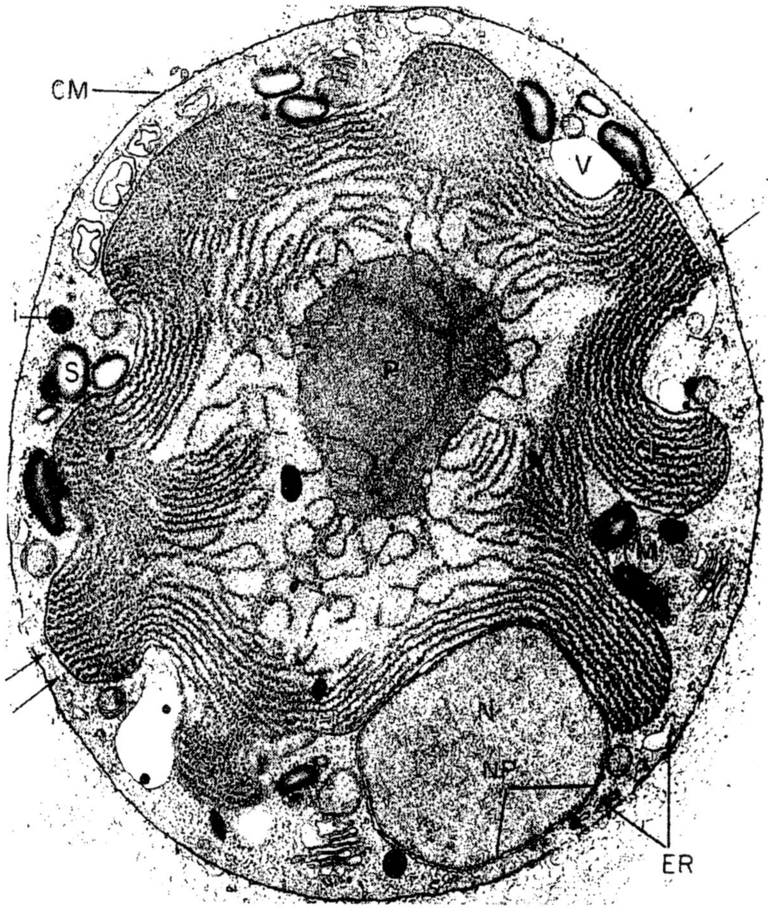

Fig. 6. Ultrastructure of *Porphyridium cruentum* (after Gantt and Conti, 1965).

(CM, cell membrane; CL, chloroplast lamellae; ER, endoplasmic reticulum; G, golgi bodies; M, mitochondria; N, nucleus; NP, nuclear pores; P, pyrenoid; S, starch grains; V, vacuoles).

Electron micrograph of vegetative cell showing a large chloroplast, with opaque pyrenoid, occupying the central portion. Double membrane lamellae do not fuse with the chloroplast membrane.

Michael *et al.*, (1990) through immuno-electron microscopy studies revealed that phycoerythrin, which is an important structural component of the phycobilisomes of red algae is absent from the pyrenoid of *Porphyridium cruentum* and thus the thylakoid lamellae traversing the pyrenoid of this alga also lack phycobilisomes.

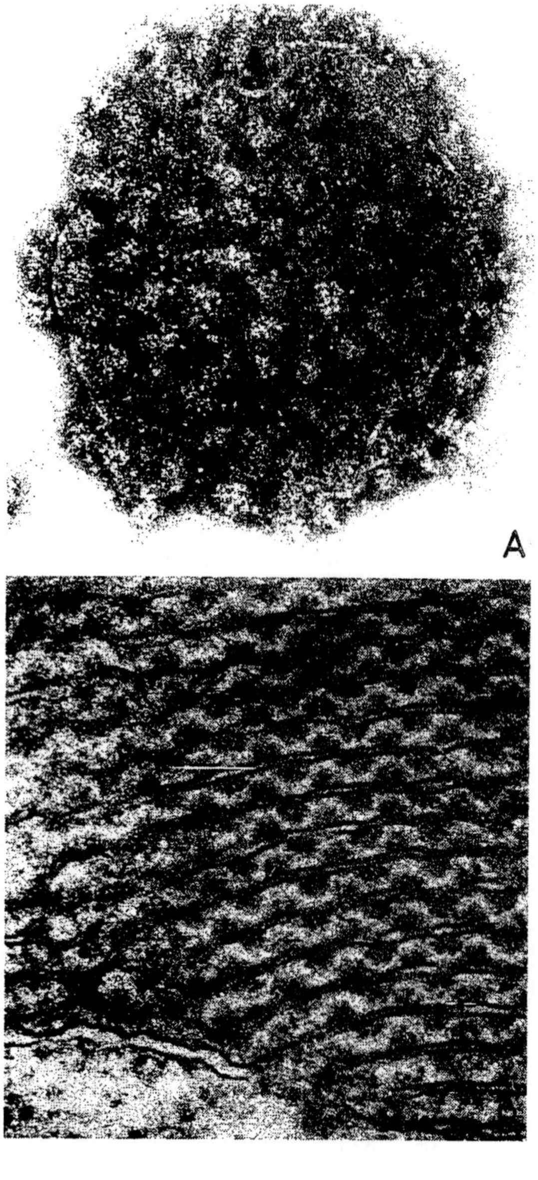

Fig. 7 A,B Ultrastructure of chloroplast of *Porphyridium cruentum*
(after Gantt and Conti, 1966)

A Section of a chloroplast. Granules with small rounded projections are attached to lamellae.

B Same, showing small distinct subunits. Membrane pairs of chloroplast lamellae stick together even after fragmentation of chloroplast.

Lange *et al.*, (1990), using Freeze-fracture techniques conducted extensive studies on the supramolecular structure of Photosystem II-Phycobilisome complexes of *Porphyridium cruentum* and concluded that the hemiellipsoidal phycobilisomes found here are 20 nm thick and center to center distance is 30-40 nm. in contrast to the hemidiscoidal phycobilisomes of red algae. These phycobilisomes are cleaved by a central faint furrow, parallel to the long axis from top to base. Freeze-fracture studies revealed the phycobilisomes as 18-22 nm thick, 35-40 nm high, 55-65 nm long structures.

Negatively stained isolated phycobilisomes in face view show a semi-circular outline, and three cylindrical units of 10-12 nm diameter joined to form a central triangular core through which biliprotein rods radiate in semicircular pattern. Each rod has three or four hexameric biliprotein aggregates.

Freeze-fractured thylakoid membranes of *Porphyridium cruentum* chloroplast reveals two types of fractured faces; Exoplasmic fracture face (EF) and Protoplasmic fracture face (PF).

The PF freeze-fracture particles are densly packed 8.5 - 0.9 and 10.5 nm structures arranged in rows on the thylakoids. These rows are separated by particle free grooves. The EF particle rows fit into these spaces in the native membrane. These EF particles are 10 nm in size but may get aggregated with their long face resulting in particles of size 10x20 nm. Four such 10 nm particles aggregate to form a package which can be resolved (Fig. 8). EF particles are correlated with PS II complexes and a 1 : 1 ratio of the tetrameric particle packages and phycobilisomes is inferred. Each phycobilisome is thus supposed to bind four PS II particles (Lange *et al.*, 1990)

Porphyra umbilicalis shows both ellipsoidal and hemidiscoidal phycobilisomes that are well-coupled, Negative staining revealed the PBSI and PBS2 phycobilisomes. PBS2 particles appear ellipsoidally shaped 60 nm long and 30 nm wide with poor structural details. The PBS1 particles are made of several radial projections with stacked discs approximately 12 nm in diameter and 6 nm thick associated to the central portion (Fig. 9 A,B; Algarra *et al.*, 1990).

Freeze-fracture technique used to study the organization of photosynthetic membranes of thallus stage of *Porphyra leucosticta* show single lobed chloroplast, lacking peripheral thylakoid and the paralled thylakoids terminating close to chloroplast envelope (Fig. 9C; Tsekos and Reiss, 1990). The two fracture faces (PF and EF) of thylakoid membrane are structured asymmetrically as is the case with red algae, but the density of PF particles of *P. leucosticta* differs from that of *P. umbilicalis*.

Both PF and EF particles fall in two size classes namely, a major size class (10-11 nm) and a minor size class (14-15 nm). These two particle sizes may exist due to variable sizes of PS 1 and PS II complexes in thylakoid membranes of red algae as the PF and EF particles are considered equivalent to Photosystem 1 and Photosystem II, respectively (Tsekos and Reiss, 1990).

Compsopogon coeruleus shows the presence of a single, complex, interconnected parietal chloroplast which appears as numerous discoid or ribbon shaped structures. The older cells have widely spaced chloroplasts but are still interconnected by narrow extensions that contain only a single continuous encircling thylakoid simulating a proplastid. The chloroplasts central stromal region lack thylakoids but show electron-transparent areas that could be interpreted as containing DNA (Fig. 10 A, B; Gantt *et al.*, 1986). In *Compsopogon coeruleus* phycobilisomes appear hemidiscoidal and closely spaced.

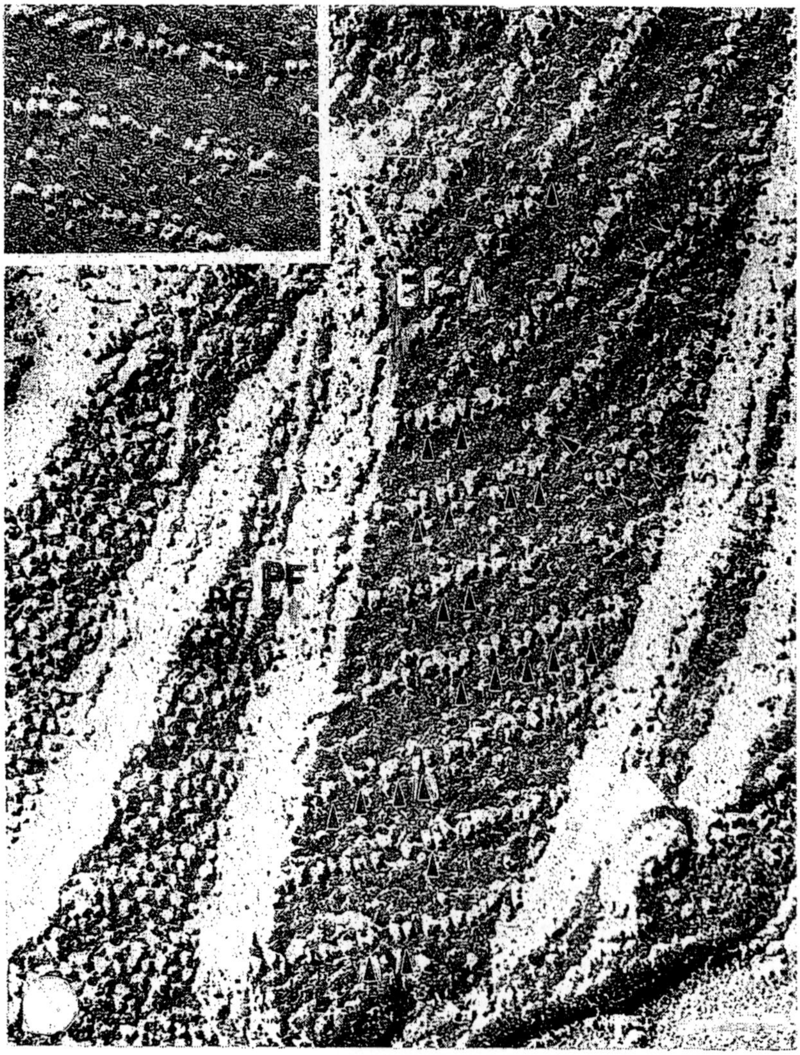

Fig. 8. Ultrastructure of chloroplast of *Porphyridium cruentum* (after Lange *et al.*, 1990)

PF and EF, Freeze-fracture particles arranged in rows on the thylakoids. The particles may be arranged in aggregations of two 10 x 20 nm particles (arrowheads). Each particle is resolved into two 10 nm particles (arrows). A few rows show conditions of slight dissociation with particle clusters containing mainly three and four 10 nm particles (inset)

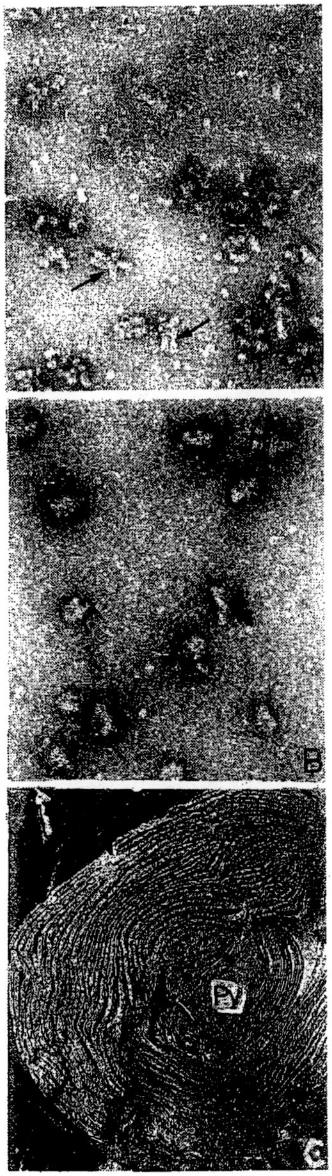

Fig. 9. A-C. Ultrastructure of chloroplast of *Porphyra*

(PY, Pyrenoid)

A,B *P. umbilicalis* showing PBSI, and PBS2 types of phycobilisome particles (negative staining). The former particles show radial projections attached to central core (arrows) (after Algarra *et al.*, 1990).

C Freeze-fracture chloroplast of *P.leucosticta* showing thylakoid membranes covering most of the pyrenoid body as well as transecting the body of pyrenoid (after Tsekos and Reiss, 1990).

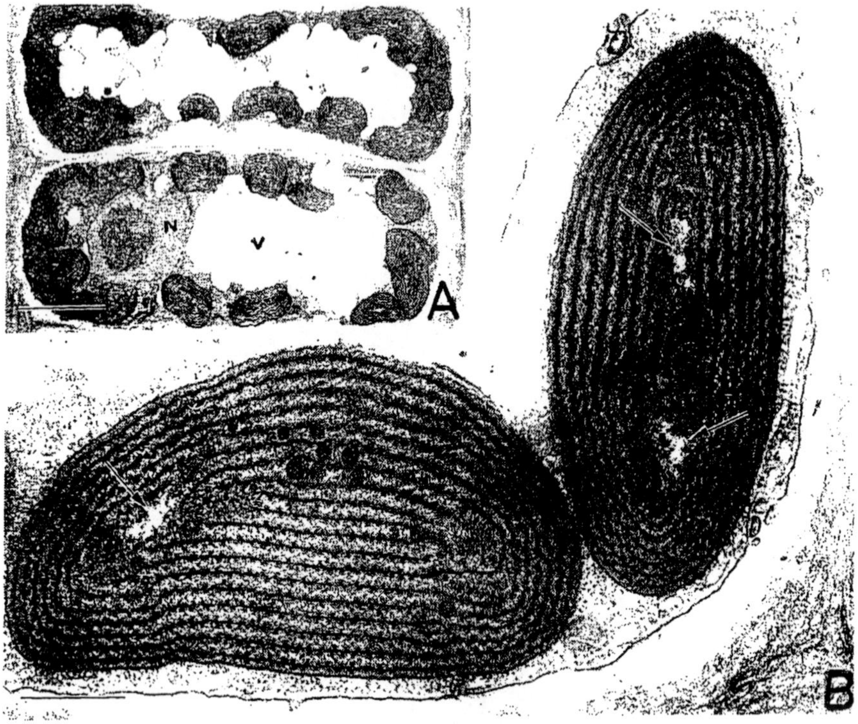

Fig. 10 A,B. Electron micrographs of *Compsopogon coeruleus* (after Gantt et al., 1986).

A Uncorticated axial cells containing at least eleven separate chloroplasts.

B Chloroplasts showing thylakoid membrane. The central stromal region, lacks thylakoids, which could be the genophore (arrows). Distinct phycobilisomes are seen on the thylakoid membranes.

Bangia, reveals structural differentiation of the fracture-faces of inner and outer chloroplast envelope membrane indicating separate evolutionary origin of the two membranes (Pueschel, 1990).

Cole and Sheath (1980) studied spermatiogenesis in *Bangia,* and reported the sequential changes that occur in the chloroplast. A large lobed structure with parallel, single, evenly spaced thylakoids and a central pyrenoid is converted into a small, discoid structure with a few distorted, dialated thylakoids with many platoglobuli but lacks pyrenoid. These changes are generally associated with degeneration of aging. Mature spermatia lack floridean starch, but have reduced chloroplasts containing many plastoglobuli. Those cells that differentiate into speramatia show thylakoid reticulation during chloroplast transformation. This change is initiated within the pyrenoid and ultimately extends to rest of chloroplast.

The unicellular alga *Rhodella* was distinguished from other unicellular genera by presence of an irregularly shaped pyrenoid not penetrated by thylakoids and also not separated from plastid envelope by thylakoids. In *R. reticulata* aggregations of plastoglobuli that occur at the periphery of plastid lobes is believed to be identical to the orange spots visible in living cell. These aggregations are called "eye spots" (Pueschel, 1990).

MITOCHONDRIA

Investigations on the red algal genera reveal that the number and morphology of the mitochondria show a great variability from one cell type to another and the number of mitochondria actually present in a cell is often less than expected.

Broody and Vatter (1959) reported the absence of mitochondria in the unicellular alga *Porphyridium cruentum*. Later in this taxon all the eukaryotic organelles including the mitochondria was reported which, led to further studies (Gantt and Conti, 1965). Each mitochondrion is a long, slender, double membranous structure, which may be branched. Tubular cristae arise from the inner membrane.Mitochondria are postitioned between the starch granules, chloroplast extensions and vacuoles. Due to this characteristic positioning and slender shape of the mitochondria they are hard to identify and can be easily over looked in sections (Fig. 11 A).

Studies regarding the organization of the mitochondrial system in *Rhodella reticulata* revealed that mitochondria are generally present in outer regions of the cell, though a few branches may sometimes be found in the cell interior. The mitochondrion varies greatly in width where a few areas are about 50 nm wide.

In *Rhodella reticulata,* serial sectioning and three dimensional reconstruction of mitochondrial system revealed one large, anastomosing mitochondrion located at the cell periphery. Some cells contain a few small mitochondria that are not connected to the main body of mitochondria system. In one cell, one to three large branched mitochondria may, however, be present (Fig. 11 B; Broadwater and Scott, 1986).

The *Bangia* mitochondria show characteristic tubular cristae, and are situated close to the dictyosomes. During the process of spermatium differentiation the spermatium shows many large mitochondria, that are often present near the chloroplast (Cole and Sheath, 1980).

The above studies clearly indicate that all the Bangiophycean genera show the presence of mitochondria identical to those encountered in the Floridiophycean taxa. The effect of a particular stage in cell cycle on the number and size of mitochondria is, however, not very clear in Bangiophycideae.

GOLGI APPARATUS/DICTYOSOMES

In Bangioideae Golgi bodies are generally composed of 4 to 15 stacked cisternae that are arranged parallel to the cell membrane surrounding the stacked Golgi cisternae are tubules which are circular to elliptical in profile with diameters ranging from 450 to 550 A^0.These tubules have a smaller average diameter than the inflated Golgi cisternae and are arranged at right angles to the cisternae. They may be components or derived from Golgi bodies as they are concentrated more around the Golgi apparatus.

The peripherally situated Golgi bodies (at the forming face) are characteristically associated with an arc–shaped endoplasmic reticulum. This endoplasmic reticulum appears to produce many electron-dense small vesicles which lie close to the cisternae at the forming and lateral faces of the Golgi body. At the maturing face, the cisternae become distendend and contain fine fibrillar

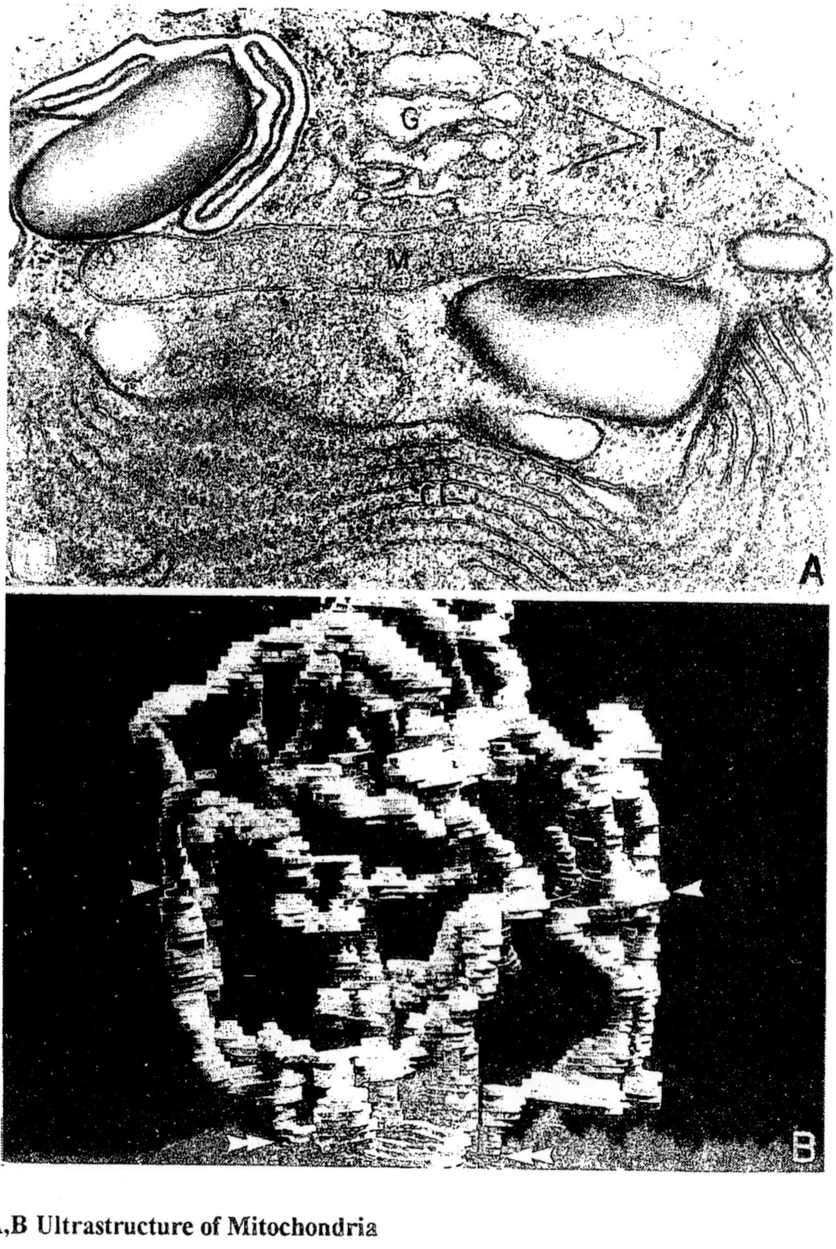

Fig. 11 A,B Ultrastructure of Mitochondria

(CL, Chloroplast lamellae; G, Golgi body; M, Mitochondria; T, Tubular elements)

A Longisection of double membranous mitochondrion of *Porphyridium cruentum*. The inner membrane is connected to the tubular cristae (after Gantt and Conti, 1965).

B Lateral view of reconstructed mitochondrial system of *Rhodella reticulata* (after Broadwater and Scott, 1986).

materials. These cisternae are cut off from the maturing face and they round off to give membrane bound vesicles which are seen near the Golgi surface and between Golgi bodies and plasmalemma (Fig. 12 A,B; Evans et al., 1974).

In Bangia, the Golgi bodies, at the cis face are generally found to coexist with other organelles like endoplasmic reticulum and mitochondria. *Porphyridium* is also known to show Golgi cisternae-dictyosome-mitochondria associations (Gantt and Conti, 1965).

Rhodella cyanea and *R. reticulata* are exceptions in that they show cis Golgi associated closely with the nucleus. There is no endoplasmic reticulum involvement (Deason *et al.*, 1983).

In *Rhodella maculata* (Evans, 1970), *R. viloacea* (Cole and Sheath, 1990), *Smithora* (McBride and Cole, 1971), *Rhodochaete* and *Compsopogon* (Pueschel and Magne, 1987) only endoplasmic reticulum is associated with Golgi bodies. Ultrastructural autoradiography in *Rhodella maculata* showed that sulfation of mucilage occurs in Golgi apparatus and by radioactive labels it was calculated that it took less than 2 hours.

Similar studies were performed in *Porphyridium* by incorporation of carbon-label where the process was completed within 30 minutes (Ramus, 1972).

In *Porphyridium,* morphometric analysis of Golgi apparatus in stationary phase versus log phase cells revealed that the number and the size of dictyosomes were approximately three times greater in log phase cells than the stationary phase cells. Contradictory observations between the synthesis and the secretion rates were obtained. The amount of secretion was less when Golgi volume was high. This was mainly due to the accumulation of product synthesized during log phase which was not secreted until the stationary phase. Golgi bodies are common in active vegetative cells, but they are specially prominent during sporogenesis and mucilage formation.

MICROBODIES

True microbodies have not been reported in Rhodophyceae but membrane bound peroxidase has been reported in *Cystoclonium* (Murphy and Heocha, 1973). Oakley and Dodge (1973) reported in *Porphyridium purpureum,* microbody-like organelles, that are 0.2 - 0.4 μ in diameter, spherical to slightly ovoid in shape, about 6-7 nm in width, and bound by a single membrane. These microbody-like organelles, however, lack crystalline inclusions usually associated with microbodies. In *Porphyridium,* the microbodies when incubated with gluteraldehyde and DAB show intense staining which may be due to the high peroxidase concentration. Cole and Sheath (1980) reported in *Bangia,* that during spermatia differentiation, there is occurrence of microbody-like organelles near the nucleus. Similar structures are known near the mitotic spindle poles of *Porphyridium.*

GERMINATING CARPOSPORES

Ultrastructural studies on the developmental changes occuring in germination of carpospores to produce the conchocelis phase were carried out in *Porphyra* (Pueschel and Cole, 1985). Vegetative cells of *P. variegata* containing a single stellate chloroplast with a large central pyrenoid show packets of 8-16 interspersed carposporangia. Each carpospore possesses abundant starch grains and a single chloroplast with arms that radiate from the region of the central pyrenoid. Thylakoids penetrate the pyrenoid but peripheral encircling thylakoids are rare (Fig. 13 A).

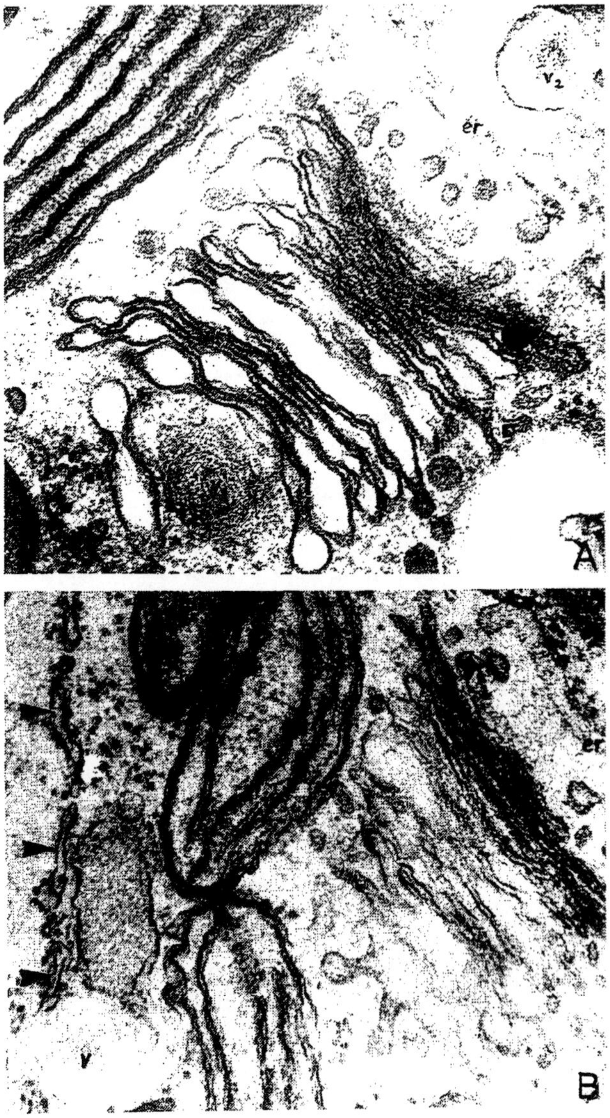

Fig. 12 A,B. Ultrastructure of Golgi apparatus in *Rhodella maculata*
(after Evans et al., 1974)
(er, endoplasmic reticulum; v, vesicle)

A Section through a Golgi body showing fibrillar material in cisternae at the maturing face. A mature mucilage vesicle (V_1) is closely pressed to the maturing face and another (V_2) lies free in the cytoplasm. Many small vesicles occur in forming and lateral faces of the Golgi body.

B Part of the periphery of a cell showing a Golgi body with associated endoplasmic reticulum and a vesicle with fibrillar content passing between the subplasmalemmal network of endoplasmic reticulum (arrowheads) in transit to the plasmalemma.

During carpospore differentiation, two distinct types of vesicles with electron-dense fibrils in an electron transparent matrix are seen. The small vesicles are 0.3 µm in diameter, abundant, and spherical with dense cores. The large vesicles fuse among themselves which result in variable sizes and shapes (Fig. 13B). Similar fibrillar vesicles are also reported in *Porphyra pseudolinearis* (Kito, 1978).

Emergence of a unipolar germ-tube marked the begining of carpospore germination. A few spores, in addition to the germ-tube, also formed small protrusions, which contained extracellular material. The vesicles, which were abundant in mature unattached spores, were absent in attached spores (Fig. 13B) and sporelings. Kito (1978), however, found large fibrillar vesicles still present in the germinating spores of *P. pseudolinearis*. The surface of the attached spores and the germ-tube was coated with a layer of moderately electron-dense material which could act as the adhesive mucilage (Fig. 14). It was suggested that when spores settle, the vesicles discharge a substance which fix them firmly to the substrate.

The adhesive material ocassionally breaks and curves due to the emergence of germ-tube. The germ-tube wall material is thicker but the tip has relatively thin wall.

Conchocelis phase has the presence of pit-plugs whose formation is concomittant with the septum formation which when formed is contiguous with the spore wall. The septum may be formed either prior to or after the emergence of germ-tube. The variability of septum formation is far greater in *Porphyra variegata* as compared to *P. schizophylla* where evidence of a septa in spore bodies is lacking (Boney, 1978). The cleavage of spore body was roughly perpendicular to the germ-tube axis. After germ-tube formation, the spores remain either spherical or become elongate.

In contrast to the unattached spores the germinating spore cytoplasm showed many vaculoes which push the other organelles toward periphery. During spore germination, there is a decrease in starch content and most of it was retained in spore body. The released carpospore chloroplast showed, in addition to a single large pyrenoid, a few, new pyrenoids (Fig. 15). The sporelings revealed electron-dense lipids and their muclei had extremely irregular outline (Fig. 15).

SPERMATIOGENESIS IN *PORPHYRA* AND *BANGIA*

In Bangiophyceae, the ultrastructural details on the process of spermatiogenesis are limited to two genera namely *Porphyra* (Hawkes, 1978) and *Bangia* (Cole and Sheath, 1980).

In *Porphyra gardneri*, the marginal vegetative cells produce a new wall layer before undergoing division to form the spermatium. The vegetative cell contains a stellate chloroplast with centrally located pyrenoid and peripheral nucleus. The immature spermatia do not differ much from the vegetative cells, retain well developed chloroplast and floridean starch. In the later stages, a few large and small fibrous vesicles appear. The smaller vesicles contribute to the formation of large vesicles. The chloroplast becomes reduced in size and gives a degenerate appearance. Mitochondria are present and nuclei have nuclear membranes. Before sperm liberation, the large fibrous vesicles become extracytoplasmic and nuclear membrane begins to degenerate. Mature spermatium contains highly condensed nucleus bereft of nuclear membrane and has abundant small fibrous vesicles and reduced chloroplast. The spermatium, at the time of release, is 3-5 µm in diameter; lacks a cell wall, is surrounded by a plasmamembrane and contains both large and small fibrous vesicles.

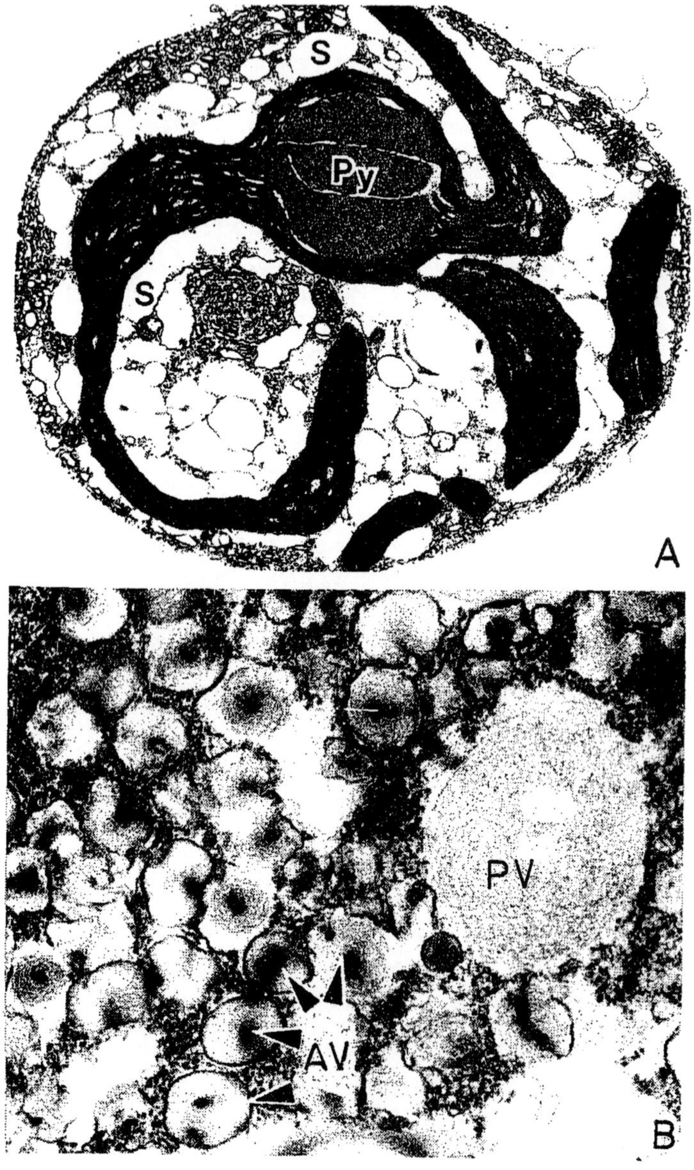

Fig. 13 A,B. Ultrastructure of carpospore in *Porphyra variegata*.
(after Pueschel and Cole, 1985)

(AV, Adhesive vesicles; FV, Fibrillar vesicles; Py, Pyrenoid; S, Starch grains).

A A prematurely released carpospore showing stellate plastid, single pyrenoid and numerous starch grains.

B Magnified view of unattached mature carpospore showing small adhesive vesicles and large fibrillar vesicles.)

Fig. 14. Ultrastructure of an attached carpospore of *Porphyra variegata*
(after Pueschel and Cole, 1985)

(Mu, Mucilage)

Carpospore surrounded by a thin wall and mucilage. Transections show a plasmalemmal invagination filled with extracellular mucilage. Encircling peripheral thylakoids are absent from the ends of the chloroplast arm (arrows).

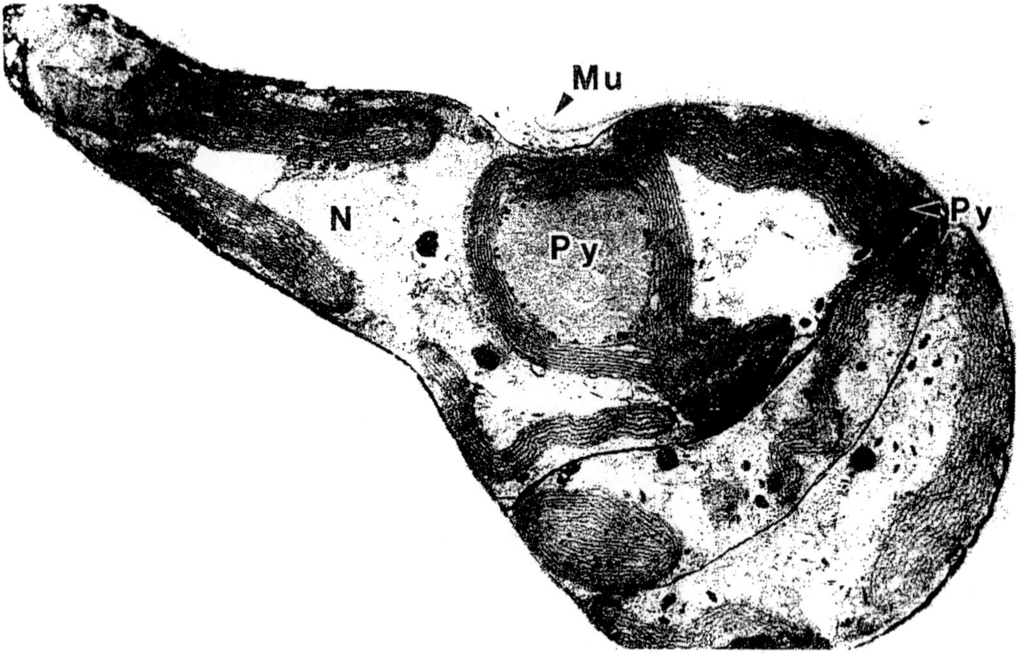

Fig. 15. Ultrastructure of germinating carpospore in *Porphyra variegata*
(after Pueshel and Cole, 1985).

(Mu. Mucilage layer; N, nucleus; Py, Pyrenoid).

Germinating carpospore showing a broad germ-tube and twice divided original spore-body. Large original pyrenoid and smaller secondary pyrenoid at the end of chloroplast-arm can be seen. Mucilage layer coats only the spore-body but not the germ-tube.

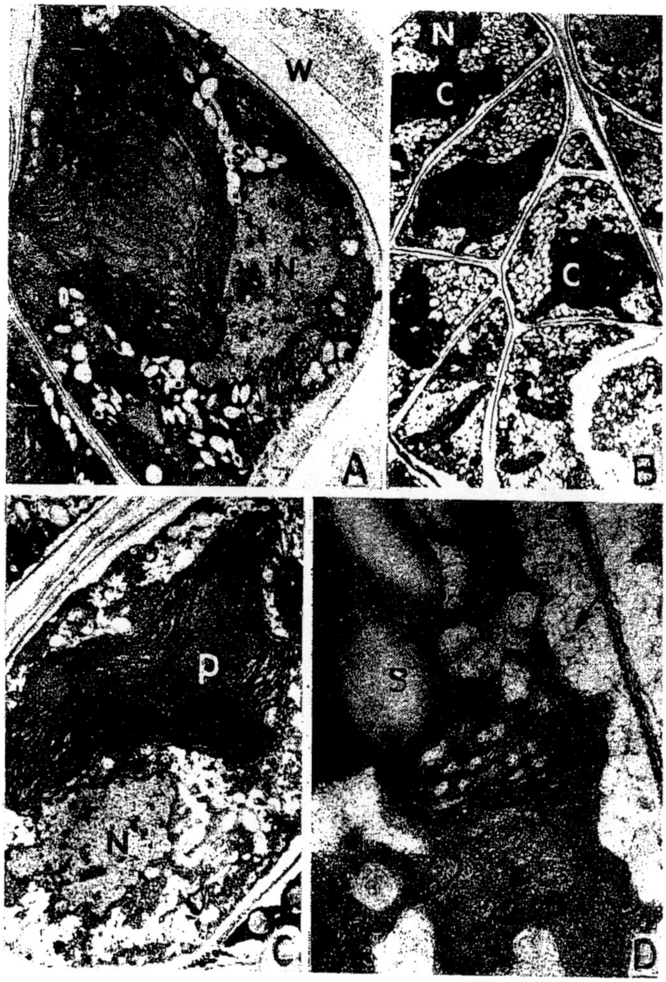

Fig. 16 A-D. Ultrastructure of spermatiogenesis in *Bangia atropurpurea*
(after Cole and Sheath, 1980)

(C, chloroplast; D, dictyosome; M, mitochondria; N, nucleus; P, pyrenoid; S, starch grains; CW, cell wall)

A A vegetative cell showing the large lobed chloroplast with central pyrenoid and a number of plastoglobuli; large nucleus adjacent to the chloroplast contains heterochromatic bodies; a number of floridean starch granules; several mitochondria and a thick cell wall.

B-D Changes during the transformation of a vegetative cell into spermatangia. A packet of cells is formed which shows associaton of nucleus and chloroplasts (B). Each cell in the packet shows a reduced chloroplast with large pyrenoid. Chloroplast envelope begins to separate from the nucleus and starch granules and vesicles are located between them at one side (C) and dictyosome-mitochondria association. Cytoplasm shows large starch granules and fibrous vesicles. Fibrillar material (large arrow) is present outside the plasmamembrane and endoplasmic reticulum is parallel to plasmamembrane (small arrows) (D).

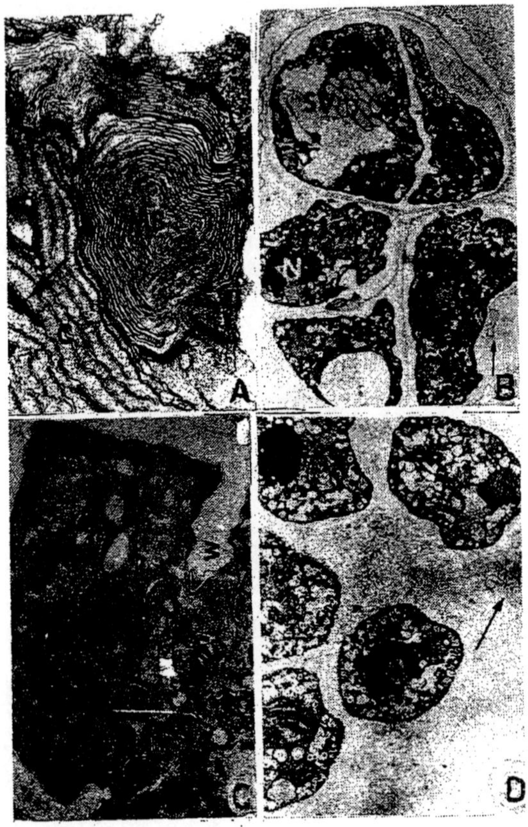

Fig. 17 A-D. Ultrastructure of spermatiogenesis in *Bangia atropurpurea*
(after Cole and Sheath, 1980).

(C, chloroplast; MB, multilamellar body; M, mitochondria; N, nucleus; SV, spermatial vesicle; W, wall material)

A Large multilamellar body adjacent to the chloroplast is formed in the cell packets, during spermatangia formation. A few multilamellar bodies (large arrow) may also be associated with thylakoids.

B After the formation of spermatangial packets there is marked differentiation of cell organelles in the major differentiation phase and results in cytokinesis in differentiating cell. Wall material is centripetally deposited. Double membrane septum is formed across the cell. Large number of multilamellar bodies are present near the septum. Mitochondria and number of microbody-like structures can be seen near plasmamembrane.

C Spermatangial packets show large spermatial vesicles. In the lower right, cell vesicle has released fibrous contents (arrow). The cells have small nuclei and are bereft of floridean starch.

D Several irregularly shaped spermatia are released as intercellular wall breaks down. Some fibrillar material from spermatial vesicle can be seen between the cells (arrow). A few cell structures like, discoid chloroplast, nucleus and a number of small vesicles are evident.

Similar ultrastructural transformations in the major organelles were also reported for *Bangia atropurpurea*. The process of spermatiogenesis occurs in three stages and each stage has specific ultrastructural features. The vegetative cell divides to form spermatangia. This is followed by a period of major differentiation within the cells and spermatia are produced. In the final stage the walls around the spermatia dissolve, the outer wall around the packet breaks and the spermatia are released (Figs. 16 A-D, 17 A-D ; Cole and Sheath, 1980).

The mature spermatium of *Bangia* contains very little endoplasmic reticulum and lacks dictyosomes. However, during certain phases of spermatiogenesis multilamellar bodies and microbody-like structures occur within the cells.

METABOLIC BYPRODUCTS

POLYSACCHARIDES

Two main classes of polysaccharides have been identified in cell walls of red algae (a) hot water soluble polymers which consist of galactose and/or modified galactose units. (b) the insoluble, structural polymers that are fibrillar in nature and consist of 1, 3-linked xylans and 1,4-linked mannans. Many marine Rhodophycean taxa accumulate, in their cell walls or intercellular regions, large amounts of sulphated polysaccharides that have the same basic structure namely a chain of galactose units linked alternately at α - 1,3 and β-1,4 positions.

The information available on cell wall polysaccharides of the Bangiophyceae is less and most of the reports emanate for *Porphyra* and *Bangia*.

Anderson *et al.*, (1965) showed that the polysaccharide Porphyran (a sulphated galactan) isolated from *Porphyra* sp. is structurally related to agarose. In agarose 1,4-linked galactose unit occurs as 3,6-anhydride of the L-sugar, while the 1,3-linked galactose occurs as unmodified D-isomer. Porphyran also has L-galactose units linked by 1,4-linkage but here only a part of this sugar occurs as 3,6-anhydride and the 1,3-linked units are partially 6-0 methylated (Fig. 18). Porphyran can be converted into methylated agarose by a series of reactions.

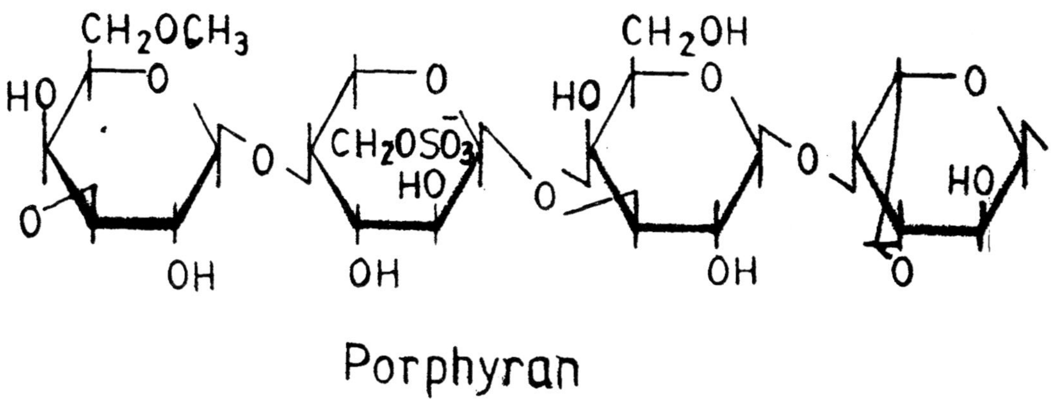

Fig. 18. Chemical structure of Porphyran, a sulphated galactan, isolated from *Porphyra* sp. shows L-galactose units linked by 1,4-linkage.

Xylose is a known component of marine algal polysaccharides that has at least two distinct structures (a) in the form of a separate xylan or combined in heteropolysaccharide, neither of which is a skeletal material, and in which both 1,3 and 1,4 -linkages occur (b) as a linear polysaccharide in the cell wall in which it is either completely 1,3-linked, or completely 1,4-linked.

In *Porphyra umbilicalis* cell wall alkali extractable xylan microfibrils gave an X-ray diffraction pattern of β-1,3, linked xylan whch is almost equivalent to a pure xylan except for traces of a glucan as contaminant (Turvey and Williams, 1970).

Methylation analysis of cell walls of conchocelis and thallus phases of *Porphyra tenera* confirmed the presence of cellulose microfibrils in conchocelis phase in contrast to xylan microfibrils in the thallus (Mukai *et al.*, 1981).

In *Bangia,* the cell walls of the conchocelis phase differ from the gametophyte in that they contain cellulose component. The sulphated galactans of gametophytic phase of *B. fuscopurpurea* are similar in structure to porphyran and that the major structural polysaccharides resemble those of *Porphyra* (Gretz *et al.*, 1982).

Cellulose was later demonstrated in conchocclis phases of *Porphyra leucosticta, P. umbilicalis,* and *Bangia atropurpurea* (Gretz *et al.*, 1980, 82). The quantities reported were 3.8%, 5.5% and 11.8% of the purified cell wall fractions for *B. atropurpurea, P. leucosticta* and *P. tenera,* respectively. No cellulose was, however, reported in generic phases of either *P. tenera* or *Bangia atropurpurea.*

Cuticle of *Bangia atropurpurea* filaments showed the presence of neutral polysaccharides. Acidic polysaccharides were present in the outer wall materials (Cole *et al.*, 1985).

There is difference in cell wall proteins and polysaccharides of conchocelis and thallus phases of dimorphic alga *Porphyra.* This may be as an adaptation to the rather unusual ecological niche of the conchocelis stage, where it is able to bore into shells and calcareous rocks, in comparison to the leafy, *Porphyra* thallus.

The low molecular weight alcohol-soluble carbohydrates namely O-a-D-galactopyranosyl - (1- - >2) - glycerol ('floridoside') and 0-a-D-Galactopyranosyl - (1 - ->1) - glycerol ('isofloridoside'). have been reported in *Porphyra purpurea.* Floridoside is not only a constitutive photosynthetic product but may also be involved in osmotic adaptation (Reed *et al.*, 1980).

Unicellular red alga *Rhodella maculata* possesses a thick mucilaginous envelope outside the plasmalemma which is similar in composition to that of *Porphyridium aeruginium* (Ramus, 1972). The mucilage in *Rhodella maculata* comprises about 50% carbohydrate, 16% protein and 10% sulphate. The major sugars are xylose, uronic acid, a small amount of galactose, traces of glucose and in addition two reducing substances (Evans *et al.*, 1974).

Glycoproteins contain a group of polysaccharides attached to their protein parts and these have been reported rarely from the red algae. Glycoprotein was extracted from the extracellular mucilage of *Porphyridium purpureum* (Craigie, 1990).

Kaska *et al.*, (1988) fractionated marker glycoprotein from differentiated regions of *Porphyra perforata* thallus. The study indicated that areas of the thallus which differ in morphology, susceptibility to enzyme digestion and regeneration potential differ also in their glycoprotein composition.

The cell wall polysaccharides of other members of Porphyridiales differ significantly from those of *Porphyridium* and *Rhodella* in having less uronic acid and sulphate besides insignificant methoxyl contents. Rest of the basic composition may vary as follows :

Chroodactylon ornata (Asterocystis) produces a (1-->3) - linked galactan that lacks anhydrogalactose; *Stylonema alsidii* (as *Goniotrichum elegans*) yields an alkali soluble galactan having (1-->3), (1--->4) and (1--->6) linkages, and contains anhydrogalactose but lacks sulphate. *Rhodosorus marinus* contains an unsulphated (1-->4) linked xylan with terminal galactose groups.

LIPIDS AND ESSENTIAL OILS

The unicellular marine red alga *Porphyridium cruentum* contains arachidonic acid and eicosapentaenoic acid (Nichols and Appelby, 1969; Neyberg, 1985) both of which are relatively rare polyunsaturated fatty acids. Relative amounts of the lipids and fatty acids in *P. cruentum* was found to be a function of light intensity, temperature, pH and salinity (Cohen *et al.*, 1988).

Cultivated conchocelis filaments of *Porphyra tenera,* used as food in Japan, yield an essential oil with major volatile compounds : cubenol (4.2%), phytol (19.2%), palmitic acid (13.2%) and long-chain fatty aldehydes (16.9%), tetradecanal (TD), pentadecanal (PD), (8z, 11Z) - heptadecadienal (HDD) and (8Z) heptadecanal (HD). (Kajiwara *et al,* 1990)

ENZYMES AND GROWTH SUBSTANCES

Floridoside (a-D.galactopyranosyl 1---->2' - glycerol) is the major photosynthetic product in most red algae. The key enzyme in its biosynthesis, floridoside phosphate synthase (FPS, UDP - galactose, sn-3-glycerol phosphate : 1---->2' galactosyltransferase) was purified from *Porphyra perforata* (Meng and Srivastava, 1991). This enzyme has a high substrate specificity, only UDP-galactose and sn-3-glycerol phosphate could be used as substrates, no cofactor or activator was found for the reaction.

Two types of cysteine synthase were purified from *Porphyra yezonensis* (Fujimori *et al.*, 1991).

Cytokinins which have important physiological roles in flowering plants have only a limited information on their function in algae. Cytokinins have been shown to promote growth in conchocelis stage of *Porphyra* (Iwasaki, 1965). The thalli of *Porphyra,* possessing mature spermatangia and carpogonia, yielded 0.6 ug.kg^{-1} fresh weight of isopentyladenosine. This cytokinin may be originating from tRNA and is present at very low levels (Zhang *et al.*, 1991).

IN VITRO STUDIES

PROTOPLAST ISOLATION AND REGENERATION

Protoplasts, plant cells stripped of their walls, isolated from different tissues of higher plants when induced to promote callus are used as tools both in physiological and cytological studies besides research in protoplast fusion to produce somatic hybrids. In algae the report of protoplast fusion and regeneration date back to 1981 when a few multicellular algae yielded protoplasts that produced multicellular thalli. However, protoplast isolation and fusion from seaweeds is of late gaining momentum.

In *Porphyra,* during blade differentiation, four distinct areas are encountered. Regeneration rates of the isolated protoplasts from these areas depended upon their morphological types, and are found to be the highest in vegetative and reproductive regions that are away from the holdfast. This study helped to identify the area of maximum regeneration potential, which could be economically useful and commercially exploited for farming (Polne-Fuller and Gibor, 1984). This induced other workers like Fujita and Migita (1987) to fuse protoplasts isolated from wild-type and green-type thalli of *Porphyra yezoensis* and trace the further development of fusion products that differentiated into callus–like cell aggregates. These chimeral thalli are irregularly variegated with greenish and reddish-purple cell groups and upon maturity produced both carpogonia and spermatia. Carpospores grew into greenish-conchocelis-phase. Conchospores released from either conchosporangia of wild type or from greenish type could be distinguished by their hue. In their study they obtained only distinct green plants.

Protoplasts isolated from spermatangia and carposporangia of *Porphyra haitanensis* could not regenerate thalli but they developed spermatia and carpospores, respectively. The protoplasts from vegetative area, however, could regenerate thalli but a small part of them formed carpogonia or spermatia. The older parent thalli regenerated increased number of spermatia or carpogonia as compared to vegetative thalli (Dai *et al.*, 1988).

Freshwater *et al.*, (1989) isolated spheroplast from *Porphyra carolinensis* and reported methods for vegetative propagation under nonaxenic conditions. Isolated cells demonstrated a high degree of survival and totipotency regenerating leafy-blade phase.

Studies were also undertaken to fuse protoplasts of several species of *Porphyra* to generate somatic hybids. Developmental morphology of heteroplasmic fusion products of *P. yezoensis* and *P. pseudolinearis* showed that heterokaryons produced are similar to one of the other of the parental types. This might have been due to gradual elimination of one nucleus and chloroplast of one of the parents during the course of development to adult plant. A few chimeral thalli may, however, result and the dual chromosome count in them suggests the coexistence of both nuclei at initial stages, which later might have in the same thallus produced to different chromosome counts (Fujito and saito, 1990).

Viable protoplasts, capable of cell wall formation, sustained cell division and differentiation, are isolated from epiphytic marine alga *Porphyra nereocystis* (Fig. 19 A,B; Waaland *et al.*, 1990). Under uniform culture conditions, regenerates from these protoplasts exhibited three different pathways : (a) directly from filaments resembling the conchocelis phase (Fig. 19 C,D). (b) from calli with relatively thick walled, pigmented cells; and (c) indirectly form blades from the edges of these calli.

Tissue and cell culture studies on *Porphyra,* to develop micropropagation techniques, were conducted (Xue–Wu and Gordon, 1987). The callus cells could be maintained long enough to show early stages of regeneration, some cells even producing rhizoids and a few dividing equitorially into two cells. Except *Porphyra,* there have been no reports on protoplast fusion and regeneration in any other member of Bangioideae. Culture studies of many unicellular members like *Porphyridium* and *Rhodella* have been conducted.

Nori is extensively cultured and cultivated in Japan for food. Its cultivation was affected by pathogenic microorganisms and therefore, it was desired that strains of cultivated *Porphyra* be improved through genetic engineering. Thus, the knowledge and technique to isolate seaweed

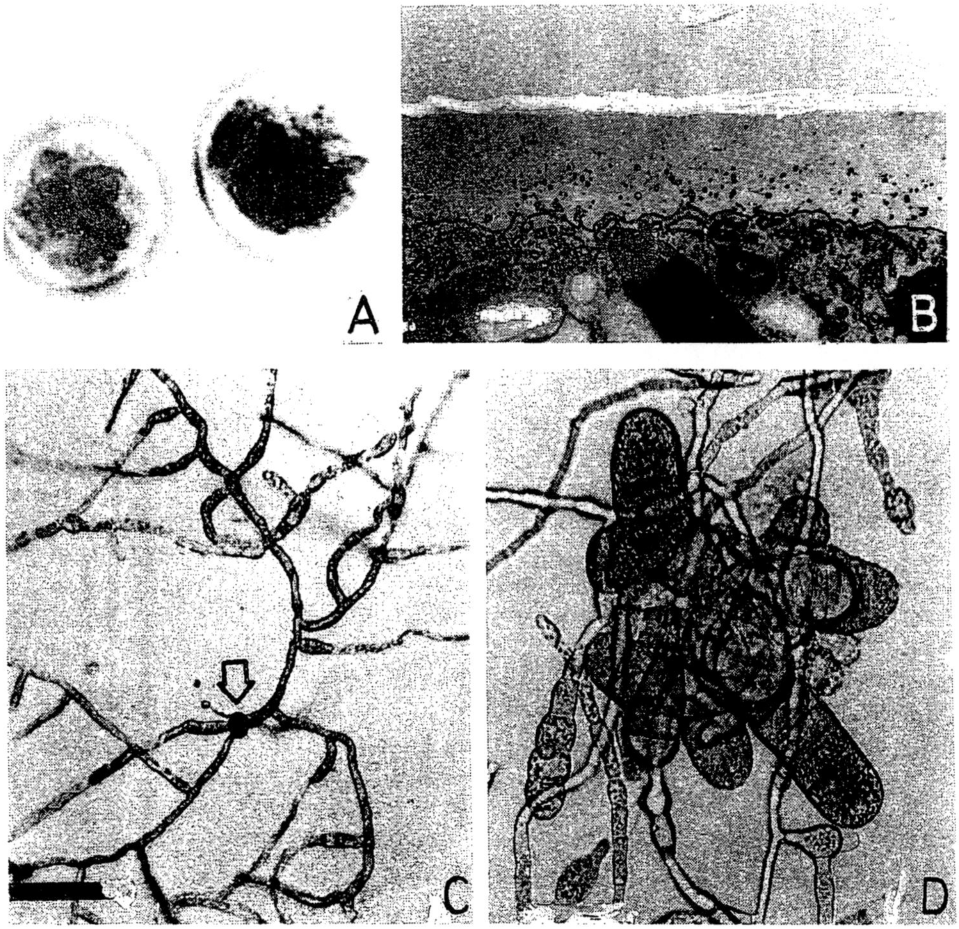

Fig. 19 A-D. Protoplast regeneration in *Porphyra nereocystis*
(after Waaland *et al.*, 1990)

A Isolated protoplasts, after washing, are free of wall debris.

B TEM of 20 h protoplast showing fibrillar elements of newly synthesized cell wall.

C Protoplasts (arrow) after 10-14 days of isolation showing multiple germination-tubes which after subsequent development produce conchocelis-like filaments.

D Conchosporangia produced by protoplast-derived conchocelis-like filaments.

protoplasts and regenerate them opens a wider range of possibilities. It allows for an easy selection of both natural and induced mutations, and permits investigations on their somatic genetics. As an example, the protoplasts of *P. nereocystis,* could offer a useful and promising material to explore alternative propagation methods for 'nori' products. This will include studies of cell wall regeneration, cell division, and development and differentiation of multicellular thalli. If successful, such studies are useful in somatic selection, somatic hybridization and gene transfer.

2 FLORIDEOPHYCIDEAE

Florideophycideae are characterized by apical growth, pit-connection and pseudoparenchymatous thallus. The growth is affected by an apical cell which is commonly dome-shaped. The adjacent cells are connected to one another by pit-connections which are occupied by broad cytoplasmic strands. The thallus exhibits varied degree of compactness and is invariably pseudoparenchymatous, composed of dense aggregates of the branch-system of either one (uniaxial) or more (multiaxial) richly branched filaments. Sometimes foliaceous development of thallus, as seen in Gigartinaceae and Delesseriaceae, also occurs in contrast to the common cylindrical type.

Some members of Florideophycideae show iridescence when exposed to strong light. This is due to reflection of shorter wavelengths by light reflecting bodies which mostly consist of protein and are located in the vacuoles. The phenomenon is shown by *Chondria, Callithamnion, Laurencia, Nitophyllum, Champia, Chylocladia and Gastroclonium*. Some members bear vesicular cells especially on the younger parts of the thallus. In certain species these vesicular cells may contain iodine. Iodine containing vesicle cells are common in Bonnemaisoniaceae whereas non-iodine containing vesiculate cells are seen in Rhodymeniales.

Tendril-like structures are seen in diverse Florideophycideae for example- *Asparagopsis armata, Cystoclonium purpureum, Hypnea musciformis, Spyridia aculeata and Laurencia virgata.* These structures not only help in attachment but also in vegetative propogation. Wound healing and regeneration is also a feature of Florideophycideae posessing compact thallus.

Sexual reproduction is oogamous. The female sex organ (carpogonium) bears a special receptive process (trichogyne) and retains a single nucleus. The male sex organ (spermatangium) produces a single, non motile male cell (spermatium). The zygote may undergoes meiosis (Nemaliales) but in most of the Florideophycideae reduction divisions are postponed to the time of tetraspore formation. The gonimoblast filaments may originate from the fertilized carpogonium or the hypogynous cell. The gonimoblast filaments produce carposporangia. Each carposporangium forms a single carpospore.. The carpospores germinate to give rise to gametophytic plants so that the life-cycle becomes haplobiontic (Nemaliales).

In the remaining Florideophycideae, the fertilized carpogonium produces connecting-filament which establishes contact with certain specialized cells called auxiliary cells. When auxiliary cells and carpogonia form parts of the same branch-system, the plant is called as procarpic. In Gigartinales and Cryptonemiales the auxiliary cells are widely separated from the carpogonia so that connecting filaments are of considerable length. Such plants are called as non-procarpic. In the more specialized Florideophycideae the connecting filaments are extremely short. The fusion of connecting filament with auxiliary cells induces the formation of gonimoblast filaments from the auxiliary cell. The carpospores give rise to a diploid plant called tetrasporophyte which bears the tetrasporangia. The meiosis occurs in tetrasporangium resulting in formation of tetraspores which may show various types of arrangements eg. zonate (in many Cryptonemiales and Gigartinales) ; cruciate, (in Rhodymeniales) or tetrahedral (in Ceramiales). Thus the life cycle in most of the Florideophycideae is diplobiontic.

3 NEMALIALES

CLASSIFICATION

NEMALIONALES VS NEMALIALES

Bertoloni (1818) created the genus *Nemalion* and indicated it to be synonymous to *Fucus*. Since then this generic name has become valid (see Nicolson and Norris, 1983). The controversy regarding the ordinal orthography revolves around grammar where this genus is treated as the neuter or as masculine gender. The derivations of *Nemalion* that are considered are as follows (Nicolson and Norris, 1983):

a) the first part nem - is from the neuter Greek word *nêma, nêmatos* (that which is spun, meaning thread)

b) the final element in neuter, either as noun or in adjective form, ending in short - on (masculine - os; feminine - a in adjectives) and not a long - lion.

According to Nicolson and Norris (1983) the usage involving stem of *Nemalion* is inconsistent and the most accepted version is where *Nemalion* is regarded as Second Declension with a genitive singular *Nemalii* and the stem *Nemalio*. Hence the subclass Nemaliophycidae (instead of Nemalionophycideae), order Nemaliales (instead of Nemalionales), suborder Nemaliineae (instead of Nemalionineae) and family Nemaliaceae (instead of Nemalionaceae) has been accepted.

Most of the older classification of Nemaliales are in accordance with the Kylin's (1956) classification, albeit there exists many minor differences. The present classification of the Florideophyceae is derived essentially from that proposed by Schmitz 1889; in Schmitz and Hauptfleishch 1896-7. He recognized four orders Nemaliales (Nemalionales), Gigartinales, Rhodymeniales and Cryptonemiales based on the development of the carposporophyte with special emphasis on the position of the auxiliary cell, if present, or its absence. This classification, thus, resulted in several misplaced families and orders which were later brought to their correct place.

VARIOUS PROPOSED CLASSIFICATIONS: According to Fritsch (1945) the Nemalionales (Nemaliales) are divided into uniaxial and multiaxial forms.

UNIAXIAL FORMS

1. Acrochaetiaceae - *Acrochaetium, Kylinia, Rhodochorton.*
2. Batrachospermaceae - *Batrachospermum, Gulsonia, Nothocladus, Sirodotia.*
3. Lemaneaceae - *Lemanea, Tuomeya.*
4. Naccariaceae - *Atractophora, Naccaria.*
5. Bonnemaisoniaceae - *Asparagopsis, Bonnemaisonia.*

MULTIAXIAL FORMS

6 Thoreaceae - *Nemalionopsis, Thorea.*

7. Helminthocladiaceae - *Cumagloea, Dermonema, Dorella, Helminthocladia, Helminthora, Liagora, Nemalion, Trichogloea.*

8. Chaetangiaceae - *Chaetangium, Galaxaura, Scinaia.*

Bold and Wynne (1978) divided Nemalionales into nine families.

1. Acrochaetiaceae - *Audouinella.*
2. Nemalionaceae - *Nemalion, Trichogloea, Trichogloeopsis.*
3. Helminthocladiaceae - *Liagora, Ganonema, Helminthocladia, Helminthora, Izziella.*
4. Dermonemataceae - *Dotyophycus, Cumagloea, Yamadella, Liagoraopsis.*
5. Bonnemaisoniaceae - *Asparagopsis, Bonnemaisonia.*
6. Naccariaceae - *Neoardissonia, Naccaria, Atractophora.*
7. Chaetangiaceae - *Galaxaura, Actinotrichia, Whidbeyella, Scinaia,*
8. Gelidiaceae - *Gelidium, Pterocladia.*
9. Gelidiellaceae - *Gelidiella.*

Dixon (1982), proposed another classification and included thirteen families which are as follows:

1. Acrochaetiaceae - *Acrochaetium, Rhodochorton, Audouinella.*
2. Batrachospermaceae - *Batrachospermum.*
3. Lemaneaceae - *Lemanea.*
4. Thoreaceae - *Thorea, Nemalionopsis.*
5. Helminthocladiaceae - *Helminthocladia, Helminthora, Liagora.*
6. Nemaliaceae - *Nemalion.*
7. Dermonemataceae - *Dotyphycus.*
8. Chaetangiaceae - *Scinaia, Galaxaura, Chaetangium.*
9. Naccariaceae - *Naccaria, Atractophora,*
10. Bonnemaisoniaceae - *Asparagopsis, Bonnemaisonia.*
11. Gelidiaceae - *Gelidium, Pterocladia.*
12. Gelidiellaceae -*Gelidiella.*
13. Wurdemanniaceae - *Wurdemannia.*

CHAETANGIACEAE Vs GALAXAURACEAE

According to Parkinson (1983) the generic name *Chaetangium* (family Chaetangiaceae) was published based on the type specimen *Fucus ornatus*. Later, it was found to be synonymous to *Suhuria vittata* J. Agarth ex Endlicher (1843). Since *Suhria* was published earlier hence it was conserved over the name *Chaetangium* and *Chaetangium* became invalid. Parkinson (1983), included the six species described previously under the name *Chaetangium* readopted the old name *Nothogenia* and formed combination nova. So all the species of *Chaetangium* were transferred to *Nothogenia* and comb. nov. were formed.

Since *Chaetangium* was synonymous to *Suhuria vittata* and later became invalid, there was a need to form a new family name to accommodate the seven genera which were previously placed under Chaetangiaceae. *Suhuria vittata* belonged to Gelidiaceae and so the seven genera had to be given a new family name. Parkinson (1983) selected the genus *Galaxaura*, which enjoyed priority over other genera as the type genus and changed the name of the family to Galaxauraceae.

MORPHOLOGY

The members of the order Nemaliales show wide variation in their morphology. The thalli are generally erect, filamentous and if pseudoparenchymatous, plants may be uniaxial or multiaxial in construction. The filaments may be aggregated loosely or compactly. Occasionally the thallus reveals calcification (Fig. 20 A,B).

The filamentous habit is found in *Acrochaetium* and *Rhodochorton* (Fig. 21 A). Most species of *Acrochaetium* are erect attached by a basal cell but some species also show heterotrichous habit with branches terminating in hyaline hairs. The freshwater species of *Acrochaetium* are sometimes referred to a separate genus *Audouinella*.

Batrachospermum is a simple uniaxial red alga (Fig. 20 D) called commonly as "Frog spawn". The thalli appears as chains of delicate beads, showing monopodial branching. The "beads" are actually densely branched laterals of limited growth, all of about the same length.

Lemanea shows modification of ordinary uniaxial type. The thallus is in the form of stiff, olive green or greenish-black bristles which are simple or branched in the mature parts. The axial threads and its whorled branches thus become evident. It is further distinguished by the parenchymatous structure of the tips and thus constitutes a transition from the uniaxial forms with a loose construction to the pseudoparenchymatous forms.

Nemalion shows a simple multiaxial structure. The thalli appears outwardly as soft gelatinous cylinders reaching a length of 25 cm and often shows some dichotomy. In *Helminthocladia* the plants are attached to the substratum by a basal disc. The fronds are brick-red or purple; gelatinous or cartilaginous; tufted; branched; cylindrical or compressed (Fig. 20 C).

Many species of *Liagora* show variation in their morphology. *L. erecta* (Fig. 20 B) is strongly calcified; profusely and pinnately branched . The lower branches are of considerable length. *L. amplectans* shows pyramidal, pinnately to bipinnately branched highly lubricous and slightly calcified thallus which is dioecious. The cystocarpic plants (Fig. 21 E) are more regularly pinnately branched than the spermatangial plants (Fig. 21 F) In culture the tetraspore of *L. harveyana* on germination produces a filamentous thallus (Fig. 21 D) instead of densely tufted, dichotomously branched and calcified thallus.

Thorea also shows multiaxial, branched, mucilaginous thalli which reach a length of one meter. The thalli are densely covered with a felt of deeply pigmented hairs which abscise in the mature thalli (Fig. 20 F). The bulk of the thalli consists of irregularly interwined hyaline threads which are oriented in longitudinal direction.

The gametophytes of *Nothogenia ovalis* are erect, saccate, hollow bladders which reache at maturity, 20-40 mm in length. Those of *N. erinacea* even attain a length of 300 mm and in *N. fastigiata* are blade-like, simple or divided once or twice and are covered with papillae which bear the reproductive organs (Fig. 21 C).

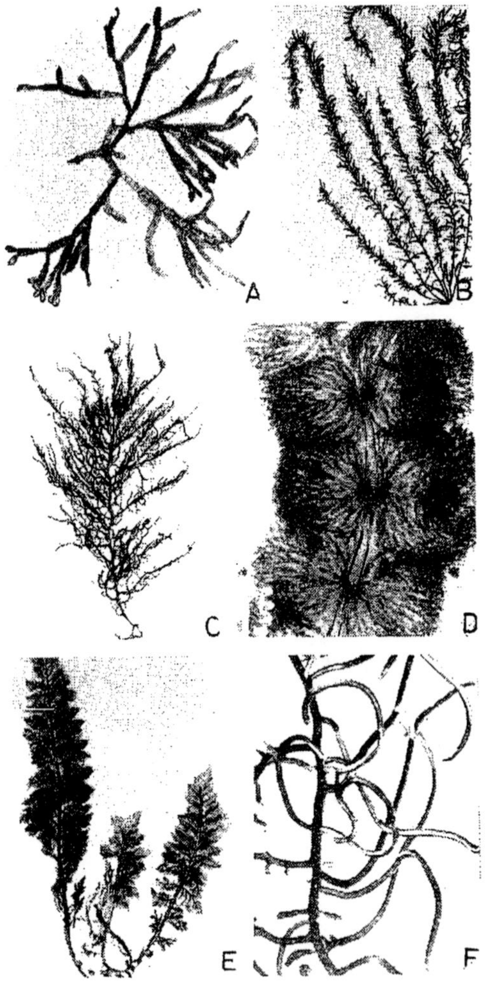

Fig. 20. A-F. Morphology

A *Galaxaura.* sp., showing slight calcification of the thallus which shows joint-formation (after Bold and Wynne, 1978).

B *Liagora erecta*, showing profuse pinnate branching, strong calcification and lateral branches ramified like main axis (after Srinivasan, 1969).

C *Helminthocladia clavadosii*, showing tufted cylindrical fronds, robust main stem covered with closely set branches (after Srinivasan, 1969).

D *Batrachospermum carpocontortum*, showing main axis with lateral whorls of a female plant containing large peripheral carposporophyte (arrow) (after Sheath *et. al.*, 1986).

E *Asparagopsis taxiformis*, showing erect plant with a much branched entangling rhizome and bushy shoots lacking branches in the lower portion (after Srinivasan, 1969).

F *Thorea okadai*, showing branched mucilaginous thallus lacking hairs (after Yoshizaki, 1986).

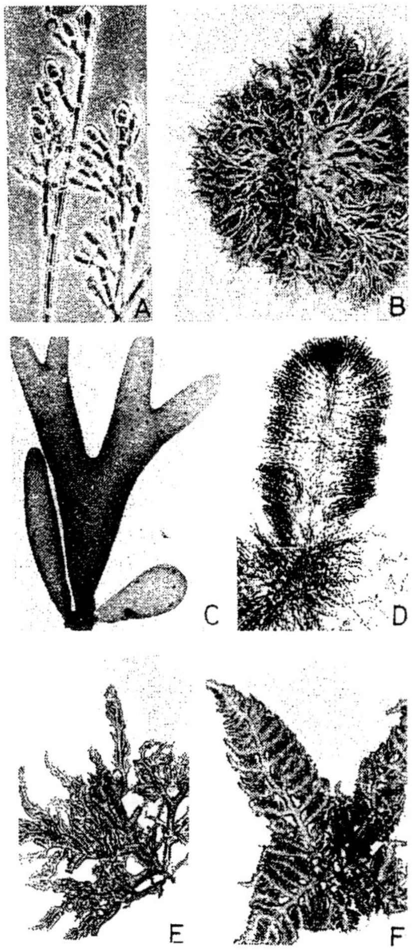

Fig. 21 A-F. Morphology

A *Rhodochorton purpureum*, showing small tufted filaments bearing clustered tetrasporangia at the top of the filaments (after Woelkerling, 1971).

B *Galaxaura*. sp., showing calcified tufted and dichotomously branched thallus (after South and Whittick, 1987).

C *Nothogenia fastigiata*. The thallus is blade like, simple, divided and covered with papillae which bear the reproductive organs (after Huisman and Womersley, 1992).

D *Liagora harveyana*, showing erect thallus produced *in vitro* from the tetraspore (after Guiry, 1990b).

E *L. amplectans*, female plant has lubricous, slightly calcified, pinnate to bipinnate branches (after Abbott, 1984).

F *L. amplectans*, male plant has lubricous calcified, pyramidal branches (after Abbott, 1984).

The plants of *Gloiophloea* are 18 cm tall dichotomously branched with terete thalli, 3-6 mm wide, flat at the base of the branches and with slight constriction at the nodes. Branch apices have distinct apical pits.

Among the multiaxial forms, *Scinaia* shows greater specialization. *S. forcellata* (Fig. 22 A,D) has a discoid holdfast; is gelatinous and shows a compact tufted habit wth abundant dichotomous branching. The branch apices have pits. *S. turgida* (Fig. 22 C) is similar to *S. forcellata* and can be differentiated only by the microscopic examination of the surface cells. *S. monliformis* (Fig. 22 B) is also attached by a discoid holdfast but the thallus is many times dichotomously branched, with regular constrictions which divide it into oblong segments. In *S. australis* the thallus is again 10-11 times dichotomously branched and each furcation shows a branch pair equal in length with distinctly pointed apices (Fig. 22 F). The plants of *Pseudoscinaia* are 12-20 cm in length, unconstricted and 9-13 times dichotomously branched (Fig. 22 E). Internal construction is identical to *S. forcellata*. The different species of *Galaxaura* show same type of branching pattern as in *Scinaia*, but the cortex is usually strongly calcified. Some species also show jointed appearance of the thallus (Figs. 20A; 21B).

In Nemaliales the specialized uniaxial forms are *Naccaria, Bonnemaisonia* and *Asparagopsis*. *Asparagopsis* plants have a much branched, rhizome-like creeping portion, that produces a few rhizoids. In well developed plants, the shoot is erect, bushy, generally 20 cm or more in height with terete stem (Fig. 20 E).

In *Gelidium pteridifolium* one to several primary axes arises from a branched, stoloniferous holdfast. Each axis bears up to four orders of pinnately branched, indeterminate vegetative laterals that are largely denuded below and beset with alternate to subopposite, distichously, arranged rows of determinate fertile branchlets above. The general outline of major axis is pyramidal.

REPRODUCTION

MONOSPORANGIA

Monosporangia are single and uninucleate spores. Occasionally, they occur in groups and are produced after mitotic divisions either by the gametophyte or by the tetrasporophyte.

Many taxa of Florideae reproduce only through monosporangia. In *Acrochaetium botryocarpum* both mono–, and tetra–, sporangia occur in plants grown in nature and in culture (Guiry *et al.*, 1988; Woelkerling, 1970; see also Cole and Sheath 1990). Woelkerling (1970) found in *A. botryocarpum* such monosporangia throughout the year. Fritsch (1945) regarded the tetrasporangia occurring with monosporangia as divided monosporangia.

In *Rhodochorton purpureum* (Fig. 21A), the origin, cytology and spore release of the unbranched chain of sporangia are unknown (Kalvestad, 1957). Lee (1985), however, suggests that this taxon lacks monosporangia both in nature and in culture.

In *Batrachospermum* species, the monosporangia are formed both in the erect and in the chantransia phases but generally they are confined to the latter phase. Further, both lateral and spherical monosporangia are produced in considerable number for the perpetuation of pseudochantransia stage (Dixon, 1973; Balakrishnan and Chaugule, 1980). Yoshida (1959) found that monospores are formed singly or in pairs on the tip of side branches of chantransia stage in *Batrachospermum*. These spores are indistinguishable upon release.

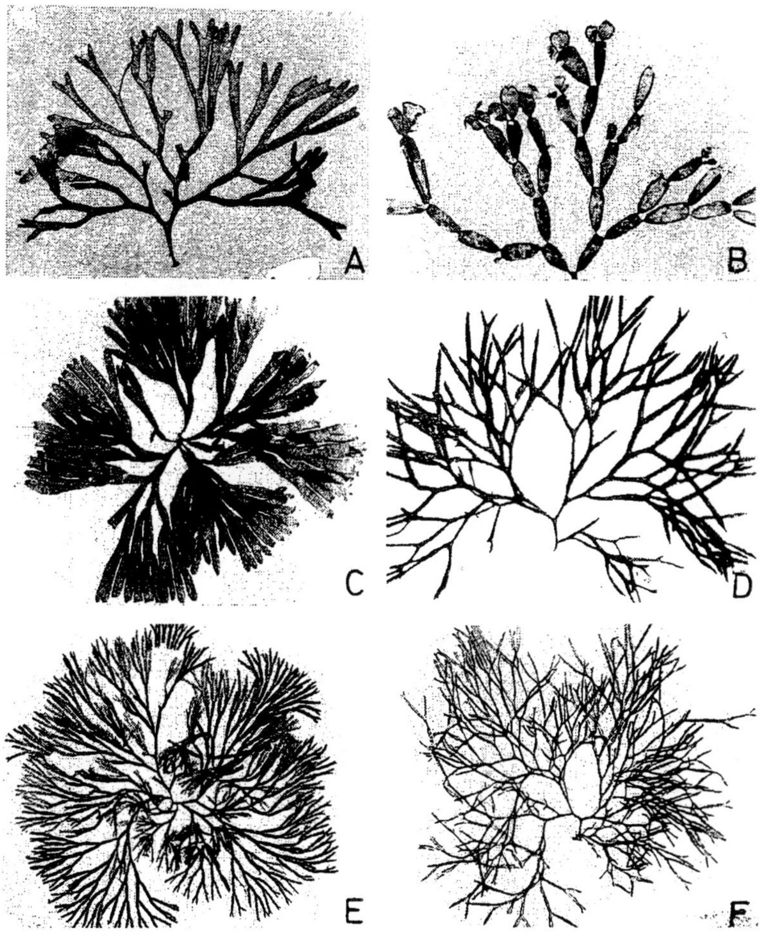

Fig. 22 A-F Morphology

A *Scinaia forcellata* from Ireland. The thallus is compact, tufted and gelatinous (after Huisman, 1985a).

B *S. moniliformis.* Mature plant showing profuse dichotomy with regular constrictions which divide the plant into oblong segments (after Huisman, 1986).

C *S. turgida*, showing rosette–like thallus structure due to repeated dichotomy (after Maggs and Guiry, 1982).

D *S. forcellata,* The thallus is compact, tufted and dichotomously branched (after Maggs and Guiry, 1982).

E *Pseudoscinaia snyderae*, Mature plant showing cylindrical unconstricted and dichotomous branches (after Huisman, 1985a).

F *Scinaia australis,* Habit of a mature plant showing discoid holdfast, dichotomous branching and branches almost equal in length (after Huisman, 1986).

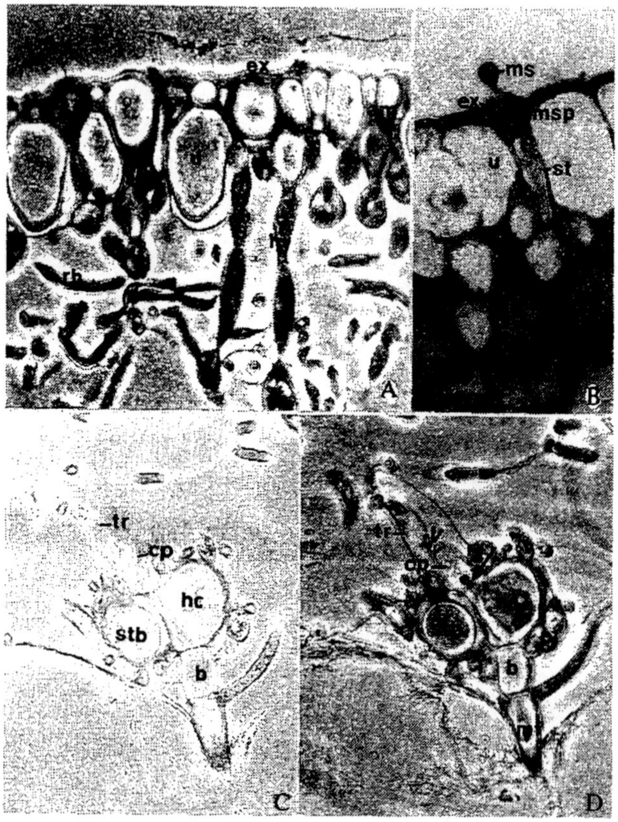

Fig. 23 *A-D Scinaia pseudocrispa*

(*am*, assimilatory cell ; *b*, basal cell ; *cp*, carpogonium ; *ex*, extracellular mucilage ; *hc*, hypogynous cell ; *hy*, hypodermis ; *mf* , medullary filament ; *ms*, monospore, *msp*, monosporangium ; *pi*, pericarp initial ; *rh*, rhizoidal filament ; *st*, stalk-cell ; *stb*, sterile branch of hypogynous cell ; *tr*, trichogyne, *u*, utricle).

A. A portion of thallus enlarged to show epidermis consisting of utricles and elongate assimilatory cells. The epidermis is covered by a thick mucilage. Hypodermal cells show dense cytoplasmic contents whereas utricles are devoid of such contents. The unbranched rhizoidal filaments are thick walled with narrow lumen. (PAS, phase contrast) (after Vijayaraghavan and Bhatia, 1996).

B. Transverse section of thallus to show monospore release. The stalk cell elongates. In the monosporangium, the peripheral layer of polysacccharides extrudes through the extra cellular mucilage and aids monospore release. The monospore cytoplasm is naked and stains blue (TBO) (after Bhatia and Vijayaraghavan, 1996).

C. Mature carpogonial branch to show carpogonium with trichogyne, enlarged hypogynous cell, sterile branch of hypogynous cell and the basal cell. The cells of the carpogonial branch are rich in wall polysaccharides. The cytoplasm of hypogynous cell and its sterile branch are bereft of floridean starch (PAS) (after Vijayaraghavan and Bhatia 1996).

D Same, showing transfer of zygote nucleus (arrow) to the hypogynous cell. The trichogyne is isolated by the deposition of plug materials (arrowhead) (PAS, phase contrast; after Bhatia and Vijayaraghavan, 1995a).

Hymes and Cole (1983) studied the ultrastructure and development of monosporangia in *Audouinella hermanii* and agree that the carposporophyte represents a monosporangial generation (see also Guiry, 1978). In *A. hermanii* the monosporogenesis is identical to Bangiophycideae whereas the carposporogenesis and tetrasporogenesis resemble the Florideophycideae (Hymes and Cole, 1983). They also found that monosporangia in this taxon are similar to the young vegetative cells containing a central nucleus, a parietal, lobed, chloroplast, a few scattered starch grains, mitochondria, dictyosomes and endoplasmic reticulum. Monospore differentiation is a three step process:

Step I : There is an increase in the number of dictyosomes whose electron-transparent vesicles coalesce to form large fibrous vesicles which fuse subsequently with the plasmamembrane. Floridean starch and electron-dense, membrane-bound, lipid deposits are known. Peripheral endoplasmic reticulum filled with electron-translucent materials coexist. The dictyosome vesicles and dilated peripheral endoplasmic reticulum fuse with plasmalemma and release electron transparent materials inside the monosporangial wall (Fig. 24). The plasmalemma is also convoluted.

Step II : During this stage the dictyosomes become large and are concentrated in the peripheral cytoplasm, their cisternae are hypertrophied and bud off vesicles. These vesicles, contain electron-dense materials and called "cored vesicles", fuse with plasmalemma accentuating the lybyrinths (Fig. 25). The peripheral endoplasmic reticulum are not evident and the chloroplasts divide by constriction.

Step III : In the final stage, both the dictyosome activity and mitochondrial number decrease. Fibrous vesicles disappear and the cytoplasm is gorged with floridean starch and cored vesicles. The chloroplast continue to divide (Fig. 26). Prior to monospore release, the vegetative cell plasmalemma and cytoplasm retract in the area of the pit-plug and the monosporangial inner wall becomes more electron-dense.

Monosporangia are an important means of dispersal mechanism in *Thorea* and *Nemalionopsis* (Sheath, 1984) as these taxa are bereft of diploid tetrasporangial phases (Sheath and Cole, 1990). Kajimura (1988) found that in *Scinaia tokidae*, *S. okiensis* and *S. flabellata*, the monosporangia are formed singly and terminally on the assimilatory cells. In *Scinaia* species the monosporangia are also formed in groups on the outer cortical cells of the gametophyte (Dixon, 1973; Huisman, 1985; Maggs and Guiry, 1982) In *S. confusa* monospore recycle tetrasporangial plants (Ramus, 1969).

Monosporangia have been found in the tetrasporophyte of *Nemalion, Liagora, Helminthocladia* and *Cumagloea* (Cole and Sheath, 1990). Monosporangia-like bodies have also been found in *Helminthocladia calvadosii* and *H. pacifica* (Umezaki, 1972). These are short chains of round cells that superficially resemble tetrasporangia and detach easily forming new filaments. There was no spore release. Bhatia and Vijayaraghavan (1996) in *Scinaia pseudocrispa* (= *S. forcellata*) reported monospore release. During the spore release the stalk cell elongates. In the monosporangium, the peripheral layer of polysaccharides extrudes through the extracellular mucilage and aids in monospore release (Fig . 23 B).

SPERMATANGIA

Acrochaetiaceae : The spermatangia are borne in pairs on the supporting cells of the carpogonia in *Audouinella alariae* (Lee and Kurogi, 1983). In *Chromastrum alariae* spermatangia occur on

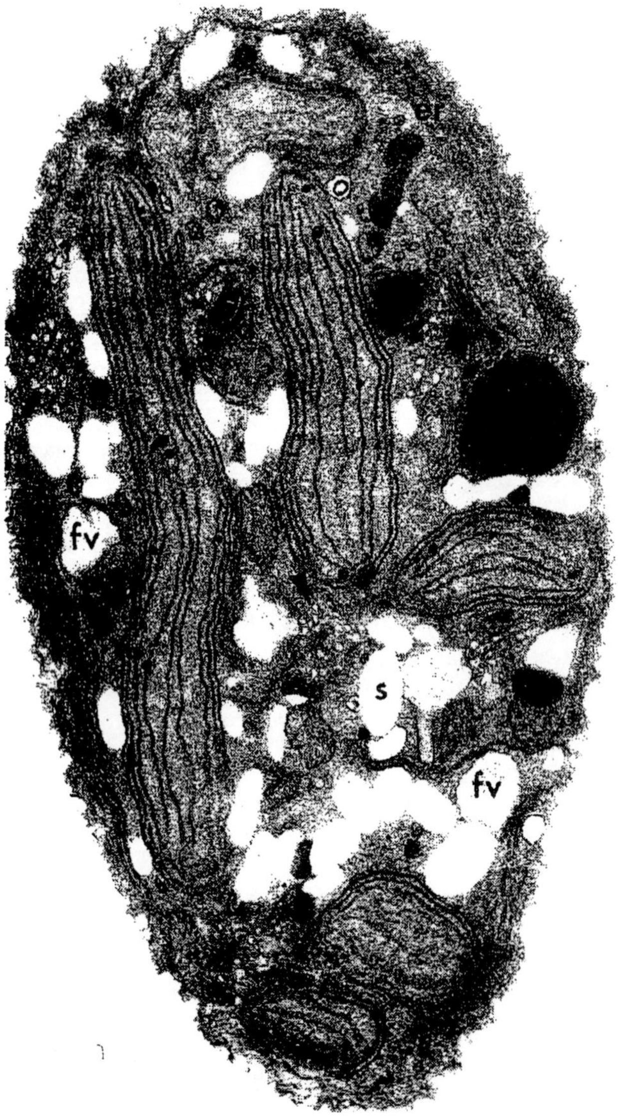

Fig. 24. Ultrastructure of monosporangia of *Audouinella hermannii*.

(after Hymes and Cole, 1983)

(d, dictyosome; er, endoplasmic reticulum; fv, fibrous vesicle; s, floridean starch)

Longitudinal section through a developing monosporangium to show first stage of monosporogenesis. The monosporangium shows the presence of floridean starch, layers of peripheral endoplasmic reticulum, production of electron-translucent vesicles (large arrow heads) released from the dictyosomes and fusion of these smaller vesicles into the larger fibrous vesicles which are released at plasmalemma (small arrow head).

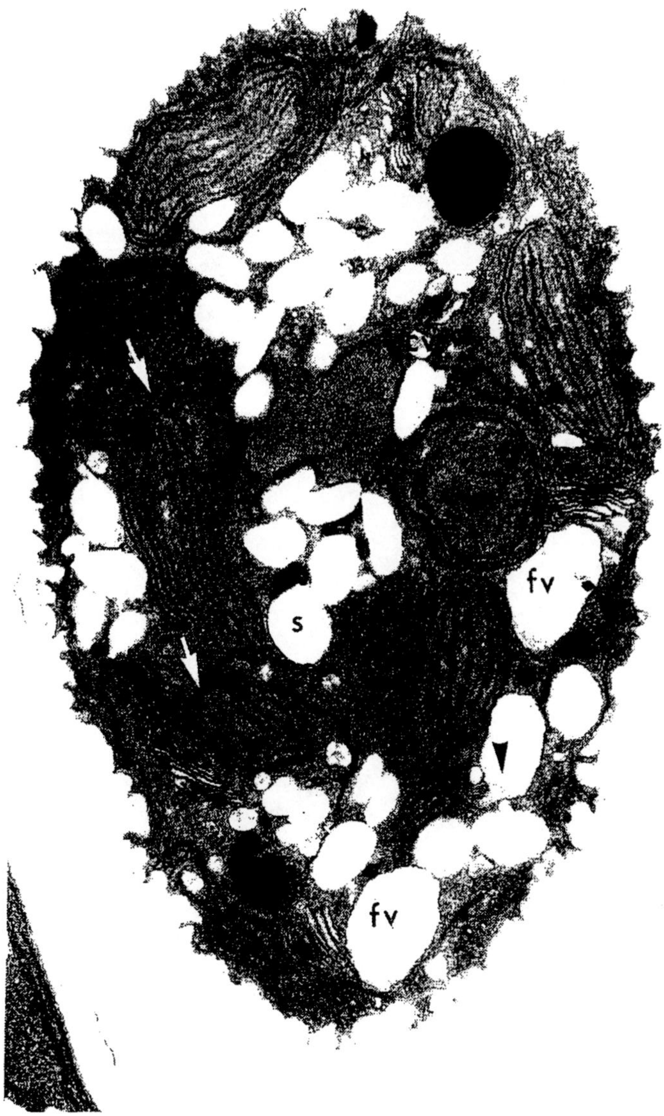

Fig. 25. Ultrastructure of monosporangia of *Audouinella hermannii.*

(after Hymes and Cole, 1983)

(cv, cored vesicle; d, dictyosome; fv, fibrous vesicle; n, nucleus; s, floridean starch)

Longitudinal section of a developing monosporangium showing second stage of monosporogenensis with dictyosome and cored vesicles. A few cored vesicles fuse with the large fibrous vesicle and deposit electron-dense substance. The chloroplast show begining of division. The monosporangium shows central nucleus, bilayered sporangial wall and increase in floridean starch.

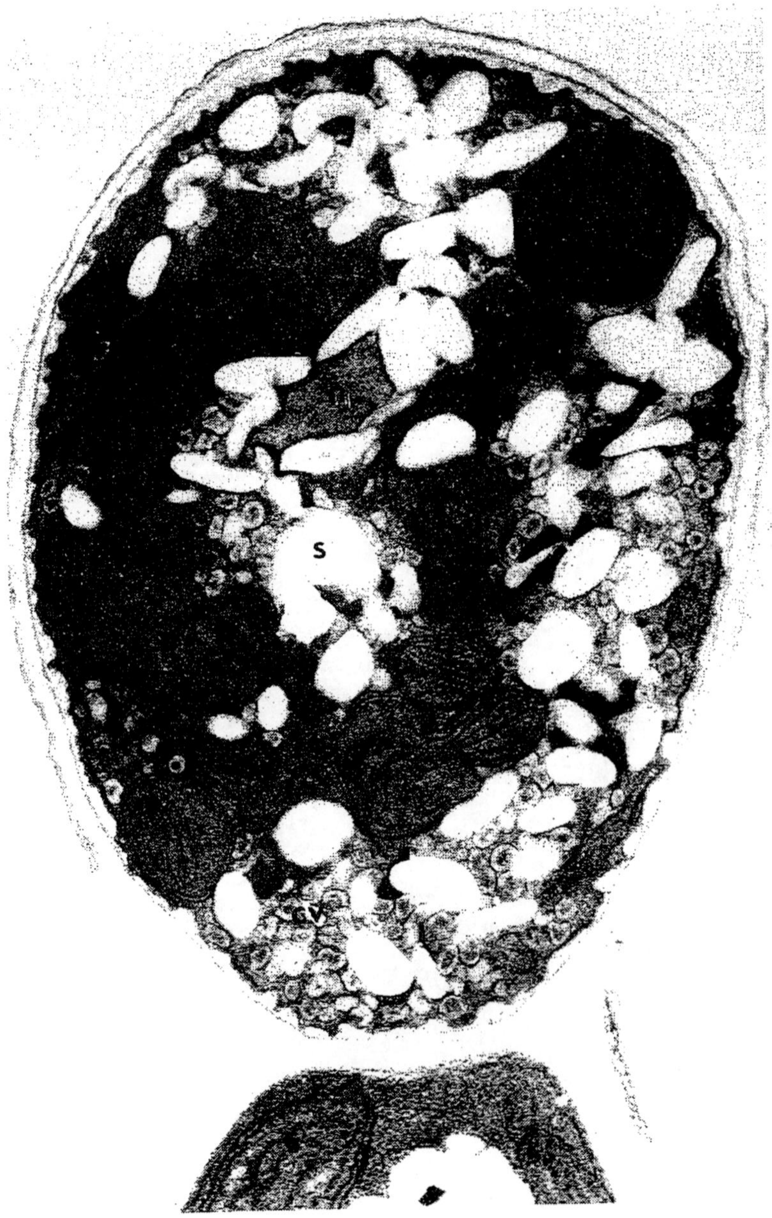

Fig. 26. Ultrastructure of monosporangia of *Audouinella hermannii*.

(after Hymes and Cole, 1983)

(cv, cored vesicle; n, nucleus; s, floridean starch).

Longitudinal section of the developing monosporangium to show third stage of monosporogenesis. There is abundance of floridean starch, numerous cored vesicles, a few dictyosomes, dividing chloroplast, lipid body, central nucleus and electron-dense mature wall.

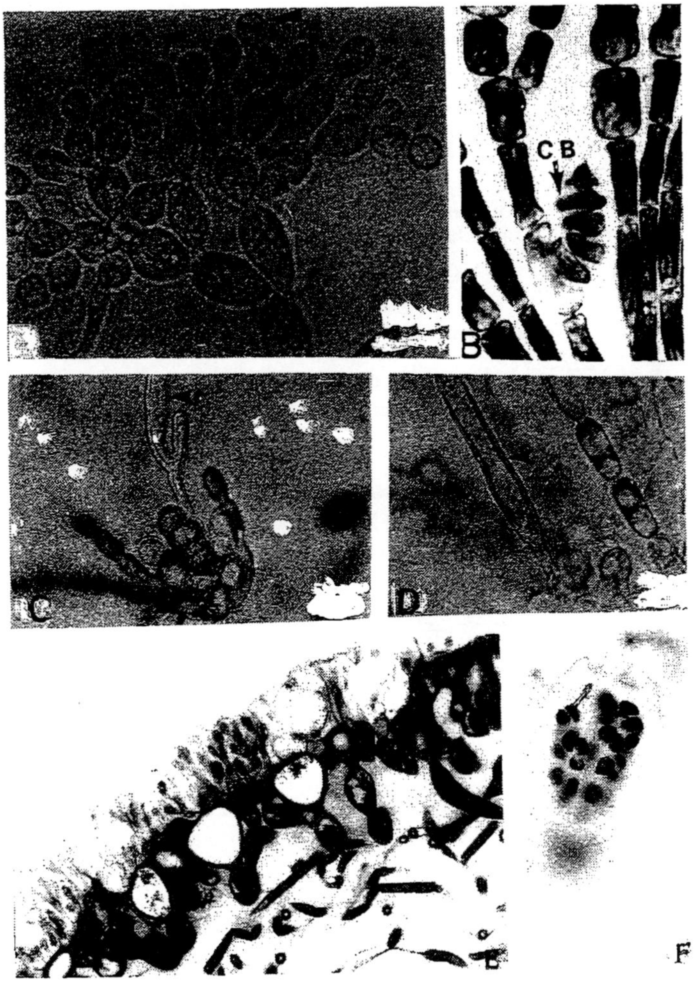

Fig. 27 A-F Reproductive structures (Carpogonial branch and Spermatangia).

(cb, carpogonial branch)

A *Batrachospermum carpocontortum*. Spherical to ovoid spermatangia at the tip of fasicles (after Sheath *et al.*, 1986).

B *Liagora norrisiae*. Five celled carpogonial branch (after Abbott, 1984).

C *Batrachospermum coerulescens*, showing carpogonia with stalked, branched and cylindrical trichogyne (after Sheath *et al.*, 1986).

D *B. virgatum*. Same with a cylindrical and pedicillate trichogyne (after Sheath *et al.* 1986).

E *Scinaia pseudocrispa*. Spermatangial sorus showing spermatangial filaments interspersed with utricles. The outer tangential wall of utricles lyse and facilitate release of polysacha ridic materials into the sorus (after Bhatia and Vijayaraghavan, 1995a)

F *Asparagopsis taxiformis*. Showing three-celled carpogonial branch with club shaped trichogyne, (after Jha, 1996).

any cell of the carpogonial branch (Kuiper, 1983). Lee (1985) found that the population of *Rhodochorton purpureum* from Hokkaido islands lack these male reproductive structures, whereas those from Muroran islands have them.

Batrachospermaceae : *Batrachospermum mahabaleshwarensis* is monoecious, producing globular spermatangia on both the primary as well as the secondary laterals (Balakrishnan and Chaugule, 1980). In contrast, the dioecious *Batrachospermum* species from Northern Kyushu Japan, have less male plants than the female (Yoshida, 1959). In male plants, spermatangia are formed at the tip of the laterals which are richly branched in the upper parts (Fig. 27A). The mature spermatium is spherical, 6 or 7 μm in diameter and lacks chromatophore. The mature spermatium during release is bereft of membrane but soon develops one when it attaches to the trichogyne. In *B. carpocontortum*, spermatangia are numerous at the tip of fasicles (Fig. 27A; Sheath et al., 1986).

Thoreaceae: The plants *of Thorea okadai* are dioecious and the ovoid spermatangia are borne in small, terminal clusters on short assimilatory branches (Yoshizaki, 1986). The liberated spermatia are spherical and 7-9 μm in diameter.

Helminthocladiaceae: In *Helminthocladia calvodosii. f. comorinensis,* the spermatangia are borne in dense terminal clusters on noninflated terminal cells of short laterals, which are one to a few celled, unbranched and usually originate from cells immediately below the end cell. They may be dichotomously branched and according to Chiang (1971) their development follows the Helminthocladia-type (see Krishnamurthy and Sundararajan, 1985). In *Helminthocladia nizamuddinii,* the inflated cells are absent and the spermatangial branches develop directly on the second cell of secondary assimilatory branches (Husain and Shameel, 1990). The spermatangial branches are branched upto five times and when young they develop in heads which upon maturation are embedded in mucilage. Martin (1939) in *Helminthocladia papenfusii* reported that spermatangia are formed in dense clusters on the tip of short laterals which are differentiated from the assimilatory filaments by short and narrow cells. Chang (1971) observed supporting cell in *H. papenfusii.* In *Helminthora furcellata,* the male reproductive organs are either directly produced from the ultimate cells of the assimilatory filaments, as described for *H. divericata* (Kylin, 1956) or from the supporting cells which originate from the ultimate cells of the assimilatory filaments (Chang, 1971). Both the developmental stages coexist.

In *Liagora harveyana,* the spermatangia develop in digitate manner on the outer cortical cells (Guiry, 1990), whereas in *L. amplectans,* they are laterally attached in small clusters on the assimilatory filaments and later cover the terminal cell. They are present on different filament than those bearing carpogonia (Abbott, 1984). In *L. norrisiae* both the spermatangia and the carpogonia are produced in the same plant. The spermatangial heads are always lateral to the penultimate and lower cells of assimilatory filaments. Spermatangia are spherical, attached directly on surfaces of spermatangial mother cells.

Chaetangiaceae: In *Galaxaura oblongata,* the spermatangial conceptacle intiates, with the enlargement of a terminal cell of the filament in the apical-pit. The terminal cell divides and the resulting apical cell.continues division. The lower cell, form lateral filamentous, branches which results in the conceptacle lined by the filaments that produces spermatangia (Magruder, 1984). In *G. fastigiata,* the conceptacle has stoloniferous spreading branches, (Krishnamurthy and Sundara rajan, 1986).

In *Scinaia furcata*, the spermatangia occur in large patches. The spermatangial branches develop from cells in the hypodermal layer which function as support cells and produces one or two stalk-cells of the spermatangial mother cell, that protrudes upward between the utricle and produces 3-6 spermatangial mother cells (Zablackis, 1987). It has also been found true for *S. proliferata* (Huisman, 1986).

In *Scinaia tokidae*, the utricle is absent in the spermatangial sori. Spermatangial mother cells are formed in both the hypodermal and the epidermal cells (Kajimura, 1988). In *S. okiensis*, the spermatangial sori are formed in a belt between dichotomies and utricles are present in the spermatangial sori (Kajimura, 1988). In *S. flabellata*, the spermatangial sori are scattered on thallus surface and utricles are present (Kajimura, 1988).

In *Scinaia indica*, spermatangia are produced in the sori and are stalked. The stalks are produced from the pigmented end cells of the assimilatory filaments. A few spermatangia are borne at the tip of these stalks. In *S. bengalica* spermatangia are produced in definite sori. A hypodermal cell gives rise to 2-4 spermatangial mother cells. In *S. fascicularis*, the spermatangia occur in irregular sori and the spermatangia are borne on one or a few celled narrow filaments produced from the peripheral cells (Krishnamurthy and Sundararajan, 1987). In *S. forcellata*, the spermatangial branch is three or four celled and arise from the hypodermal cell (Huisman, 1985). In *Scinaia pseudocrispa* (= *S. forcellata*) spermatangia are borne in sori and are scattered superficially on the thallus surface. The spermatangial filaments are interspersed by utricles. The spermatangia are borne at the tips of 3 or 4 celled spermatangial filaments that project beyond the epidermal cells (Fig. 27 E). One of the hypodermal cells, cuts off one or two cells, each of which produces one or two spermatangial mother cells. The lowest cell acts as the stalk cell. Two or three spermatangia are produced from each spermatangial mother cell. The mature spermatangial sori are veneered by a thin mucilage layer. The released spermatium is round and bound by a distinct polysaccharide spermatium coat (Bhatia and Vijayaraghavan, 1995a ; see also Vijayaraghavan and Bhatia, 1995 b)

Bonnemaisoniaceae: In *Delisea fimbriata*, the spermatangia are formed in groups covering almost both upper and under surfaces of the long branches. Any cortical cell can be the progenitor of spermatangial mother cell, that are arranged in palisade-like manner. These ultimately produce the spermatangia (Chihara, 1961). Whereas in *Asparagopsis armata* axial cells of the male branch cuts off four periaxial cells, all at the same level. These in turn segment several layers of medullary cells that give rise to spermatangial mother cells. The spermatangia producing branches are borne on opposite determinate branches and occur throughout the male plant. The male branches are usually long, narrrow, unbranched and covered entirely by spermatangia, except near the base, which appear as stalk (Bonin and Hawkes, 1987).

Gelidiaceae : In Gelidiaceae the male plants are indistinguishable from cystocarpic or tetrasporic plants. The sex organs are borne on special branches called ramuli . In *Gelidium japonicum* the male fertile ramuli closely remembles the tetrasporangia, branch profusely in pinnate manner and thus crowd the terminal part of the main and lateral axis of vegetative thallus (Akatsuka, 1979). In *Gelidium latifolium* the spermatangial fronds are small, and indistinguishable from sterile fronds (Dixon, 1959). The spermataniga occur superficially as irregular patches on the main axes and lateral branches of the frond. The apical cells of the lateral branches of the limited growth, enlarge to form the spermatangial mother cells. Each of these gives rise to 2 or 3 spermatangia. As the thallus ages, three or four times the normal thickness of the mucilage is seen over the spermatangial sorus (Dixon, 1959).

CARPOGONIAL BRANCH

Acrochaetiaceae: In *Audouinella alariae* the terminal carpogonia, with short lateral trichogyne, occur on short branches (Lee and Kurogi, 1983). In *Chromastrum alariae,* the carpogonium has similar morphology, but the trichogyne is always directed towards spermatangia (Kuiper, 1983). Lee (1986) found that *Rhodochorton purpureum,* occurring in Hokkaido islands, lacks carpogonia whereas those from Muroran islands produces them.

Batrachospermaceae: In *Batrachospermum mahabaleshwarensis,* the carpogonial branch is spirally twisted and arises from the basal cell (Balakrishnan and Chaugule, 1980). The trichogyne is distinctly stalked, conical and compressed. The hypogynous cell in *B. macrosporum* (Kumano and Orlando, 1990) and in *B. hypogynum* (Kumano and Ratnasabapathy, 1982), forms rosette-like laterals, and there is pronounced protoplasmic connection between the carpogonium and the hypogynous nutritive cell. In yet another species of *Batrachospermum* from Japan, the carpogonial branch in early stages is not distinguishable from primary laterals and carpogonia have ellipsoidal trichogyne (Yoshida, 1959). In *Batrachospermum carpocontortum* the carpogonial branch is 3-6 celled, straight, and proximal to the fascicles. The carpogonium has trichogyne with protrusion and bends. Involucral branches are long and extend well beyond the trichogyne tip (Sheath et al., 1986). In *B. ectocarpum,* the carpogonial branch is differentiated after fertilization. Initially these branches are indistinguishable from the vegetative branches. Kumano *et al* (1970) consider it as a primitive character. However, in *B. cayennense* the carpogonial branch is different from the vegetative fascicles both in size and shape. This is considered as an advanced feature (Kumano and Ratnasabapathy, 1982).

In *Batrachospermum moniliforme, B. coerulescens and B. boryanum,* a distinct nucleus was observed both in the carpogonium as well as in the trichogyne (Kumano and Ratnasabapathy, 1982). Schmidle (1899) and Kylin (1917), however, did not observe a trichogyne nucleus in *B. moniliforme.* The trichogyne also show variations; it is stalked and branched in *B. coerulescens* (Fig. 27 C); cylindrical and pedicellate in *B. virgatum* (Fig. 27 D); sessile, elongate and obconical in *B. keratophytum;* clavate in *B. boryanum* and ellipsoidal and sessile in *B. moniliforme.*

Various species of *Sirodotia* were divided into advanced and primitive forms on the basis of origin of the carpogonial branch (Kumano, 1982). Besides, *S. sinica* is regarded as the most primitive taxon since the carpogonial branch arises from the intercalary cell of the cortical filament and further the cell that bears the carpogonial branch is not fixed. In *S. suecica* and *S. delicata,* however, the carpogonial branches are short and moderately developed, from the basal cells of the primary branchlets (Kumano, 1982).

Thoreaceae: One-celled carpogonium is the characteristic feature of *Thorea.* In *T. okadai,* the carpogonia are formed in abundance in the younger parts of the thallus either laterally or terminally. They are one-celled, urceolate in shape and have straight and cylindrical trichogynes (Yoshizaki, 1986).

Helminthocladiaceae: In *Helminthocladia calvadosii f. comorinensis,* the carpogonia are terminally situated usually on 3-celled and rarely 4-celled carpogonial branch (Krishnamurthy and Sundararajan, 1985). The carpogonial branch is straight, is borne in the peripheral part of the assimilatory filaments and arise laterally on the ultimate forking of the assimilatory filament or even deeper down on the distal two or three forkings. In *H. nizamuddinii,* the carpogonial branch consists of 2-4 cells, which are borne laterally on the upper-half of the fourth or fifth cell of an assimilatory branch. When more than two branches develop on the same assimilatory branch they appear nonfunctional and sterile (Husain and Shameel, 1990).

In *Helminthocladia calvadosii,* the carpogonia are conical with a long trichogyne that extend beyond the end cells of the assimilatory filaments (Krishnamurthy and Sundararajan, 1985). In *H. nizamuddinii,* the carpogonia are also conical but the trichogynes are much short (Husain and Shameel, 1990).

In *Liagora harveyana,* the carpogonial branch is curved and 4-celled which is formed on the sixth or seventh cell from the apex of the cortical fasicle and only one carpogonial branch is formed by the supporting cell *per se* (Guiry, 1990). In *L. valida,* the carpogonial branch is relatively straight and arises from the supporting cell (Abbott and Mokoto, 1982). However, in *L. amplectans* up to 4 carpogonial branches arise from the supporting cell (Abbott, 1984). In *L. norrisiae,* the carpogonial branch is 5-celled (Fig. 27 B; Abbott, 1984). In *L. indica* the carpogonial branch arises laterally from a cell above, the middle of an assimilatory filament and is 5-celled (Krishnamurthy and Sundararajan, 1985). In *L. mannarensis,* the carpogonial branch is a special accessory lateral branch of 2-4 cells produced from the middle or most often from the lower part of an assimilatory filament and carpogonial branch is slightly curved. The supporting cell is distinguished from the rest of the cells and rarely two carpogonial branches may arise from the same supporting cell (Krishnamurthy and Sundararajan, 1985). Similar structures were observed in *L. harveyana* (Guiry, 1990) and in *L. indica.*

Dermonemataceae : In *Dotyophycus corymbosus,* the carpogonial branch is terminal and arises straight from the medullary filament or from the lowermost cell of the assimilatory filament. The carpogonial branch is long and has 9-13 cells. Of these, the lower 4 or 5 cells are elongate and cylindrical. The length of upper cells gradually decreases and becomes spherical or subspherical (Krishnamurthy and Sundararajan, 1985). In *D. pacificum* the carpogonial branch has 8 or 9 cells, that are borne terminally on special unbranched filaments which arise directly from the medullary filaments and subtended by 3 or 4 vegetative cells. A total of 13-15 cells occur in a carpogonial branch (Abbott, 1976).

In *Dotyophycus corymbosus,* the carpogonium is the terminal cell of the carpogonial branch. It is conical and has a terminal, thick trichogyne that gradually thickens toward the apex. Occasionally, the upper cells of the carpogonial branch give out 1-3 short, lateral, 2-6 celled carpogonial branches that result in a corymbose cluster (Krishnamurthy and Sundararajan, 1985).

Chaetangiaceae : In *Galaxaura oblongata,* the carpogonial branch is formed by the enlargement of the terminal cell of a filament within the apical-pit. The carpogonial branch is 3-celled with a terminal carpogonium, a hypogynous cell and a basal cell. The hypogynous cell produces four short filamentous branches of 4-9 cells and the basal cell also produces four branched filaments of 4-12 cells (Magruder, 1984). The branched filaments are also observed in *G. rugosa, G. cylindrica, G. marginata* and *G. obtusa* (Huisman and Borowitzka, 1990) and also in *G. fastigiata* (Krishnamurthy and Sundararajan, 1986).

In *Scinaia pseudocrispa* (= *S. forcellata*) the carpogonial branches arise from the subcortical cells near the apex of the plant and are 3-celled. Before fertilization, the basal cell of the carpogonial branch produces several short branches which eventually forms the pericarp. The hypogynous cell also cuts off on either side one or two celled sterile branches (Huisman, 1985; Bhatia and Vijayaraghavan, 1995 b). At maturity the hypogynous cell enlarges. The cells of carpogonial branch are rich in wall polysaccharides. The cytoplasm of hypogynous cell and its sterile branch are bereft of floridean starch grains (Fig. 23 C; Bhatia and Vijayaraghavan, 1995 b). The same has been reported for *S. furcata* (Zablackis, 1987) and *S. proliferata* (Huisman, 1986). In *S. tokidae, S. okiensis* and *S. flabellata,* the mature carpogonial branch is however, 3-celled. Hypogynous cell divides longitudinally to form 1-3 sterile cells. The first cell of the carpogonial branch gives rise to pericarpial filaments (Kajimura, 1988).

In *Scinaia indica,* the carpogonium is the terminal cell of a 3-celled carpogonial branch. The hypogynous cell undergoes division to form a whorl of four, cruciately arranged cells which assume a spherical shape. In *S. bengalica; S. fascicularis; S. complanata* and *S. carnosa,* it was found that the hypogynous cell of the 3-celled carpogonial branch is the largest of the three cells (Krishnamurthy and Sundararajan, 1987).

In *Gloiophloea scinaioides,* the carpogonial branch is also 3-celled but the hypogynous cell produces three sterile branches. The carpogonia and the carpogonial branch in *Pseudogloiophloea berggrenii* and *Pseudoscinaia snyderae* are similar to *S. forcellata* (Huisman, 1985).

Bonnemaisoniaceae : In *Asparagopsis armata* and *A. taxiformis* the gametophytes are dioecious. In the female branch the 9th, 10th or 11th axial cell cuts off a periaxial cell which becomes the supporting cell of the three-celled carpogonial branch (Fig. 27 F). The mature carpogonial branch is surrounded by pericarp cells derived from the upper axial cells and the periaxial cells of the supporting branch. A long trichogyne extending from the carpogonium passes through an apical or subapical opening in the pericarp. The hypogynous cells cuts off a nutritive tuft which consists of several filaments made up of irregularly shaped cells that branch freely (Bonin and Hawkes, 1987; Jha, 1996).

Gelidiaceae : The carpogonia of both *Gelidium pulchellum* and *G. latifolium* are formed in the apical regions of the short and lateral axes whose growth is inhibited by the development of the carpogonia. The resulting cystocarp is always terminal. The development of the carpogonium is initiated by the formation of a lobe on a cell of a lateral branch of a limited growth, a short distance behind the apical cell of an axis. The lobe elongates rapidly and then gets cut off by a wall to form a sessile carpogonium which continues to elongate. The apical prolongation penetrates the surface of the thallus to form the trichogyne. There is a marked constriction at the surface of the thallus between the trichogyne and the basal part of the carpogonium. In contrast to the Bonnemaisoniaceae the nutritive tissue development commences only after fertilization. Due to late initiation, the nutritive tissue is not fully developed until after the development of the carposporophyte has commenced (Dixon, 1959). The nutritive filaments are branched and short-celled that develop primarily from the basal cells of vegetative filaments of the third order in the fertile area (Hommersand *et al.*, 1988).

CARPOSPOROPHYTE

Acrochaetiaceae : In *Audouinella alariae,* the fertilized carpogonium produces a single carposporophyte which cuts off terminally or laterally, 3 or 4 carposporangia (Lee and Kurogi, 1983). In *Chromastrum alariae* also, carposporophyte is simple with 3 or 4 carposporangia (Kuiper, 1983).

Batrachospermaceae : In the fertilized carpogonium of *Batrachospermum cayennense,* the trichogyne possesses a nucleus, since the spermatium before fertilization divides into two, out of which one remains in the trichogyne. In *B. beraense,* the fertilized carpogonium divides transversely and the gonimoblast filaments develop from the divided carpogonium. The lower cell of the divided carpogonium does not fuse with the hypogynous cell (Kumano and Ratnasabapathy, 1982). Yoshida (1952) reported in *Batrachospermum* sp. from Northern Kyushu Japan, that after fertilization, carpogonial branch enlarges, and many short laterals project from these large cells, which partially surround the developing carposporophyte and occasionally give rise to an elaborate structure with rhizoidal filaments. This structure simulates the main axis. The carposporophyte thus is situated outside the periphery of the whorls. The gonimoblast filament formation in *B. mahabaleshwarensis* follows the general pattern. The zygote divides and a lateral protuberance develops from the carpogonium. One of the daughter nuclei moves into the protuberance which then gets cut off from the carpogonium. This functions as gonimoblast initial and by repeated division forms the gonimoblast filaments. The terminal cells of the filaments behave as carposporangia and produce single carpospores bereft of resting periods (Balakrishnan and Chaugule, 1980).

Thoreaceae : The fertilized carpogonium in *Thorea okadai* produces 3-5 gonimoblast initials which divide to produce gonimoblast filaments, that grow downward towards the medulla, run parallel to the surface of the thallus, penetrate among the assimilatory filaments and result in the formation of widely diffuse carposporophyte. This type of diffuse carposporophyte is an important feature in identification of the genus. The gonimoblast filaments ultimately produce many short, erect branches, which again branch forming another order, that grow among the long assimilatory filaments. Carposporangia are formed on the tips of the gonimoblast filaments.

Helminthocladiaceae : In *Helminthocladia nizamuddinii*, the trichogyne remains partly or wholly intact with the dividing carpogonium for a long time and no change in the size of any cell of the carpogonial branch is observed (Husain and Shameel, 1990). The division of the zygote is variable in different species. It is either transverse or oblique or longitudinal in *H. nizamuddinii;* longitudinal in *H. calvadosii. f. comorinensis* (Krishnamurthy and Sundararajan, 1985; Desikachary, 1957); slightly oblique in *H. australis* (Umamaheswara Rao, 1974); transverse in *H. papenfusii* (Martin, 1939). In *H. nizamuddinii,* the gonimoblast filaments are produced as a result of division and redivision of the daughter cell, resulting in round mass of cells (Husain and Shameel, 1990). In *H. calvadosii. f. comorinensis* the fertilized carpogonium divides and produces 4 cells arranged tetrahedrally. Further divisions result in the cluster of cells from which gonimoblast filaments are initiated (Krishnamurthy and Sundararajan, 1985).

The gonimoblast filaments in *Helminthocladia nizamuddinii* are broader at base and elongate and slender distally (Husain and Shameel, 1990). The branching of gonimoblast filaments is copious but not dichotomous and they form conico-round cystocarp. In *H. calvadosii. f. comorinensis* the gonimoblast filments are branched and compactly arranged and form dense spherical knots. In *Helminthocladia nizamuddinii* the involucral filaments arise from the cells above and below the bearing cell of the carpogonial branch, whereas in *H. calvadosii. f. comorinesis* the sterile filament initials are put forth from the cells of the adjacent filaments of the carpogonial branch. In *H. nizamudinii,* 1-4 distal cells of the gonimoblast filaments form the carposporangia. After the formation of carpospores, the lower vegetative cells of the gonimoblast filaments give rise to new branches whose apical parts again produce carpospores which are borne singly and terminally on the gonimoblast filaments.

In *Helminthocladia australis,* and *H. macrocephala,* the lower cell fuses with the hypogynous cell (Kumano and Ratnasabapathy, 1982) But Desikachary (1955) reported that in *Helminthora lindaueri*, the carpogonium does not fuse with the hypogynous cell, but there is pronounced widening of the protoplasmic connection between the carpogonium and the hypogynous cell.

In *Liagora harveyana,* the cytoplasm in the fertilized carpogonium retracts from the trichogyne and cleaves transversely (Guiry, 1990). In *L. indica,* the trichogyne is usually cut off after fertilization and the carpogonium undergoes a transverse division and the upper cell gives rise to the gonimoblast initial (Krishnamurthy and Sundararajan, 1985). In *L. mannarensis,* after fertilization, the trichogyne is cut off and the fertilized carpogonium divides twice transversely, resulting in the formation of 3 rows of cells, all of which contribute in the gonimoblast formation. This is in contrast to all other species. In *L. amplectans*, after fertilization, a transverse division in the carpogonium is followed by a vertical division of the distal cell (Abbott, 1984). The gonimoblasts are produced by the rapid and subsequent division of the distal cell. In *L. norrisiae,* the gonimoblasts arise in tight clusters from the distal cells of the divided zygote (Abbott, 1984).

In *Liagora harveyana,* after fertilization, the supporting cells branch irregularly to form sterile filaments that become profuse, branch dichotomously and curve around the carpogonial branch (Guiry, 1990). In *L. amplectens* the sterile filaments are at first uniseriate but later at the time of carposporangia formation branch and become inflated. In *L. norrisiae,* the cystocarp generally lacks sterile filaments. However, adjoining assimilatory filaments produce 1-3 celled, sterile, uniseriate filaments which do not surround either the carpogonial branch or the carpospore mass (Abbott, 1984). In *L. valida,* the cystocarp is surrounded by an involucre of sterile filaments, whose cell shape and size are similar to those of vegetative filaments (Abbott and Mokoto, 1982). In *L. indica,* a pair of characteristic horizontal gonimoblast filaments arise on top of the carpogonial branch, by the division of the fertilized carpogonium (Krishnamurthy and Sundararajan, 1985). In *L. harveyana,* the carposporangia form 4 cruciately, decussately or irregularly arranged spores. This is the only Australasian taxon that produces multispored carposporangia which generally develop laterally from the cell below an existing terminal carposporangium (Guiry, 1990).

Nemaliaceae : In *Nemalion helminthoides,* following fertilization, the carposporophyte differentiates into 4 distinct regions namely the fusion cell, the sterile gonimoblast cell, the carposporangial mother cell and the carposporangia. The gonimoblast is formed by apically dividing, monopodial filaments of limited growth which may later become pseudodichotomous. The carposporangia differentiates terminally on gonimoblast filaments which may continue to grow sympodially. A single carposporangial mother cell produces carposporangia in several different directions as well as proliferates successive carposporangia within the sporangial wall that remains intact after carpospore liberation. As the carposporophyte matures, the gonimoblast initial, the stalk cell, the hypogynous and subhypogynous cells fuse (Anderson and Wetherbee, 1982).

Dermonemataceae : In *Dotyophycus corymbosus* the carpogonium divides transversely and has well developed diffused gonimoblasts. These consist of a basal, horizontally spreading gonimoblast fillments, that give rise to many erect, branched and a few celled filaments. The terminal cell of these filaments produce carposporangia (Krishnamurthy and Sundararajan, 1985). In *D. pacificum,* the fertilized carpogonium divides and produces initials from the lateral faces. These initials spread horizontally and develop into diffused gonimoblast (Abbott and Mokoto, 1981). Further, the terminal and lateral divisions of the carpogonium gives rise to the cystocarp. At maturity, the first formed cells of the gonimoblast fuse with the terminal cells of the carpogonial branch leaving an irregularly branched, darkly stained mass at the base of the cystocarp (Abbott, 1976). *D. corymbosus* is bereft of sterile involucral filaments (Krishnamurthy and Sundararajan, 1985).

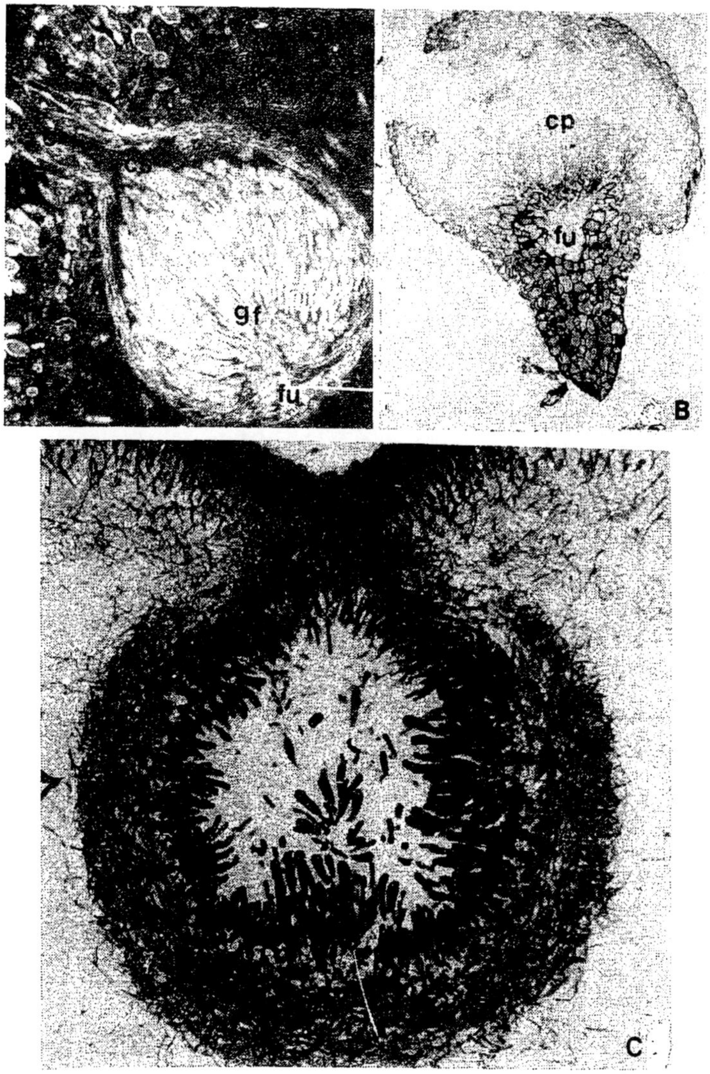

Fig. 28. A-C Carposporophyte (cystocarp).

A *Scinaia pseudocrispa*. Mature cystocarp with well developed ostiole. The carposporangia, carpospores, gonimoblast filaments and pericarp cells contain autofluorescent materials (after Bhatia and Vijayaragavan 1995b).

B *Asparagopsis taxiformis*. Longitudinal section of the mature cystocarp showing spore producing region and the stalk. Spores appear to arise from tree-trunk shaped fusion tissue. The carpospores are clavate with prominent nucleus. The stalk is rich in sulphated and carboxylated polysaccharides (after Jha, 1996).

C. *Nothogenia fastigiata*. Showing mature cystocarp with gonimoblast filaments lining the pericarp and projecting into the cystocarpic cavity, producing terminal carposporangia (after Huisman and Womersley, 1992).

Chaetangiaceae : In *Galaxaura oblongata,* a thick pericarp arises from the basal cell and surrounds the hypogynous cell and the carpogonium (Magruder, 1984). The carpogonium divides transversely and both cells produce gonimoblast initials. In *G. rugosa, G. obtusa* and *G. marginata,* a clear pericarp is not formed and the gonimoblast filaments grow laterally, curve distally and form the wall of the cystocarp (Huisman and Borowitzka, 1990). Further, in *G. cylindrica* and *G. oblongata,* the pericarp filaments project into the cystocarp cavity and to a varying degree intermix with the gonimoblast filaments (Magruder, 1984; Huisman and Borowitzka, 1990). In *G. oblongata,* the carpogonium, hypogynous and the basal cells fuse to form a large fusion cell (Magruder, 1984).

Tricelocarpa oblongata produces a cystocarp in which all the gonimoblast filaments line partially at the base of the pericarp. The carposporangia, thus, are present only in that region. In *T. cylindrica* the gonimoblast filaments line completely the pericarp and the carposporangia arise from all faces of the cystocarp cavity (Huisman and Borowitzka, 1990).

In *Nothogenia fastigiata,* the fertilized carpogonium loses trichogyne and even before the gonimoblasts arise, the pericarp is well developed. The first division of the fertilized carpogonium is transverse, with the gonimoblasts arising from both the resultant cells. The gonimoblasts are initially a mass of compact cells but subsequently they line the already established pericarp wall (Fig. 28 C).

In *Scinaia tokidae, S. okiensis* and *S. flabellata,* the gonimoblast initials arise either laterally, subterminally or terminally from the fertilized carpogonium (Kajimura, 1988). The upper half of the fully developed gonimoblasts consists of free filaments, all or several cells of which become carposporangia. In *S. furcata,* a single gonimoblast initial is cut off laterally at the base of the carpogonium (Zablackis, 1987). The gonimoblast filaments are decussately branched and the branches terminate in chain of 3-6 carposporangia. In contrast to *S. furcata,* a single gonimoblast initial appears to arise from the upper end of one of the hypogynous cells in *S. indica, S. bengalica, S. fascicularis and S. complanata* (Krishnamurthy and Sundararajan, 1987). However, contrary to the earlier reports, Bhatia and Vijayaraghavan (1995b) in *Scinaia pseudocrispa,* showed transfer of zygote nucleus from the carpogonium to the hypogynous cell through a pit-connection and further the origin of gonimoblast initial from the hypogynous cell (Fig. 23 D). Three or four gonimoblast initials arise from the hypogynous cell. At this stage extensive pericarp surrounds the developing carposporophyte. The gonimoblast initials fork repeatedly and differentiate into gonimoblast filaments.

Scinaia shows characteristic fusion cell formation. In *S. proliferata,* fusion cell is composed of the hypogynous cell and the carpogonium, but not the sterile branches on the hypogynous cell (Huisman, 1986). In *S. furcata,* the fusion cell comprises the gonimoblast initials, carpogonium, the hypogynous cell and the sterile branches. In contrast to *S. furcata,* the carpogonium in *S. pseudocrispa* does not take part in the fusion cell formation. The fusion cell consists of hypogynous cell, its sterile branches and the basal cell. The mature cystocarp opens through a well developed ostiole (Bhatia and Vijayaraghavan, 1995b; Fig. 28 A). In *S. tokidae, S. flabellata* and *S. okiensis,* this cell is made up of gonimoblasts, fertilized carpogonium, hypogynous cell and the first cell of the carpogonial branch, leaving sterile branch cell distinct (Kajimura, 1988). Fusion cell is not formed in *S. indica, S. bengalica, S. carnosa* and *S. complanata* (Krishnamurthy and Sundararajan, 1987), but in *S. fascicularis,* a fusion cell is formed due to coalescence of carpogonium and two or three of the hypogynous cells. A single gonimoblast initial arises from this fusion cell.

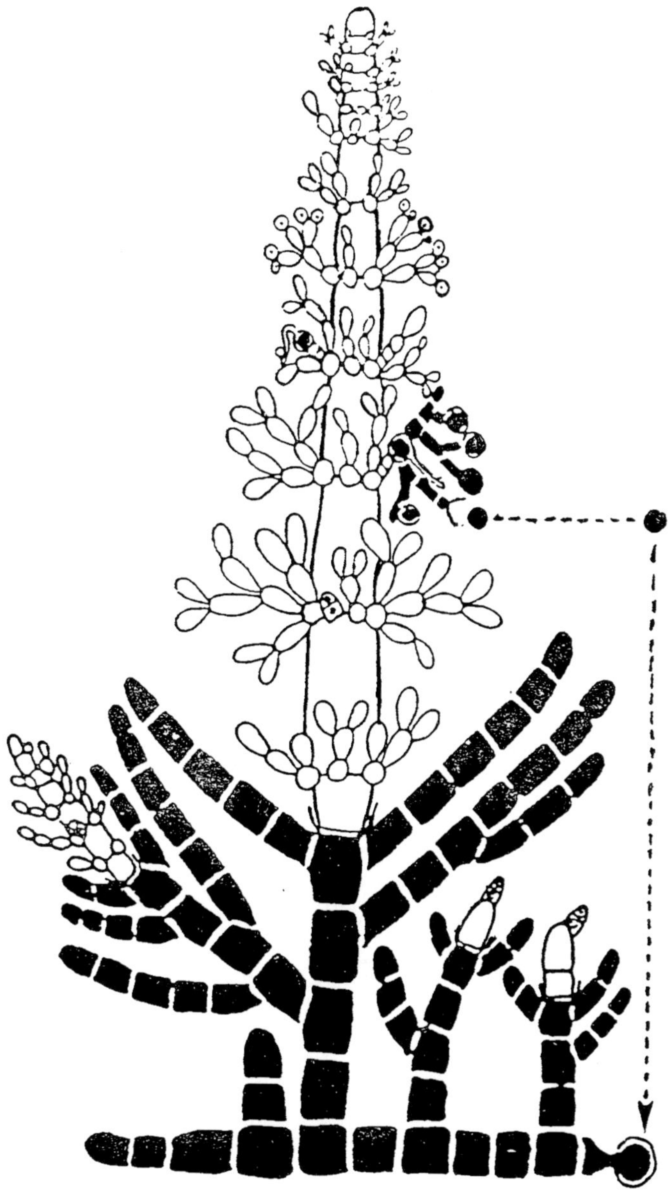

Fig. 29. Life-cycle of *Batrachospermum* (after Balakrishnan and Chaugule, 1980).

Schematic representation to bring out the telescoping of the three phases into a single composite entity. (solid black indicates diploid condition).

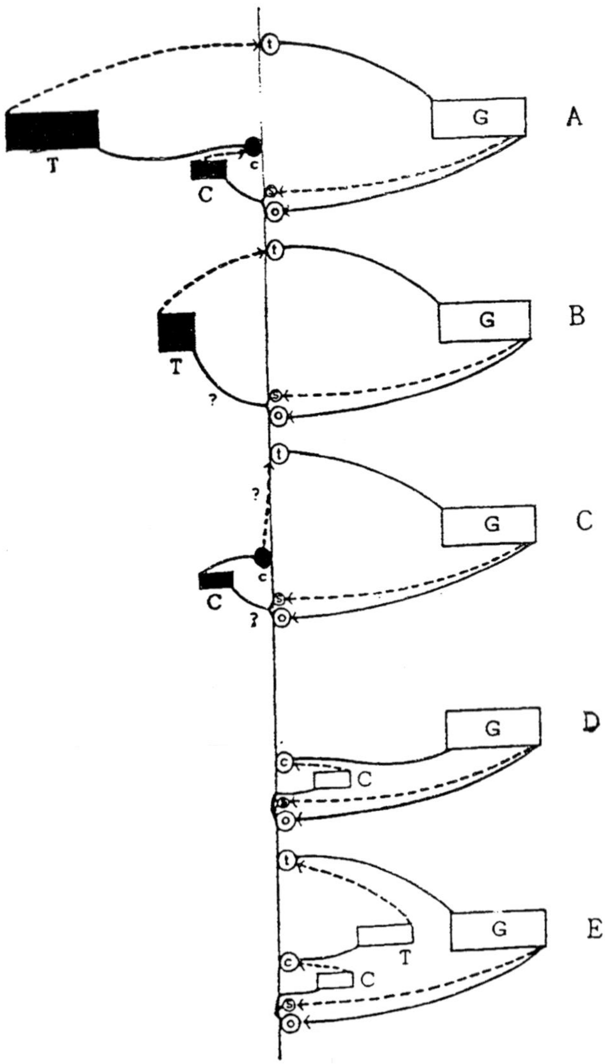

Fig. 30. Diagrammatic representation of the types of life-cycle found in the Florideae
(after Chihara, 1962)

(C, carposporphyte ; c, carpospore ; G, gametophyte; O, oosphere ; S. spermatium ; T, tetrasporophyte; t, tetraspore).

A. **Polysiphonia – type**

B. **Phyllophora brodiali – type**

C. **Liagora tetrasporifera – type**

D. **Nemalion – type**

E. **Asparagopsis – type**

(Solid area diploid ; hollow area represent haploid).

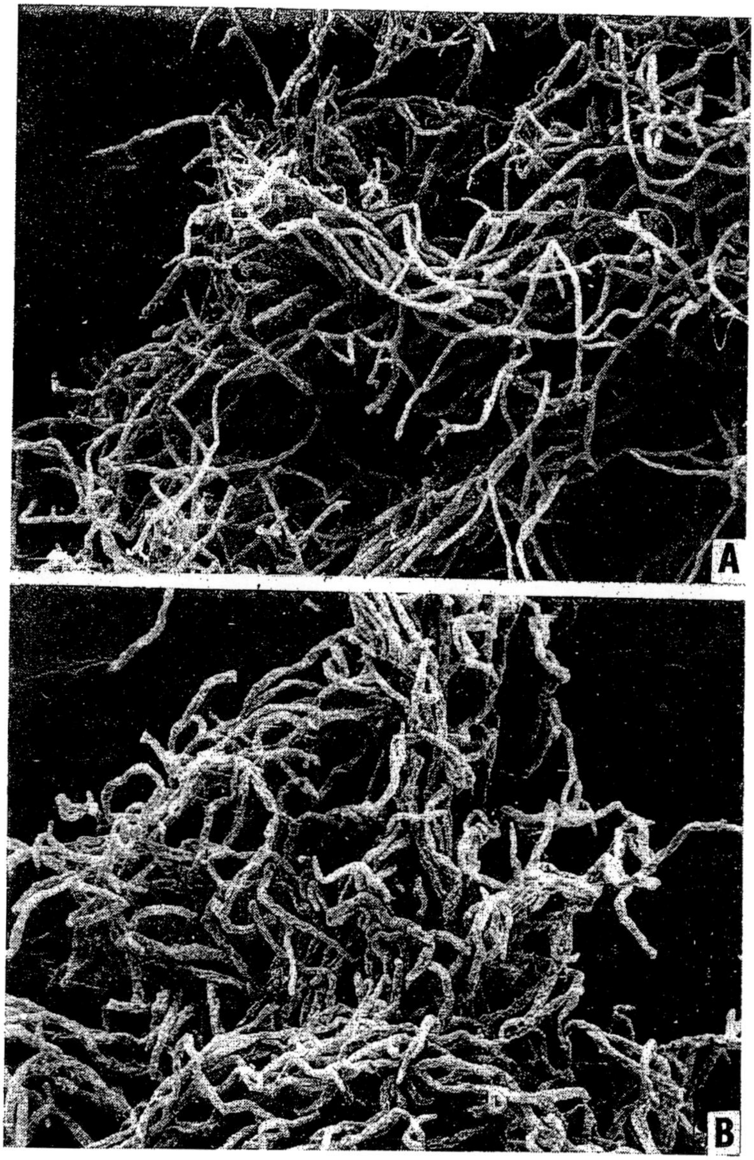

Fig. 31. A, B Scanning electron micrographs of *Falkenbergia sp.* (after Jha, 1996)

A. *F. rufolanosa*. The tetrasoporophyte has a relatively straight thallus and is not curled. The thallus does not shows dark bulges and thin constricted areas. The marking on the surface is parallel.

B. *F. hillebrandii*. The tetrasporophyte shows curled and twisted thallus, The thallus has a beaded appearence. Dark bulges are followed by light areas. The marking on the surface is reticulate.

In *Gloiophloea scinaiodes, S. forcellata Pseudoscinaia snyderae* and *Pseudogoiophloea berggrenii* fusion cell is formed by the fusion of carpogonium, hypogynous cell and basal cell of the carpogonial branch. The cystocarp in *S. furcata,* is broadly spherical to pyriform (Zablackis, 1987); *S. indica,* spherical or sub-spherical with a short or long terminal neck, opening through an ostiole or carpostome (Krishnamurthy and Sundararajan, 1987); in *S. bengalica* is flask shaped with an ostiole; in *S. complanata, S. carnosa* and *S. fascicularis,* urn-shaped; and in *S. tokidae, S. okiensis* and *S. flabellata*, pear-shaped with an ostiole (Kajimura, 1988).

Bonnemaisoniaceae: Svedelius (1933) in *Asparagopsis* sp. reported that the gonimoblast emanates from the hypogynous cell (Bonin and Hawkes, 1987). In *A. armata* the cystocarps are stalked and are usually found in pairs at the base of major determinate or inderminate branches. Within the mature cystocarp, is large fusion cell that resembles a large tree-trunk and possesses many gonimoblast branches emanating from its apex. Carposporangia are ovoid , pigmented and located at the base of the gonimoblast filaments. They are released through apical or subapical pore. Identical cystocarp construction is seen in *A. taxiformis* (Fig. 28 B; Jha, 1996).

Gelidiaceae: In *Gelidium* sp., the carpogonium after fertilization enlarges rapidly and becomes multinucleate. The trichogyne breaks off at the constriction and the carpogonium fuses with the hypogynous cell or other vegetative cell. Later long, nonseptate process develops from all parts of the multinucleate fusion cell and these ramify on both sides of the axial plate. Nuclei continue to increase in number, by division, in the fusion cell. Small primordia, containing usually a single nucleus, are formed at the tips of the processes. These primordia function as apical cells and divide to form a filament of elongate cells which constitutes the gonimoblast. Lateral branches of the gonimoblast are also formed resulting in a dense mass of interwoven filaments. As the gonimoblast develops the nutritive tissue gradually disappears .

The lateral branches of limited growth, which range between 4 and 8 cells in length when the carpogonia are formed, continue to elongate during carposporophyte development. They eventually reach a length from 12 to 16 cells. The lowermost cell of this branch elongates, bcomes thin and disintegrates, so that the outer parts of the cortex lyse from the axial plate to from the bilocular cystocarp. In *Gelidum pulchellum* and *G. latifolium* no aperture but only a superficial depression is seen in the cystocarp. In both the species the entire sporangia are detached from the gonimoblasts and accumulate in the cystocarp cavity. The carpospores are liberated by the ultimate decay of the wall of the cystocarp. Kylin (1828) and Smith (1955) however, report a definite pore in the cytocarp wall of *Gelidium cartilagineum* and the carpospores are shed through this aperture (Dixon, 1959).

Fritsch (1945) in *G. pulchellum* and *G. latifolium* considers that the cystocarps are compound structure containing many fertilized carpogonia. About 10 to 30 carpogonia are formed in each fertile axis. The involvement of more than one fusion cell in the contribution of gonimoblast is not been clearly shown (Dixon, 1959).

LIFE-CYCLE STUDIES

In Nemaliales the initial nuclear division within the fertilized carpogonium is meiotic (Fritsch 1945). The gonimoblasts arising from the carpogonium directly produce the carpospores, that are haploid and give rise to plants which again bear the sex organs. The zygote, therefore constitutes the only diploid stage, while there are two haploid stages. The life-cycle is, thus

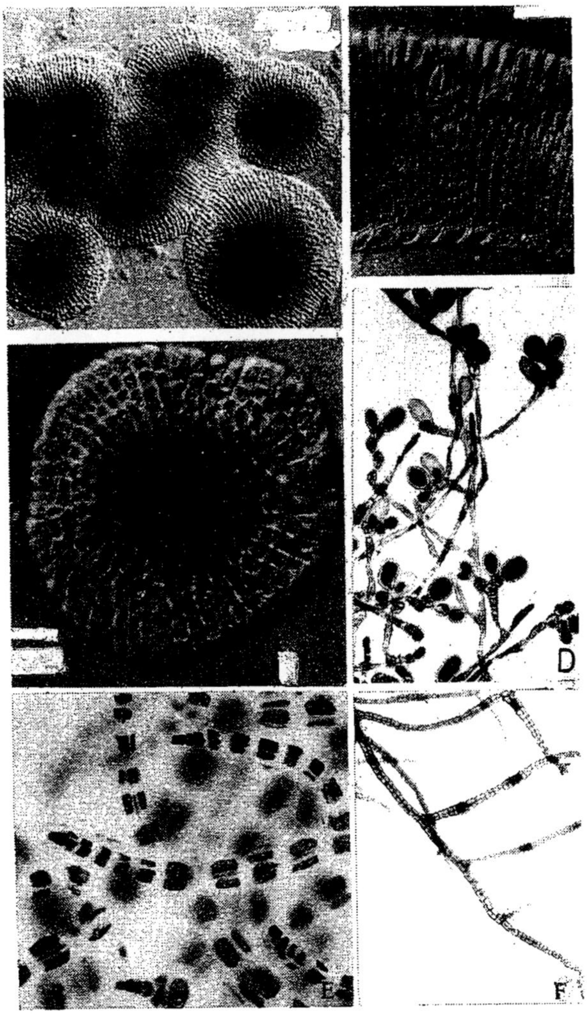

Fig.32. A-F. Tetrasporophyte (A-C after Anderson and Stegenga, 1985).

(b, basal cell; c, cohesive filaments; f. loose filaments; t, tetrasporangia)

A *Nothogenia erinacea.* Crustose tetrasporophyte

B *N. erinacea.* Same showing centrifugally radiating basal filaments.

C *N. erinacea.* Radial section of fertile tetrasporophytic crusts showing basal cell layer, cohesive filaments, loose filaments and tetrasporangium.

D *Liagora harveyana.* Formation of tetrasporangia under the long day conditions on uniaxial filaments (after Guiry, 1990b).

E *Falkenbergia hillebrandii.* Filamentous tetrasporophyte showing trisiphonous arrangement. An axial cell is surrounded by three periaxial cells (after Jha, 1996).

F *Falkenbergia rufolanosa.* Filamentous tetrasporophyte with cruciate tetraspores. The successive tiers are not separated (after Jha, 1996).

haplobiontic. Magne (1967) however, disregarded the idea of zygotic meiosis in the Nemaliales and emphasized that all Florideophycedae have diploid carposporophyte. Dixon (1973); Bold and Wynne (1978); and Balakrishnan and Chaugule, (1980) support this contention. In most of the Nemaliales, the life history is triphasic and reveals a sequence of gametangial, carposporangial and tetrasporangial phases. All the three phases are morphologically dissimilar. The carposporophyte phase develops on the gametophyte.

ACROCHAETIACEAE

The life history of *Acrochaetium* has been established as trigenetic. *A. corymbiferum* (Abdel Rahman et. al. 1990) and *A. secundatum* (Abdel-Rahman and Bidoux, 1989) have a trigenetic life-cycle with two nematothalian, heterotrichous, free living generations which are morphologically and cytologically similar. Abdel-Rahman and Magne (1983) proposed a term, sporozygote for *Acrochaetium pectinatum* where the zygote on liberation produces a tetrasporophyte. The zygote thus, is equivalent to the normally developed carposporophyte of other taxon of Rhodophyta.

In vitro studies have also been conducted to elucidate the morphogenetic response during life-cycle. In *Acrochaetium unifilum,* the tetrasporophyte formation is insensitive to photoperiod, whereas in *A. asparagopsis* the tetrasporophyte formation is a result of short-day response (Abdel-Rahman, 1981). It was shown that maximum inhibition of the response occurred with night-breaks given at approximately 24 hr intervals and coincided with the scotophilic phase while maximum stimulation was due to night-breaks in the postulated photophilic phase.

Life-cycle studies in *Audouinella botryocarpa* have shown that plants produced monosporangia throughout the year and tetrasporangia from December to May, but no gametangial plants were formed (Guiry *et al.,* 1987). In *A. alariae,* the fertilized carpogonium similarly produces rather simple carposporophyte with a few carposporangia which are cut off from the carpogonium either terminally or laterally. The gametophyte of *A. alariae* occurs throughout the year (which is in contrast to *A. botryocarpa*) ; the carposporophyte from October to February and the tetrasporophyte from December to May (as in *A. botryocarpa*) (Lee and Kurogi, 1983).

In vitro studies conducted in *Audouinella* indicate that tetrasporophyte formation is a short-day response with day length shorter than 10 hrs and temperature 8°C, 10°C and 11°C but are not formed at 6.5°C and 15°C. Plants raised from monosporangia continued to form monosporangia, but under inductive conditions formed tetrasporangia. Nightbreak of 1 hr dark period prevented tetrasporangial formation confirming its short day requirement (Guiry *et al.*, 1987).

BATRACHOSPERMACEAE

The life-cycle of *Batrachospermum mahabaleshwarensis* is triphasic with a haploid gametophyte alternating with a diploid carposporophyte and a diploid tetrasporophyte (Balakrishnan and Chaugule, 1980). Magne (1960) has suggested that the pseudochantransia stage is diploid and meiosis occurs in the pseudochantransia apical cells. Stosch and Theil (1979) have demonstrated in *Batrachospermum,* production of cladomes directly from the prostrate pseudochantransia. Balakrishnan and Chugule (1980) have suggested in *B. mahabaleshwarensis* that all the three phases are telescoped into one another as a single entity (Fig. 29). Stosch and Theil (1979) state that the plant is a chimara composed of diploid pseudochantransia with the haploid gametophyte cladome grafted on to it and after fertilization, the diploid carposporophyte grafted on the haploid cladome. Drew (1955) considers the chantransia stage of *Batrachospermum* as a somatic phase christening it as protogametophyte. It shows defined and definite morphological features and generates reproductive cell namely monospores. The differentiation of chantransia stage into the thallus is localized and sudden. Yoshida (1959) is of the opinion that chantransia should be referred to as a stage and not as a generation or somatic phase in the life cycle. He considers the number of generations or somatic phases as two, namely the thallus of *Batrachospermum* (including the chantransia stage) and the carposporophyte. *Batrachospermum* has thus a diphasic life-cycle and meiosis apparently takes place immediately after fertilization. Yoshida, (1959) reported in *B. moniliforme* that reduction division takes place in the carpogonium immediately after fertilization.

LEMANEACEAE

The life-cycle in *Lemanea* is triphasic and resembles that of *Batrachospermum mahabaleshwarensis.* The reduction division takes place in the apical cell of the diploid pseudochantransia phase that differentiates into gametophyte. The haploid gametophyte, the diploid carposporophyte and the diploid tetrasporophyte are telescoped into a single composite entity (Magne, 1967).

HELMINTHOCLADIACEAE

Liagora harveyana forms quadripartite carposporangia. The spores from the divided carposporangia germinate to form uniaxial filaments with barrel-shaped cells that directly form erect fronds, without the formation of tetrasporangia. A large number of tetrasporangia, however, were formed under long day conditions (Fig. 32 D; Guiry, 1990) with tetraspores arranged either in cruciate or decussate manner.

Dixon (1973) on the basis of cytological investigation and culture studies of *L. tetrasporifera* designated the term "Liagora tetrasporifera-type" life history where meiosis occurred in sporangia, formed on gonimoblast filaments.

NEMALIACEAE

The life-cycle in *Nemalion* is diphasic with zygote as the only diploid stage. *N. helminthoides* shows circadian rhythm in the induction of erect axis (Cunningham et al., 1989). Two isolates from west-coast of Ireland showed a heteromorphic life-history in which tetraspores, formed under short-day conditions on uniaxial prostrate tetrasporophytes gave rise to uniaxial prostrate growth, identical in morphology to the tetrasporophyte. Long day photoperiodic response induced tetraspores to produce multiaxial and erect plants.

CHAETANGIACEAE

The life-history of *Scinaia* was thought to be haplobiontic, but culture experiments on *S. forcellata* (Boillot, 1968) and *S. complanata* (Van Den Hoek and Cortel-Breeman, 1970) proved it to be diplobiontic. In these species the carpospores grow into filamentous tetrasporophyte and hence there is an alternation of a macroscopic gametophyte with a microscopic filamentous tetrasporophyte making the life-cycle heteromorphic.

In *Nothogenia erinacea* (*Chaetangium erinaceum*) instead of the filamentous tetrasporophyte, a crustose tetrasporophyte (Fig. 32 A) is produced. The carpospore from the foliose female gametophyte gives rise to crusts which are cruoriopsis-like tetrasporophytic phase consisting of centrifugally radiating basal filaments (Fig. 32 B) and supporting cohesive upright filments that later forms loose filaments. Tetrasporangia develop on the loose filaments (Fig. 32 C) and the

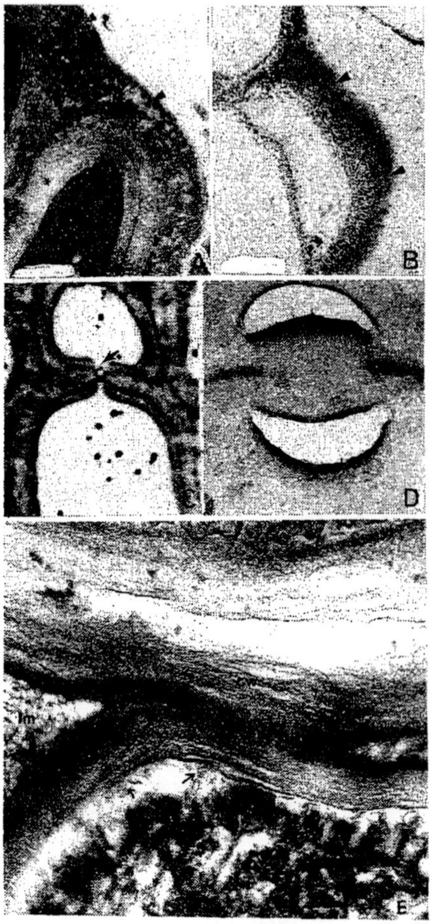

Fig. 33 A-E Histochemistry and Ultrastructure.

A *Lemanea annulata.* Portion of cell wall showing two distinct regions namely the fibrillar wall material and the amorphous region (after Gretz, *et al.*, 1991).

B *L. annulata.* Same showing CBH/ colloidal gold labelling concentrated over fibrillar appearing areas (after Gretz, *et al.*, 1991).

C *L. funica.* Longitudinal section of pit-connections showing pit-plugs (arrow) stained with PAS indicating glycoprotein nature (after Trick and Pueschel, 1991).

D *Liagora valida.* Longitudinal section of a pit-connection showing the activity of proteinase K. The thin outer cap-layer appears as a hole as a result of proteinase K digestion indicating proteinaceous nature. The hole looks like domes (after Trick and Pueschel, 1991).

E *Scinaia pseudocrispa.* Portions of adjacent hypodermal cells to show typical fibrillar and layered cell walls whose outer most layer is electron-dense. The cellulose frameworks (fibrillar materials) are present in a mucopolysaccharide matrix. Prominent plasmalemma profiles are seen near the inner wall layers (arrows) (after Bhatia, 1993).

tetraspores are either cruciate and decussate/cruciate type. The tetraspore again gives rise to the crusts. The gametophytes in *N. erinacea* and *N. ovalis* arise from the elongation of the subsurface cells of the crusts. These form the medulla consisting of many layered cortex. The gametophyte develops rhizoids and connection with the crust is obscured (Anderson and Stegenga, 1985).

Galaxaura has essentially isomorphic type of life-cycle. *G. oblongata* has a triphasic life-history with conspicuous gametophyte and small filamentous tetrasporophyte. The carpospores germinate both in unipolar and bipolar manner resulting in small filamentous thalli. Under short-day conditions cruciate tetrasporangia are produced in small clusters. Tetraspores germinate to form small filamentous thalli. Zygotic meiosis occurs (Magruder, 1984).

In nature, *Gloiophloea scinaioides* has only gametophytic plants (Huisman, 1987). *In vitro* studies have shown a triphasic heteromorphic life-history consisting of three phases namely the fleshy gametophyte which after fertilization produces the carposporophyte that produces the carpospores which after release germinate into filamentous tetrasporophyte with meiosis presumably occurring during the formation of tetraspore. The released tetraspores germinate into filamentous protonemal gametophytes which are morphologically identical with the tetrasporophytes. The fleshy gametophyte directly arises from the filamentous protonema. The species is capable of reverting to the filamentous tetrasporophyte and later as the protonemal gametophyte. The life–history resembles the related genera *Scinaia* (Boillot, 1968, 1969, 1971, 1972), *Pseudogloiphloea* (Ramus, 1969) and *Galaxaura oblongata* (Magruder, 1984).

BONNEMAISONIACEAE

The culture studies of taxa belonging to this family reveal three types of life-cycles namely Asparagopsis-type, Nemalion-type and Polysiphonia-type (Fig. 30).

(a) The Asparagopsis-type represents triphasic alternation of generation in which haploid gametophyte, haploid carposporophyte and haploid simple and small tetrasporophyte succeed one another. This type of life cycle is shown by *Asparagopsis taxiformis* (Chihara, 1961); *A. armata* (Feldmann and Feldmann, 1942) and *Bonnemaisonia hamifera* (Harder and Koch, 1949; Segawa and Chihara, 1954; Chihara, 1961).

Interestingly the carpospores of *Bonnemaisonia hamifera* gives rise to *Trailiella intricata* and later the tetraspores produced plantlets of *B. hamifera*. So *Trailiella* is now believed to be only a phase in the Asparagopsis-type life-cycle of *B. hamifera* (Chihara, 1961). The carpospores of *Asparagopsis taxiformis* similarly gives rise to the plantlets of *Falkenbergia hillebrandii* (Chihara, 1961). The tetrasporophyte is smaller than the gametophyte. The tetraspore develops into plants of *A. taxiformis*. Thus *Falkenbergia hillebrandii* is not a distinct taxon, as earlier believed, but a tetrasporophytic phase in the life-cycle of *A. taxiformis*. Jha (1996) observed that morphologically *Falkenbergia hillebrandii* (tetrasporophytic phase of *Asparagopsis taxiformis*) differs from, *F. rufolanosa* (tetrasporophytic phase of *A. armata*). *F. hillebrandii* shows, curled, twisted and moniliform thallus with dark bulges followed by light thin areas (Fig. 31 B). It reveals trisiphonous thallus with a distinct axial strand surrounded by three pericentral cells (Fig. 32 E). The successive tiers are separated by mucilage. In contrast, *F. rufolanosa* thallus is not curled and monoliform but relatively straight and further does not show dark bulges and thin constricted areas (Fig. 31 A). The successive tiers are not separated by mucilage (Fig. 32 F). It however, possesses trisiphonous thallus with a distinct axial strand surrounded by three pericentral cells.

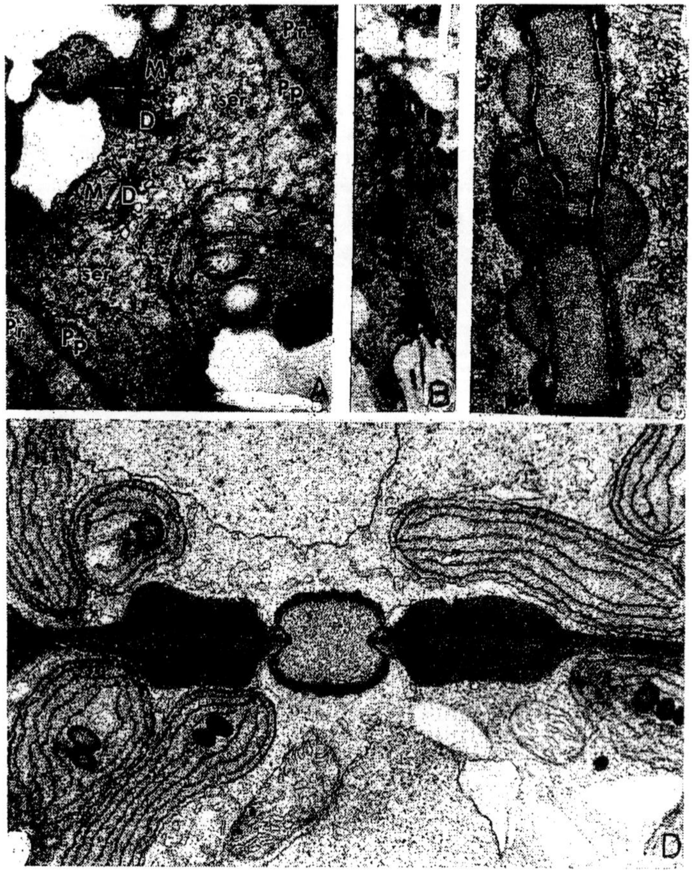

Fig. 34 A-D. Ultrastructure of pit-connection.

(d, dictyosome; m, mitochondria; pp, pit-plug; pr, pit-ring; ser, smooth endoplasmic reticulum).

A *Batrachospermum sirodotii.* Longitudinal section of a developing axial cell showing conically shaped networks adjacent to the pit-connections. Dictyosome and mitochondrial complexes are located in the centre of the cells and separate the inner elements of two ER networks (after Aghajanian and Hommersand, 1978).

B *B. sirodotii.* Median longitudinal section of the pit-connection showing electron-dense, globular structures (*) indicating areas of alkaline phosphatase activity (after Aghajanian and Hommersand, 1978).

C *B. sirodotti.* A median longitudinal section of a pit-connection between two cells of the indeterminate axis. The dough-nut-like structure of pit-plug is seen (after Aghajanian and Hommersand, 1978).

D *Pseudogloiophloea* sp. Longituidinal section through a mature pit-connection of the gametophytic filamentous phase showing septum elaboration to form a ring (after Ramus, 1969).

(b) Nemalion-type shows diphasic alternation of generation in which the haploid gametophyte alternates with the haploid carposporophyte as in *Delisea fimbriata* (Chihara, 1961), *D. elegans* (Levring, 1953) and *Ptilonia okadai. Delisea fimbriata* from Japanese coast lacks tetrasporophytic phase and follows the Nemalion-type but those from Australian coast shows Polysiphonia-type. The reason for this unusual behaviour of identical plants separated by space is unknown. In *Ptilonia okadai* and *Delisea fimbriata,* the carpospores germinate and give rise to an expanded monostromatic basal disc which closely resembles *Hymenoclonium serpens.*

(c) The Polysiphonia-type is triphasic with three generation. Haploid gametophyte, diploid carposporophyte and diploid tetrasporophyte, morphologically identical to gametophyte is seen as in *Delisea fimbriata* from Australian coast (Levring, 1953). The tetrasporophyte is diploid and morphologically identical to the gametophyte. In vegetative condition it is indistinguishable from the gametophyte whereas in Asparagopsis-type, the tetrasporophyte is haploid and morphologically simple and smaller than the gametophyte. Later it was shown that such a classification is not correct. Magne (1960) stated that the production of carpospores is not accompanied by reduction division in *Bonnemaisonia asparagoides*, but probably reduction division takes place during the tetrad formation in *Hymenoclonium serpens*. Jha (1996) showed that meiosis in *Asparagopsis armata* occurs in tetrasporangium of Falkenbergia phase (Fig. 32 F). This proves that the life-cycle of Asparagopsis-type seems to proceed through triphasic alternation of generation where haploid gametophyte, diploid carposporophyte and diploid tetrasporophyte morphologically simpler and smaller than the gametophyte follows one after another.

GELIDIACEAE

The life-cycle in Gelidiaceae is triphasic, heteromorphic with haploid gametophyte, diploid carposporophyte and diploid tetrasporophyte. Tetrasporangia are usually borne on distinct individuals. In *Gelidum sp.* the tetrasporangia occur in the flattened, inflated tips of the terminal lateral axes. The tetraspore bearing tips are similar in structure to the main axis, but they are almost transparent. The tetrasporangia develop from small, single celled primordia formed laterally on the lateral branches of limited growth. The tetrasporangia are cruciate but tetrahedral arrangement occurs in *G. latifolium* (Dixon, 1959). The tetraspores are liberated by the lyses of sporangium wall.

In *Gelidiella acerosa* the tetrasporangia are produced in an acropetal succession and are liberated from the lower end of the ramuli. New tetrasporangia continue to form at the upper ends. The termination of the tetrasporangial ramulus is swollen due to the formation of tetrasporangia but at the end of the reproductive season these fertile ramuli becomes cylindrical with young tetrasporangia at their distal ends. The tetrasporangial mother cell in *Gelidiella acerosa* develops as terminal cell of a lateral branch of the third or fourth order. The mature tetraspores are liberated through opening between the surface of the cells (Rao, 1970). The Gelidiaceae are characterized by a Polysiphonia—type of life-history in which the male, female and tetrasporangial plants are nearly identical morphologically (Kylin, 1923). Asexual reproduction is by tetrasporangia (Hommersand et al., 1988).

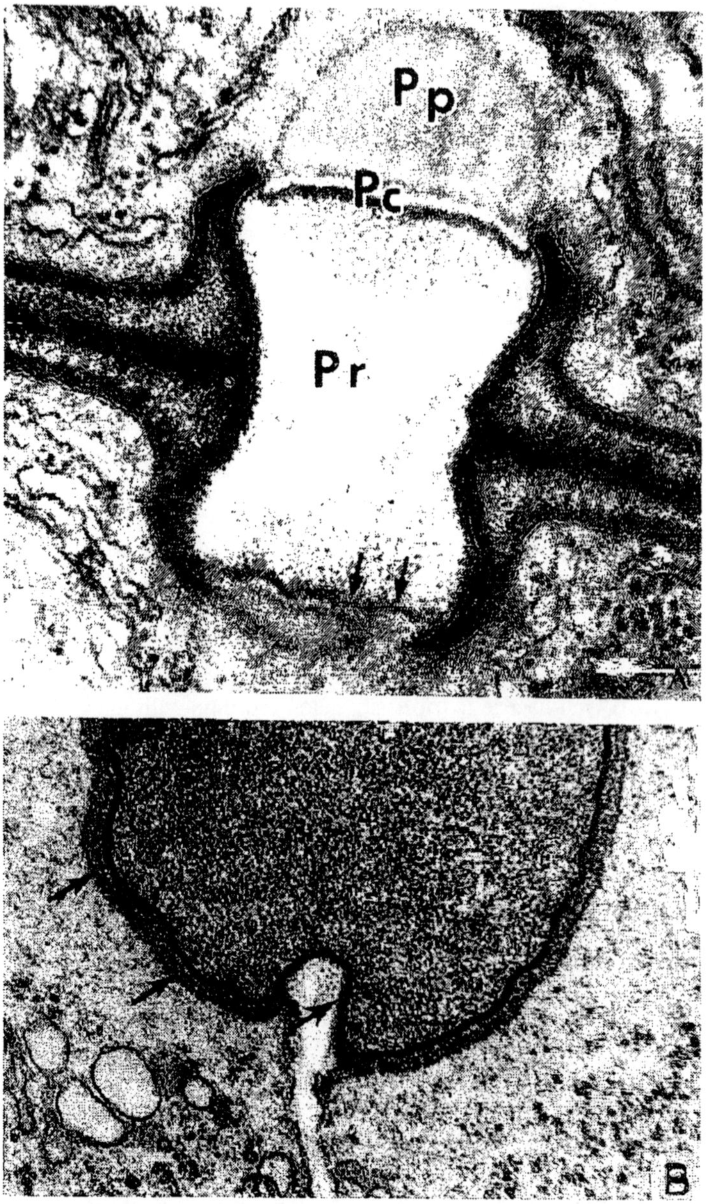

Fig. 35 A, B. Ultrastructure of pit-connection.

(pc, pit-cap; pp, pit-plug; pr, pit-ring).

A *Batrachospermum sirodotii.* Longitudinal section of pit-cap showing an electron-dense and electron-translucent layer (arrow) (after Aghajanian and Hommersand, 1978).

B *Pseudogloiophloea sp.* Longitudinal section through a mature pit-connection of the gametophytic filamentous phase showing plasmalemma around septal aperture (upper arrow) and membrane bounding the plug (lower arrow) (after Ramus, 1969).

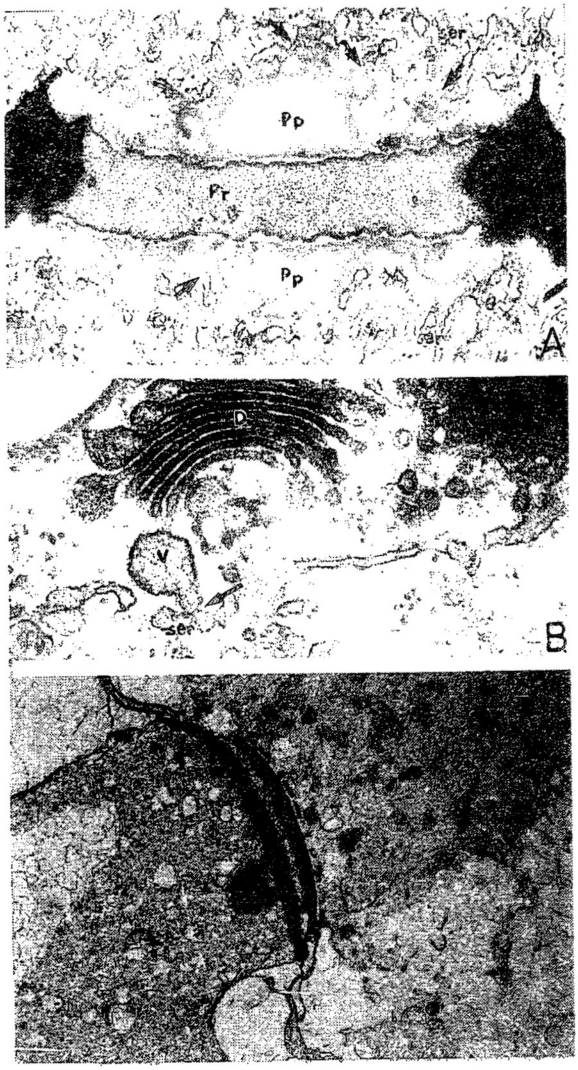

Fig. 36. A-C. **Ultrastructure of pit-connection.**

(d, dictyosome; pc, pit-cap; pp, pit-plug ; pr, pit-ring; ser, smooth endoplasmic reticulum; v, vesicle).

A. *Batrachospermum sirodotii.* Longitudinal section of the cell to show endoplasmic reticulum releasing the contents directly on the surface of the developing pit-plug (after Aghajanian and Hommersand, 1978).

B. *Batrachospermum sirodotii.* Same to show dictyosome vesicles fusing with the smooth endoplasmic reticulum (after Aghajanian and Hommersand, 1978).

C. *Asparagopsis taxiformis.* Longitudinal section of the cell to show lenticular pit-connection. The cap layer is absent and only a plug-core and cap-membrane is present. The cap membrane is continuous with the plasmalemma and the plasmalemma continues from one cell to the next along the sides of the pit-plugs (after Jha, 1996).

ULTRASTRUCTURAL AND HISTOCHEMICAL STUDIES

CELL WALL

In *Lemanea annulata* the cell wall is multilayered. In the outer portion of cortical cell wall two regions (Fig. 33 A) are distinct, namely, fibrillar wall material and an amorphous region of variable thickness. Colloidal-gold labelling of isolated cell wall preparation showed that the gold-complex deposition was concentrated over fibrillar-appearing areas but lacked such accumulation over the amorphous region (Fig. 33 B; Gretz, 1991).

Dawes *et. al.*, (1961) found that in *Helminthocladia californica* the microfibrils are not marked with any amorphous matrix and are reticulately arranged. The microfibrillar pattern of the young and the old filaments of the medulla and young cells of the cortex is essentially same. Further, when two non-related filaments are in contact, the radial walls show pit-connections.

According to Dawes *et. al.*, (1961) in the order Nemaliales, the ultrastructure of cell wall is uniform where loose reticulate pattern of microfibrils are embedded in amorphous matrix. However, in *Scinaia pseudocrispa* (*S. forcellata*) the microfibrils are parallelly arrranged. The cell wall shows multilayered, fibrillar structure and consists of an outer layer of electron-dense, mucilaginous material and is followed by an inner layer of parallel microfibrils embedded in an amorphous, mucopolysaccharide martix (Bhatia, 1993, Fig. 33 E).

PIT-CONNECTION

These are structures which are exclusive to Florideophycideae but also occur in the conchocelis-stage of the Bangiophycideae. Incomplete cytokinesis leads to holes in the cell walls which are occluded by plug-like structures called pit-connection /pit-plug (Pueschel and Cole, 1982).

Rascio et. al., (1991) found that in *Pterocladia capillacea* the pit-connections are small in the cortical cell and large in the medullary region. Pit-connections though vary in shape, show uniform pattern of organization in having a plug-core with peripheral granular region where the granules are arranged in parallel rows perpendicular to the pit-connection surface; a median region with irregularly distributed granules and a central region with fine clusters of granules and vesicles.

In contrast to *Gelidium robustum, G. pusillum, G. purpurascens* and other members of Gelidiaceae, *Pterocladia capillacea* lacks cap-membrane in the pit-connection. The plasma membrane, between the cell wall and sides of the pit-connection, does not fork to cover the pit-connection. The PTA staining was negative which also proved the lack of cap-membrane.

The PATA test for pit-connection showed it to be positive for carbohydrates and proteins which are present in both the plug-cap and the plug-core. Negative staining techniques have shown that the cap-membrane is absent in *Pterocladia capillacea* and *Gelidium pusillum* (Rascio, 1991). This shows that though pit-connections are considered uniform in the Gelidiaceae; *G. pusillum* and *P. capillacea* are exception in lacking cap-membrane.

The pit-connection in *Batrachospermum sirodotii* differ in axial and periaxial cells and between axial and juxtaposed pleuridial cells. The classical type of Rhodophycean-plug which is simple occluded type, occurs between the axial and the periaxial cells. The second type has dough-nut shaped, pit-ring with a central-pore. The pit-plug is a rivet-shaped structure which is

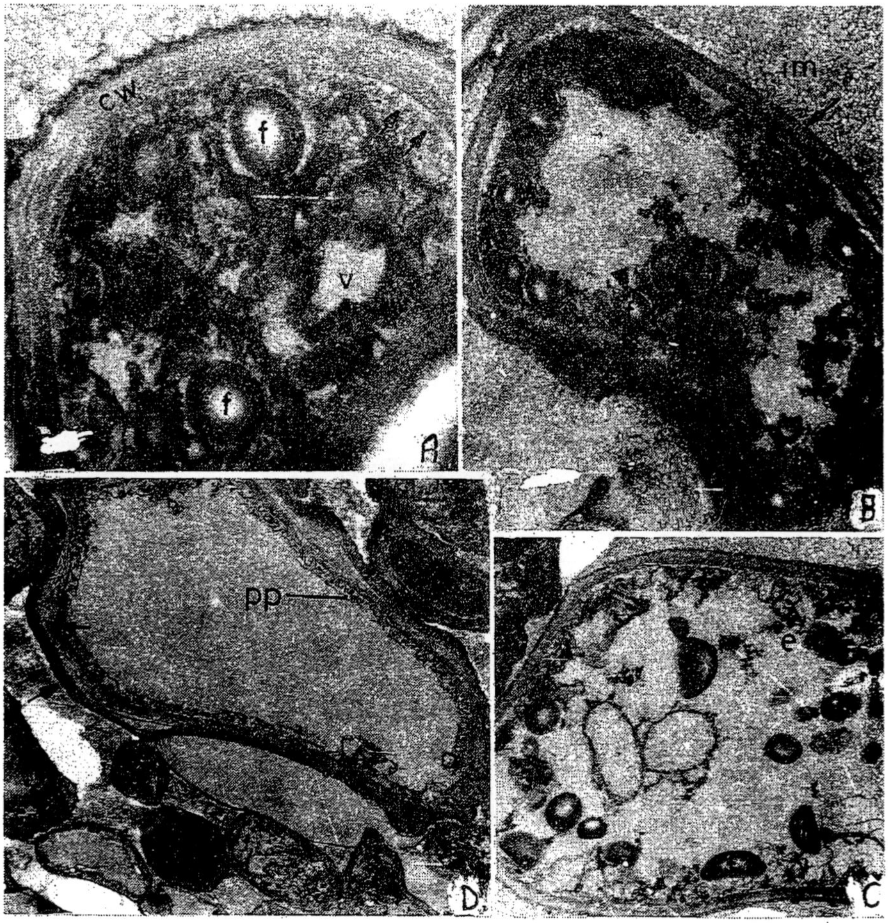

Fig. 37. A-D Scinaia pseudocrispa. Utricles (After Vijayaraghavan and Bhatia, 1995a).

(*cw*, cell wall ; *e*, electron-dense material; *f*, floridean strach; *in*, intercellular mucilage; *pp*. plasmalemma profile; *v*, vacuole)

A. An assimilatory epidermal cell showing several layers of fibrillar wall materials surrounded by extra-cellular electron-dense materials. The cytoplasm contains floridean starch, electron-dense materials and vacuoles. The plasmalemma profiles (arrows) near the cell wall are prominent.

B. A developing utricle showing vacuolate cytoplasm with dense materials and floridean starch. The starch grains are bound by membranous profiles (arrow-head). The intercellular mucilage is embellished with a fine network of honey-comb patterned fibrils.

C. A developing utricle to show a few round to kidney-shaped starch grains and electron-dense materials in the cytoplasm.

D. A developed vacuolate utricle with a multilayered fibrillar wall and peripheral layer of cytoplasm. The fibrillar wall is further surrounded by extracellular electron-dense materials (double-arrow). An additional layer of wall material inner to the cell wall accrues (arrow). Plasmalemma profiles are prominent.

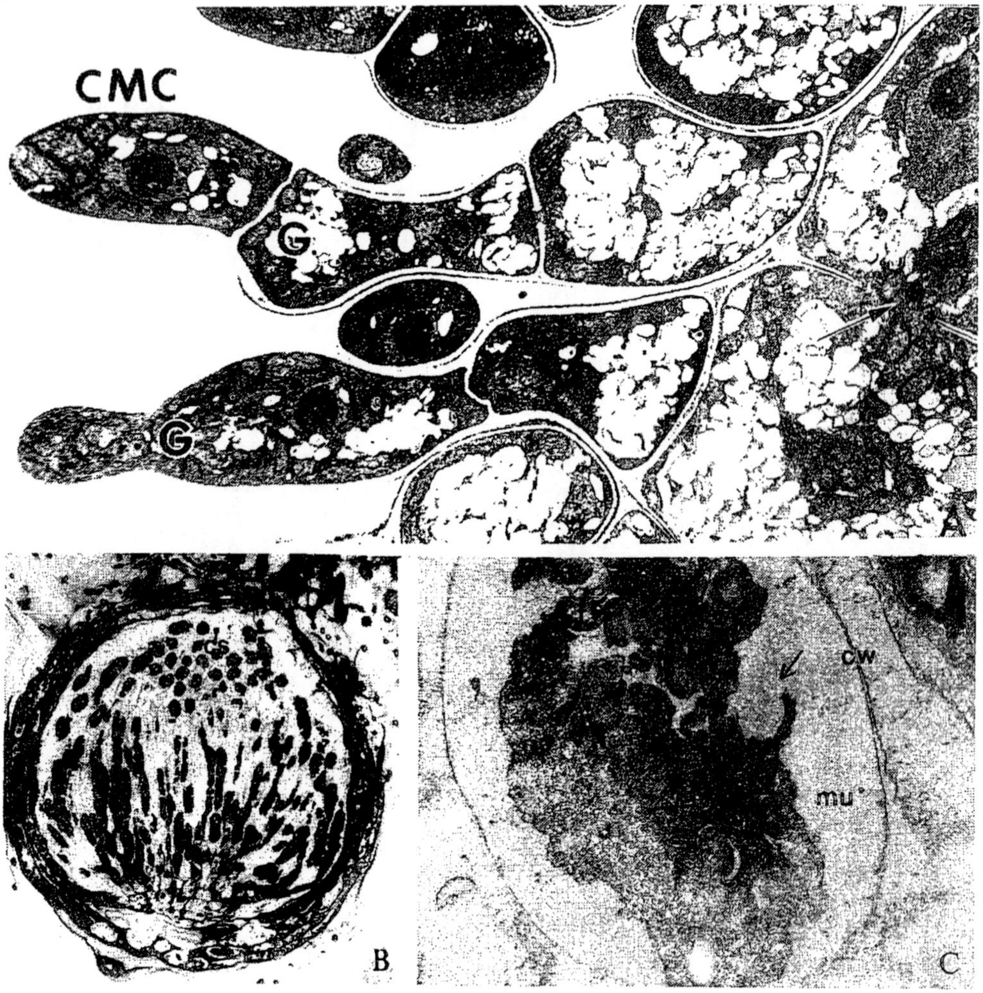

Fig. 38. A-C Cystocarp Ultrastructure.

(cmc, carposporangial mother cell; dv. distal vacuole; g, gonimoblast apical cells; pv, proximal vacuole).

A. *Batrachospermum ectocarpum* Longitudinal section of elongate gonimoblast apical cells prior to karyokinesis (after Scott, 1983).

B. *Scinaia pseudocrispa*. Longitudinal section of mature cystocarp. The carpospores are replete with floridean starch (after Bhatia and Vijayaraghavan, 1995b).

C. *Scinaia pseudocrispa*. Transmission electron micrograph of carposporangium. The cytoplasm is packed with floridean strach and vesicles. The carposporangial walls are fibrous and compact and the cytoplasm is enclosed in mucilaginous sheath. The vesicles are rich in fibrous materials that are released into the mucilaginous sheath (arrow) (after Bhatia and Vijayaraghavan 1995b).

continuous and occludes the pit-ring. This type of pit-connection is seen between the axial cells and juxtaposed pleuridial cells. The ontogeny of both types of pit-connections appears to be identical (Aghajanian and Hommersand, 1978).

There are three stages in the development of pit-plug in *Batrachospermum sirodotii* (Aghajanian and Hommersand, 1978). In the first stage, the initial polymerization of materials that form pore-plug or pit-ring occurs. The pit-ring and pore-plug are formed by centripetal polymerization of the contents of endoplasmic reticulum derived vesicles. The endoplasmic reticulum has elongate cisternae which are aligned parallel to each other but perpendicular to the plane of the septum. Within the septal-pore the microtubules are also oriented parallel to the elongate cisternae. During the second stage, the pit-cap is differentiated which is composed of an inner, 17-21 nm electron-dense and and outer 9-13 nm electron-translucent layer (Fig. 35 A). The pit-cap is membrane-bound. The third and final stage of development culminates in the formation of pit-plug or pit-domes. A central mound of material is deposited initially by the smooth endoplasmic reticulum which forms extensive network adjacent to either side of the pit-connections. The cytoplasm surrounding this ER-network is free of ribosomes and other organelles. The space occupied by the ER network is more or less conical in shape with the wider end of the cone lying towards the pit-connection and the narrow end generally associated with a dictyosome-mitochondrion complex towards the centre of the cell (Fig. 34 A). The cone appears to be physically isolated, at least in part, from the ribosomal and organellar cytoplasm by a layer of ER.

The ER network directly releases the contents on the surface of the developing pit-plug (Fig. 36 A). The material deposited on the pit-connection appears to be synthesized by the dictyosomes present in the centre of the cell. The vesicles immediately fuse with the adjacent elements of the ER (Fig. 36 B). When dictyosomes are in close association of pit-connection they directly deposit vesicles on the surface of the pit-plug.

The median longitudinal section of the pit-connection between juxtaposed axial cells reveal the pore through the pit-ring and the rivet shaped pit-plug (Fig. 34 C.) These pit-plugs undergo localized lyses, fifteen to twenty cells behind the apical cell. Cytochemical studies show alkaline phosphatase activity. This hydrolytic enzyme is involved in the pit-plug lysis (Fig. 34 B). Only the perforated occluded type of pit-plug lyse and not the simple occluded type.

The only possible explaination of the formation of such an elaborate structure and its subsequent breakdown could be that young cells require physical and physiological isolation provided by the pit-connection. But as soon as the cell attains maturity, these pit-connections are not needed by the cells, become redundant, and finally obliterate.

Ramus (1969) studied the ultrastructure of the pit-connection of *Pseudogloiophloea* and showed that the structure of the pit-connection in the filamentous, gametophytic phase is different from the fleshy, gametophytic phase thereby implying that the pit-connection is dimorphic. In the fleshy gametophytic phase the septum remains free. The plasmalemma is continuous around the rim of the aperture (Fig. 35 B). The pit-plug is bound by a dense 200 A^0 thick layer and membrane that surrounds the plug is obscured by the thick layer of amorphous material.

The plug formation in the filamentous phase is identical to that of gametophyte, but the dense material condenses on either side of the septum and is deposited differentially as large granules. The greatest deposition is around the aperture and thus a ring is formed around the aperture and the plug (Fig. 34 D). The plugs of both the phases are bound by an 80A^0 tripartite

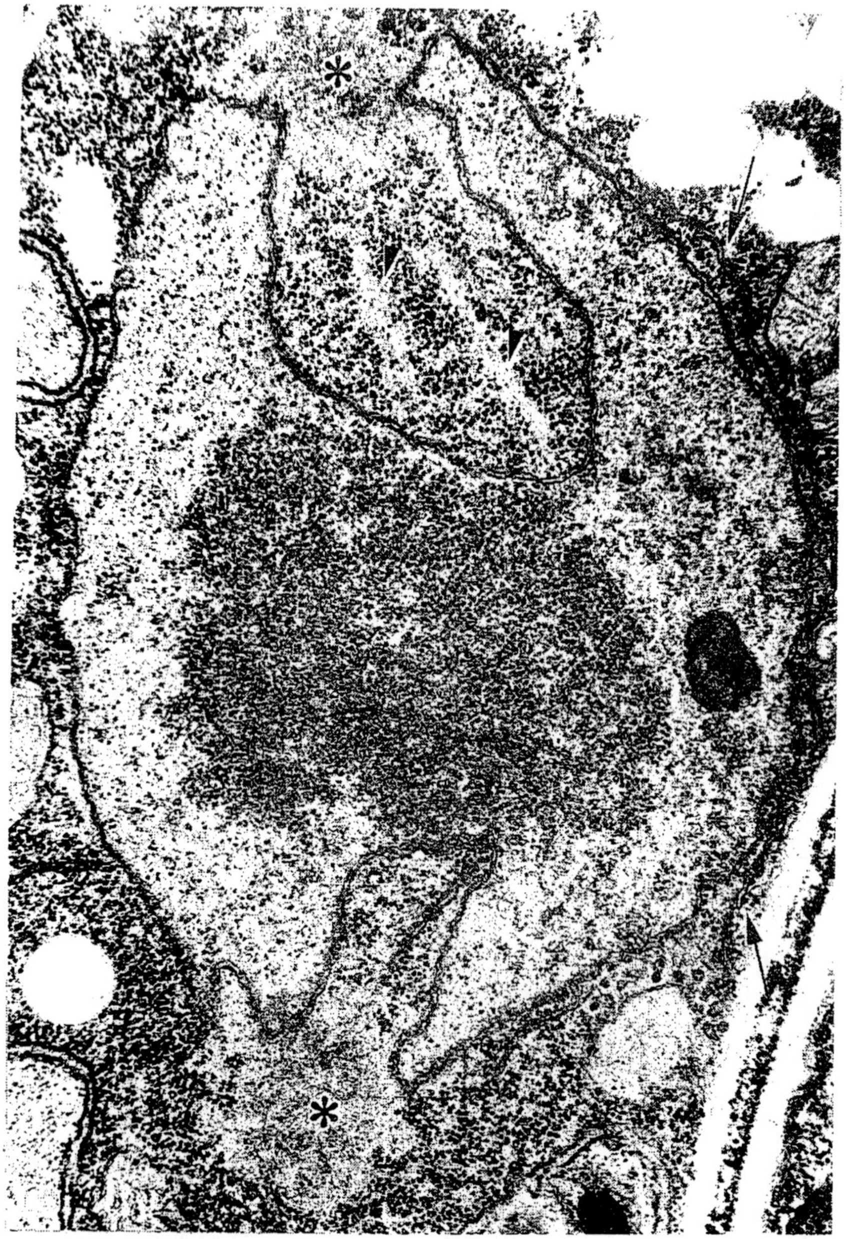

Fig. 39. Ultrastructure of cell division of *Batrachospermum ectocarpum* (after Scott, 1983).

Longitudinal section of an early prometaphase nucleus showing well developed nuclear pockets (only one of which is seen at each pole). Microtubules (arrow heads) are seen directed from the exclusion zone (asteriks) into the pockets. The nucleolus is centrally located and intact. Perinuclear endoplasmic reticulum is only partially developed.

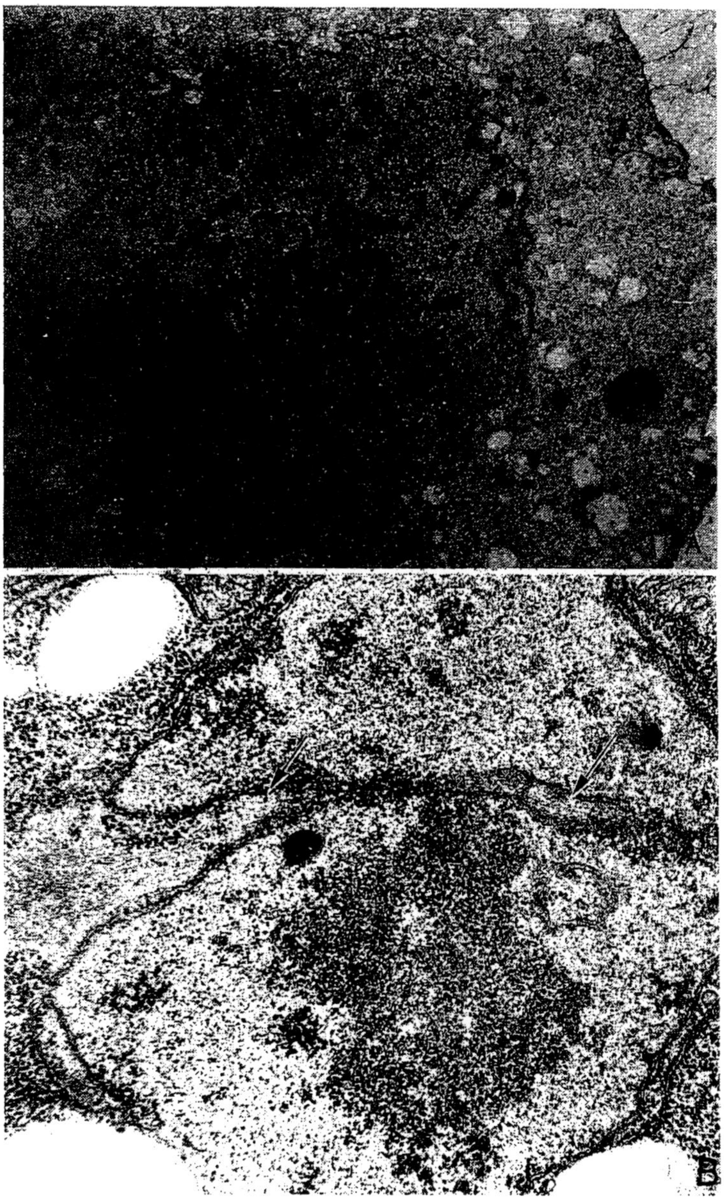

Fig. 40 A,B. Ultrastructure of nucleus

A *Asparagopsis taxiformis*. Transverse section of the prophase nucleus with convoluted nuclear envelope. The nucleoplasm shows well-differentiated euchromatin and heterochromatin areas. (after Jha, 1996).

B *Batrachospermum ectocarpum*. Longitudinal section of the mid-prometaphase nucleus showing formation of microtubules containing channels as nuclear pockets extends from pole to pole (arrows). (after Scott, 1983).

membrane. Further, the plug does not come into direct contact with the cytoplasm, nor any cytoplasmic continuity exists between the adjacent cells. Very little is known about the function of pit-connections. Fritsch (1945), however, maintains that from an evolutionary point of view, the fleshy phase is derived and consequently the simple pit, seen in this phase, should also be a derived condition.

Pit-Connection Vs Taxonomy : Pit-connection shows variation not only within the thallus, but even within the order but the number of cap-layers in pit-connection appears to be a stable character (Pueschel, 1989). The pit-plug tectonics in many heteromorphic taxa like *Asparagopsis taxiformis, Bonnemaisonia hamifera, Cumagloea andersonii, Helminthora divericata* and *Nemalion helminthoides* in both the gametophyte and the tetrasporophyte phases are identical in having a thin two-layered plug-cap except in *Asparagopsis* (Fig. 36 C; Jha, 1996) and *Bonnemaisonia* (Pueschel and Cole, 1982). Ramus (1969) reported dimorphic pit-connection in different stages of life-history of *Pseudogloiophloea* sp. Pueschel (1989), however, does not agree and feels that the differences described are features of septa and not of pit-plugs.

The most important feature of the pit-plug is the outer cap-layer and the absence of this character is the reason for the removal of Bonnemaisoniaceae (Bonnemaisoniales) and Gelidiaceae (Gelidiales) from the Nemaliales, besides the separation of three freshwater taxa namely Thoreaceae, Batrachospermaceae and Lemaneaceae from Nemaliales and transferring them to the Batrachospermales. All these three families show a dome-shaped, outer cap-layers whereas Nemaliales have a very thin outer cap-layer. On the basis of outer cap layer, which is a consistent feature, the three families previously kept in Nemaliales appear to be misfits (Pueschel and Cole, 1982).

Acrochaetiaceae is a very heterogenous group with regard to the outer cap-layer. Most of the members show thin outer cap-layer but three species show dome-shaped cap morphology. In Acrochaetiaceae, the pit-plug character varies at the species level. *Audouinella plumosa* was placed as a synonym of *A. pacifica.* Later it was found that the outer cap-layer was different in the two. *A. pulmosa* has a dome-shaped, outer cap-layer. *A. pacifica* has a thin, outer cap-layer. Similarly *A. variabilis* was considered as conspecific with *A. pacifica* and was considered as a variety of *A. plumosa.* The presence of thin, outer, cap-layers in *A. variabilis* rules out its affinity with *A. plumosa* (Pueschel, 1989).

The discovery of heterogenity in the Acrochaetiaceae has led to much taxonomic revision. This family has several genera and is placed originally under the Nemaliales. However, various differences found in the pit-plug provide additional clues. Dome-shaped, outer, cap-layers are seen in some taxa of Acrochaetiaceae but not in Nemaliales. Thus, the taxa with dome-shaped outer caps perhaps will have to be placed in a separate order regardless of their taxonomic disposition and it is expedient to segregate the Acrochaetiaceae to a separate order.

Cytochemistry of pit-plug : Ultrastructural cytochemistry has provided information about the chemical composition of the plug-caps. Pueschel (1977) found that after staining with uranyl acetate and lead citrate, plug-caps of *Nemalion* showed two layers separated by an electron-dense line presumed to be plug-cap membrane. The PTA binding affinity of the two layers of cap is also different. The layer adjacent to the cytoplasm stains intensely than the one towards the core. PTA also stained the dictyosomal membranes. In *Batrachospermum,* PTA did not stain, both the core as well as the inner cap layer whereas the dome-shaped outer cap-layer was intensely stained (Pueschel, 1980).

Cytochemical evidence is also used to study the homology between the different members of the three families of Batrachospermales namely Thoreaceae, Batrachospermaceae and Lemaneaceae. The outer cap-layer stained with PAS indicates glycoprotein composition (Fig. 33 C). Proteinase K digestion also proves this. (Trick and Pueschel, 1991). The Nemaliales are characterized by a thin outer cap-layer. Proteinase K digestion of thin sections of *Liagora valida* caused splitting along the cap-layer giving the false appearance of the domes (Fig. 33 D).

The ultrastructure of the outer cap-layer, thus, reveals diversity which implies a parallel origin rather than a single common origin in different orders. The only homology seen in all outer cap-layers is their stainability with PTA-CA. But this is not considered as a good character for homology because septal plug-caps in green alga *Smithsoniella earleae* also stain with PTA-CA (Trick and Pueschel, 1991). This will make green algae homologous to red algae which is a biological absurdity.

In all the orders of red algae the outer cap-layers stain with PAS and are also digested with proteinase K. Trick and Pueschel (1991) have suggested that the outer cap layer in all the orders of the red algae are homologous and they have a single origin, which if accepted will lead to the dome-shaped outer cap-layer to be the derived condition. If the uniformity of the relationship between inner and outer cap-layer is regarded as an advanced feature then the plate-like outer cap-layer of the Nemaliales could be the derived one as the outer cap layer completely covers the inner.

CHLOROPLAST

In *Batrachospermum* the chloroplast develops near the apex of the main axis and in the youngest cortical cells. Mature plastids are seen in one or two lower nodal cortical cells (Brown and Weier, 1968). The proplastid at an early stage of development is surrounded by two concentric, double-membrane structures that limit the chloroplast membrane and the outer most photosynthetic lamella. The outermost lamella runs approximately parallel to the chloroplast envelope and is separated by a gap of 350 A°. The first internal photosynthetic lamella appears to arise from the outermost lamella which may be connected with the innermost lamella. As the lamellae invaginate and extend inward from the outer lamella the proplastid enlarge. During further development the invaginating lamellae organize into one, two or several concentric lamellae that lie parallel to the outermost lamella. Many disc lamellae also occur in the centre of the chloroplast. The concentric lamellae by invagination produce more lamellae.

The mature chloroplasts are found in the cortical cells of the lower nodes from the apex. The maturation of the chloroplasts and the vacuolation of the cortical cells are directly related, implying that the most mature chloroplast occur in highly vacuolate cells. As the chloroplasts continue to mature, some interruptions in the lamellae occur, followed by the appearance of vesicles at the lamellar edges. The function of interruption in the photosynthetic lamellae is enigmatic. The mature chloroplast membrane system seems to be largely constructed of tubules which lie juxtaposed and separated by a narrow space forming lamellar sheet which characterises the chloroplast of highly vacuolate cells.

The chloroplasts in the cortical cells of *Scinaia forcellata* are reniform and show peripheral encircling thylakoids and many paripheral, internal thylakoids. In stroma, electron transparent zones are present. Many osmiophilic droplets lie juxtaposed to the chloroplast in the cytoplasm (Bhatia, 1993).

NUCLEUS

The interphase nucleus in *Asparagopsis taxiformis* is large and shows convoluted nuclear envelope (Fig. 40 A; Jha 1996). Perinuclear mitochondria are noteworthy. Large number of vacuoles are associated with the nuclear envelope and are seen penetrating the nucleoplasm, indicating intense nuclear-cytoplasmic interaction. The nucleoplasm is differentiated into euchromatin and heterochromatin regions. The condensation of chromatin in the centre is evident.

CARPOSPOROPHYTE

In *Nemalion helminthoides,* the gonimoblast cells, the carposporangial mother cells and the carposporangia have a characteristic organelle polarity with the proximal end of the cell occupied by a nucleus with a prominent nucleolus and the distal end containing a single, lobed, chloroplast with a pyrenoid that is surrounded by dense osmiophilic droplets. Perinuclear dictyosomes and mitochondria are present. The gonimoblast cells are characterized by the absence of vacuoles and the distinct layers of compactly arranged microfibrils. The vegetative cell wall under the same fixation is composed of three distinct layers (Ramm-Anderson and Wetherbee, 1982).

Many changes takes place in cells destined to form the fusion cell. The chloroplast thylakoid change their linear orientation and become randomly distributed, whereas the cytoplasm becomes granular and shows poor staining. In fusion cell, the nucleus lyse and numerous small osmiophilic granules contained in the remnants of chloroplasts are seen. Mitochondria are sparse and have swollen cristae (Ramm-Anderson and Wetherbee, 1982). The carpospore differentiation starts with increase in dictyosome number and activity. Straight profiled dictyosomes aggregate in groups of four or more and secrete numerous vesicles with electron-transparent contents. In the later stages, the vesicles are formed by the coalation of the entire dictyosome cisternae and they contain electron-dense core. These vesicles may also fuse to form large fibrillar vesicles which possess wall materials for deposition.

The mature carposporangia in *Scinaia pseudocrispa* show dense cytoplasmic contents and are packed with floridean starch grains, chloroplasts and vesicles (Fig. 38 B, C). The grains range from oval to kidney-shape and show transparent core being surrounded by electron-dense areas. The vesicles contain electron-translucent fibrillar contents. A thick, microfibrillar and electron-transparent mucilage sheath encloses the carposporangial cytoplasm. This sheath is produced through the release of fibrous contents of the vesicles (Bhatia and Vijayaraghavan 1995 b).

CELL DIVISION

Mitosis : Scott (1983), observed in *Batrachospermum ectocarpum,* the mitotic ultrastructure of the meristematic cells of carposporophyte generation. Prior to nuclear division, cell elongation (Fig. 38 A) results in nucleus being located at the proximal end of the cell and separated by a large central vacuole from a distal empty region. By the late interphase and early prophase, each mitotically active cell contains two regions of small vacuoles-one proximal and the other distal to the nucleus. Two "nucleus associated organelles" (NAO's) were seen to be located close to the outer membrane of the nuclear envelope. These are called polar rings (PR) (McDonald, 1972). These polar rings have a "ring-within-a-ring" configuration of the organelles. In the mid-late prophase, two nuclear pockets are formed adjacent to each PR which has microtubules, numerous ribosomes and rough ER cisternae (Fig. 39).

In the prometaphase, the spindle microtubule penetrates the nucleus but not the nucleoplasm and remains completely contained within the cytoplasmic invaginations that extends as channels (Fig. 40 B) which also contain ribosomes. The microtubules appear to arise from the polar-ring associated moderately with electron-dense material in the zone of exclusion adjacent to the entrance of the channels. The nucleolus is still intact, chromatin is not much condensed and perinuclear ER is slightly developed.

In late prometaphase, the spindle microtubule finally enters nucleoplasm by the gradual break down of the nuclear envelope binding the cytoplasmic channels. Each zone of exclusion is now in direct contact with the nucleoplasm through a single large gap in the nuclear envelope. Microtubule are now directed inwards. The perinuclear ER is more extensive and exists as a "cap" over the poles.

In metaphase, the gaps are still open. Chromosomes are arranged on the equatorial plate. Maximum development of the perinuclear ER is at metaphase. The zones of exclusion are also evident in the anaphase. Polar rings are absent.

The distal portion of the elongate nucleus pushes in between the distal vacuole and the cell wall. Further growth and breakdown of interzonal mid-piece locates the two nuclei in the central or basal central region of the future new cell. A cleavage furrow develops and separates the cell. The subapical cell has both the proximal and the distal vacuoles but the apical cell has only distal vacuole.

The nuclear channel that exists in *B. ectocarpum* are characteristic of this taxon and have never been observed in other red algal taxon. Further the studies on mitosis can serve to elucidate phylogenetic link. The variations as seen in nuclear cytokinesis can prove to be an important taxonomic character similar to the pit-plugs.

UTRICLE DEVELOPMENT

In the thallus of *Scinaia pseudocrispa* the epidermis consists of one layer of inflated utricles that are interrupted by the assimilatory cells. The utricles are either oblong obovate cells with outer convex ends, appear hyaline and lack appreciable cytoplasmic contents (Fig. 23 A). The utricles are formed by the differentiation of a few assimilatory epidermal cells. Such progenitor epidermal cells contain electron-dense materials and floridean starch grains and their cell walls have mutilayered fibrillar structures veneered by outer electron-dense layers (Fig. 37 A). During utricle development, the cell enlargement is accompanied by increased vacuolation (Fig. 37 A,B); at this stage both the floridean starch grains and electron-dense materials are present (Fig. 37 B,C). The plasmalemma profiles are prominent and are associated with electron-dense materials; many electron-dense materials fuse with plasmalemma. An additional wall layer accrues inner to the cell wall. The mature and vacuolate utricle shows multilayered, fibrillar wall and peripheral layer of cytoplasm with prominent plasmalemma profiles (Fig. 37 D) (Vijayaraghavan and Bhatia 1995 a).

METABOLIC BYPRODUCTS

POLYSACCHARIDES

Xylans : In Nemaliales, xylans are common in *Nothogenia fastigiata* (Haines *et al.*, 1990). *Nemalion vermiculare* and *Audouinella floridula* (Usov *et. al.*, 1973, 1974, 1981). In *Chaetangium fastigatum,* the cell wall after acid hydrolysis yields xylose as the main sugar and galactose and mannose as minor components. Ultracentrifugation studies reveal the presence of a single broad band characteristic of D-xylose, suggesting polydispersity.

The xylan, after methylation and acid hydrolysis, yielded 2-0-methyl-D-xylose, 2,3-di-O-methyl-D-xylose, 2,4-di-O-methyl-D-xylose and 2,3,4-tri-O-methyl-D-xylose. All these residues have β-D-configuration and the polysaccharide has random fluctuating conformation. This also suggests that (1 → 3) linkages are interspersed throughout the structure rather than grouped contiguously. The *Chaetanguim* xylan has both (1 → 3) and (1 → 4)-xylosidic linkages which occur together in branched homoxylan (Cerezo, et al., 1971). In *Lemanea annulata,* β-4-linked xylan is the major alkali soluble constituent. *Rhodochorton floridulum* contains a polysaccharide having (1 → 3) and (1 → 4) linked xylose residues which seems to contain some xylose in a heteropolysaccharide (Cerzo, et al., 1971).

The physical nature of the xylan microfibril has been investigated (Mackie and Preston, 1974), and it is believed that the β(1 → 3) xylan molecules are arranged in triads to from macromoleules. The triple helix so formed is stabilized by interchain hydrogen bonds between 0-2 of each xylose unit and by hydrogen bonds between 0-4 of one chain and the cyclic oxygen in the pyranose ring of an adjacent chain (Atkins and Parker, 1969). The helical xylan structure lacks birefrengence and, thus is different from cellulose and mannans, both of which are birefrengent in polarized light (Sheath and Cole, 1990). On hot water extraction, *Scinaia pseudocrispa* yields a polysaccharide mixture which on X-ray diffraction analysis reveals the presence of mainly xylose with galactose and mannose (after Bhatia, 1993)

Cellulose : In the freshwater species of *Lemanea annulata,* X-ray diffraction and methylation analyses have established the presence of cellulose (Gretz, 1991).

Complex Mucilages : On extraction with hot water or dilute alkali *Batrachospermum* yields a high molecular weight, viscous mucilage, that contains 3-0-methyl-L-rhamnose. The hot water soluble polymers from *Lemanea* and *Batrachospermum* share several characteristics like repeating glucuronosyl/galactosyl unit, lack of ester sulfate, low protein content and presence of xylosyl residues (Gretz, 1991).

Unlike *Batrachospermum,* where the backbone of the mucilage appears to be alternating D-glucoronic acid and D-galactose resideues (Turvey and Griffiths, 1973), members of Nemaliales produce sulfated mannans of varying complexity. The sulfated mannans are produced on extraction with water or with alkali from *Nemalion vermiculare.* The carbohydrate chain of the polysaccharide are built from α -(1 → 3) linked residues of D-mannopyranose and with sulfate groups on the 6,4 and (to a small extent) 2 carbon atoms of the anhydromannose units (Usov et al., 1975). More than 30 per cent of the dry weight of *Nemalion vermiculare* is a mannan with 15.5% of SO_3 Na, which was shown to be composed of α (1 → 3) linked D-mannopyranose, its 6-sulfate and 4-sulfate in molar ratio of 6.2:1:3:1 (Usov et al., 1973, 1974, 1975). A small quantity of D-xylose is glycosidically bonded to C-2 of the D-mannopyranose residues of the main chain.

Usov *et al.*, (1981) showed that in *Galaxaura rugosa* a novel sulfated xylogalactomannan had a linear backbone of β-(1 → 3) linked mannopyranose residues. Three, of every five mannose residue, is substituted at position 4 either with α-D-xylopyranosyl or L-galactopyranosyl residues. The sulfate hemiester was carried on C-4 of mannopyranose.

Sulfated Polysaccharides : Araki (1966) found that 6-0-methyl D-galactose is always present in Gelidiaceae but the levels are low. Some 4-linked residues can also be substituted with methoxyl at the C-2 position. Zanlungo (1979) described the same results for *Gelidium filicinum.* The sulfation of the 3 linked-D-galactose residue does not appear to be a prominent feature. The agar gel strength of *Gelidium* and *Pterocladia* is higher than that of *Gracilaria* (Doty and Santos, 1983).

4 CRYPTONEMIALES

CLASSIFICATION

The vegetative plant bodies may be filamentous or pseudoparenchymatous (blades terete or crusts); soft, rigid or fleshy. Thalli may be crustose, discoid erect, and frondose. The degree of aggregation ranges from loose to compact with uniaxial or multiaxial construction. The thalli, that bear reproductive structures, may be similar or dissimilar in their exomorphic characters.

The spermatangia are formed singly or in clusters as superficial patches in internal chains, or within the specialized branch of the female thallus. Carpogonium arises from the apical cell of a specially formed filament. After fertilization, the zygote nucleus is transferred to an auxiliary cell formed in a specialized filament. The carposporophyte varies in size and may be surrounded by sterile tissue. The life history is typically diplobiontic and isomorphic, but a few taxa, in culture, show heteromorphic pattern.

According to Fritsch (1945) the order Cryptonemiales is divided into 10 families.

Gloeosiphoniaceae	*Gloeosiphonia, Schimmelmannia, Thurtella.*
Endocladiaceae	*Endocladia, Gloeopeltis.*
Callymeniaceae	*Callocolax, Callophyllis, Callymenia, Erythrophyllum, Euthora, Nereoginkgo, Rhizopogonia.*
Grateloupiaceae	*Aeodes, Cryptonemia, Halymenia, Thamnoclonium.*
Dumontiaceae	*Acrosymphyton, Constantinea, Cryptosiphonia, Dilsea, Dudresnaya, Farlowia, Pikea, Thurtellopsis,*
Cruoriaceae	*Cruoria, Petrocelis.*
Rhizophyllidaceae	*Polyides, Rhizophyllis.*
Squamariaceae	*Coriophyllum, Cruoriopsis, Ethelia, Hildenbrandia, Peyssonelia.*
Corallinaceae	*Amphiroa, Corallina, Epilithon, Fosliella, Lithophyllum, Lithothamnion. Melobesia, Phymatolithon, Pneophyllum, Porolithon, Sporolithon.*
Choreocolaceae	*Choreocolax, Harveyella.*

Note : (i) Only selected genera are mentioned under each family. (ii) Gloeosiphoniaceae is spelt as Gloiosiphoniaceae by other authors.

Bold and Wynne (1978) give another classification in which they divided the order into six families namely Corallinaceae, Choreocolaceae, Dumontiaceae, Cryptonemiaceae, Gloiosiphoniaceae and Kallymeniaceae:

Corallinaceae *Amphiroa, Corallina, Clathromorphum, Lithophyllum, Lithothamniom, Lithothrix, Melobesia, Pneophyllum, Porolithon.*

The taxa are characterized by calcified thalli and reproductive organs are formed in conceptacles.

Choreocolaceae *Choreocolax, Dawsoniocolax, Harveyella.*

Cryptonemiaceae : *Cryptonemia, Halymenia.*

The connecting filament do not fuse with any cells of the carpogonial branch but fuse with auxiliary cells in remote auxiliary cell branches.

Dumontiaceae *Constantinea, Cryptosiphonia, Dilsea, Dudresnaya, Pikea, Weeksia.*

The carpogonia and the auxiliary cell are one and the same branch system. Prior to development of carposporophyte, the connecting filament from the fertilized carpogonium makes contact with another cell of the same carpogonial branch.

Gloiosiphoniaceae *Gloiosiphonia.*

The carpogonial branch and auxiliary cell are formed on a common supporting cell, which does not play any role in the initiation of carposporophyte.

Kallymeniaceae : *Callophyllis, Kallymenia, Pugetia.*

After fertilization the supporting cell of the carpogonial branch is involved in the formation of a fusion cell and functions as the auxiliary cell or the site of origin of the connecting filament to a remote auxiliary cell.

Dixon (1982) divided the order Cryptonemiales into 15 families.

Weeksiaceae *Constantinea, Leptocladia, Weeksia.*

Dumontiaceae *Acrosymphyton, Cryptosiphonia, Dilsea, Dudresnaya, Farlowia, Pikea.*

Choreocolacaceae *Choreocolax, Dawsoniocolax, Harveyella.*

Crptonemiaceae *Cryptonemia, Halymenia, Grateloupia.*

Corynomorphaceae *Corynomorpha.*

Pseudoanemoniaceae *Pseudoanemonia.*

Kallymeniaceae *Callocolax, Kallymenia.*

Endocladiaceae *Endocladia.*

Crossocarpaceae *Crossocarpus, Erythrophyllum, Kallymeniopsis.*

Gloiosiphoniaceae *Gloiosiphonia, Thurtella.*

Tichocarpaceae *Tichocarpus.*

Hildenbrandiaceae *Hildenbrandia.*

Pterocladiophilaceae *Pterocladiophila.*

Peyssoneliaceae *Peyssonelia*

Corallinaceae *Amphiroa, Corallina, Epilithon, Fosliella, Lithophyllum, Lithothrix, Melobesia, Phymatolithon, Pneophyllum, Porolithon.*

MORPHOLOGY

In *Constantinea* sp., (Weeksiaceae) the thalli are erect, solitary or in groups, stipe terete, branched or unbranched, terminating in one or more peltate coriaceous blades. In *Weeksia* sp. the thalli are erect, solitary, foliose, orbiculate to reniform, cleft or uncleft, with or without conspicuous veins radiating from base, with short stipe and discoid hold-fast.

The Dumontiaceae include both uniaxial and multiaxial genera, either with or without secondary pit-connections, and encompass plants ranging from foliose to filamentous types. *Dudresnaya verticillata*, ranges from shallow subtidal, and has been recorded as epiphytic, epizoic and epilithic. The thalli are terete and irregularly radially branched, and with main axes upto 2 mm in diameter with acute branch apices. The young axes are annulate. *Dudresnaya australis* occurs at depths of 2-33 m and are always epiphytic. *D. australis* superficially resembles *D. verticillata* (Fig. 41 A), being entirely terete, radially branched, and having acute branch apices and obvious annulations in younger parts. Plants reach lengths of 16 cm and main-axis width is about 3 milli meters. Branch axis are broad, often approaching 90 degree angle. Plants of *D. hawaiiensis* grow on coral rubble, encrusting calcareous algae and mollusc shells, at depths of 1.5-35 m. The thallus is dull red to pink, becomes yellow and when old even reach 18 cm in length. The higher order branches are slightly compressed and rarely terete. Margins of the thallus are rather indefinite due to the long exerted cortical filaments. The apices are obtuse. In *D. capricornica*, the branches are terete, upto 1 cm in diameter and tapers to acute apices. Colour of the plant ranges from red or light purple to light pink (Robins and Kraft, 1985).

In *Farlowia*, the terete bilateral branches are slippery in texture, and usually occur in tufts in exposed midlittoral to sublittoral sites on the pacific coast of North America (Fig. 41 B, Bold and Wynne, 1981).

This family Choreocolacaceae contains extremely reduced parasites, which form whitish or almost colourless crumpent cushions on their red algal hosts. *Choreocolax* and *Harveyella*, have been recorded from Atlantic and Pacific Coasts. *C. polysiphoniae* forms small, whitish cushions on various species of *Polysiphonia*, whereas *Harveyella* occurs as small bulbose pustules on the host genera *Odonthalia* and *Rhodomela* (Rhodomelaceae) and is an alloparasite.

The most structurally differentiated members of Cryptonemiaceae are perhaps some species of *Cryptonemia* recorded from Southern Australia (Scott *et al.*, 1982). Extensive periodic cortication of the basal parts of the thalli results in extremely hard. and massive stalks from which fleshy blades proliferate seasonally (Fig. 42 B).

Plants of *Halymenia venusta* (Fig. 42 C) are attached to the substratum through a small disc. The stipe is short, cuneate and expands abruptly into a much broad and flat thallus, which is variable in size and shape, light to deep red in hue and with age becomes dark. The frond margin is densely proliferous and divide repeatedly throughout the surface.

Halymenia asymmetrica is characterized by a planar thallus, growing from a short cuneate stipe. The blade is lobed and shows entire margin. One of the lobe is usually more expanded, giving an asymmetrical appearance to the thallus. Among the three species of *Halymenia* found in Mediterranean sea, only two are foliose (*H. floresia* and *H. latifolia*). The third *H. trigama* has a tubular thallus (Gargiulo et al., 1986).

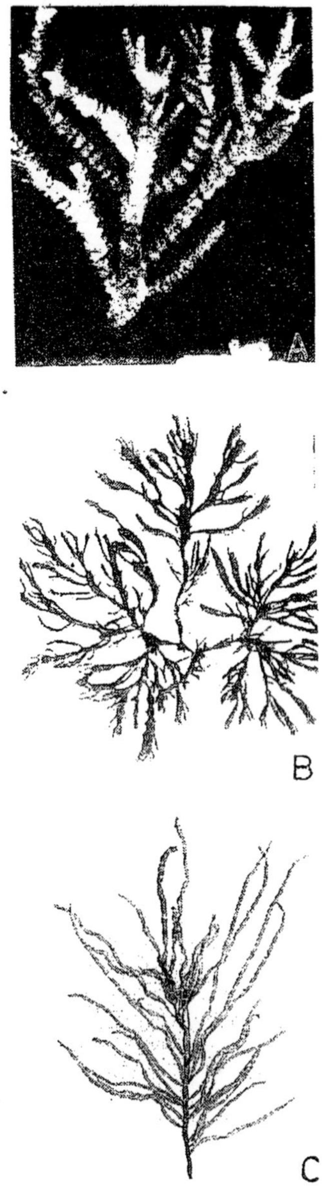

Fig. 41 A-C Morphology

A. *Dudresnaya verticillata* Showing terete and irregularly radially branched thalli (after Robins and Kraft, 1985)

B. *Farlowia compressa* Showing terete or compressed axes and bilaterally branched thallus (after Bold and Wynne, 1978)

C. *Grateloupia* sp. Showing foliose hollow and gelationous blade. (South and Whittick 1987)

In *Grateloupia sp* (Fig. 41 C) and in *G. stipitata* (Fig. 42 A), the foliose blades are hollow, stipitate and gelatinous and arise from coarse and narrow perennial stalks.

Gloiosiphonia capillaris (Fig. 42 D) (Gloiosiphoniaceae) shows uniaxial construction. The main axis bears transverse whorls of short, simple branches, the basal cells of which are large and branch out in both vertical and horizontal directions. Secondary branches give rise to progressively small branches, which ultimately unite with branches of adjacent lateral filaments to form a cortex of continuous small cells.

Members of Corallianaceae occur in tropical, temperate and subarctic sea. A few taxa parasitize only during their juvenile stages on *Amphiroa* and *Pseudolithophyllum* (Corallinaceae). *Schimitziella,* is a non-calcified genus that grows endophytically under the cuticle of *Cladphora* (Chlorophyceae).

The coralline algae generally, flourish best in cold, agitated water, but they also seem to tolerate extremes of temperature, particularly the crustose forms. They inhabit great depths of the ocean due to their capacity to tolerate low light intensity. The crustose types penetrate more deeply than the articulated types. A few taxa show noncalcified portions alternating with calcified portions of the thallus. The calcified portion is called intergenicula. On the basis of presence or absence of intergenicula the coralline algae are divided into two types (a) Articulated (Fig. 42 E) (genicula present), (b) Non-articulated (Fig. 42 F) (genicula absent). Majority of the non-articulated coralline algae have crustose habits.

REPRODUCTIVE ORGANS

SPERMATANGIA

Weeksiaceae : In *Constantinea rosa-marina* during April to June months, both upper and lower surface of the thalli, produce superficial, elongate spermatangial mother cell which bear spermatangia. In *Weeksia coccinea* spermatangia are unknown (Lindstrom and Scagel, 1987).

Dumontiaceae : The male gametophytes are usually considered to be isomorphic with female gametophytes but in many taxa they are slighly small. Spermatangia are produced singly and are either scattered over the thallus surface or may be aggregated in sori or nemathecia. They are also produced in internal chains, or aggregate in shallow depressions or line the wall of conceptacle-like depressions. The spermatangia in *Dilsea* are supereficial; in *Neodilsea natashae* and *N. borealis,* cut off superficial spermatangial mother cell; in *Dumontia contorta* and *D. simplex,* are borne on spermatangial mother cell (Lindstrom and Scagel, 1987). In *Acrosymphyton tenax* spermatangia occur on the surface cortical cells, covering entire thallus except at base and apex (Millar and Kraft, 1984).

Cryptonemiaceae : The spermatangial sori are irregularly stretched and raised on the surface of gametangial plants. In *Halymenia latifolia,* spermatangial mother cells bear pear-shaped, spermatangia which are scattered over the thallus surface, whereas in *H. asymmetrica* and *H. durvillaei,* occur in raised sori (Gargiulo and Tripodi, 1986). In *H. acuminata* male reproductive structures are found scattered on the plants bearing cystocarps. Spermatangia are cut off singly or in pairs from the outermost cortical cells (Kawaguchi, 1991).

Choreocolacaceae : In *Harveyella mirabilis,* spermatia are borne on spermatangial mother cells that form in the peripheral cell layer of the male pustules. The spermatangial mother cell undergoes a series of oblique mitotic divisions giving rise to chains of spermatia. As many as 9

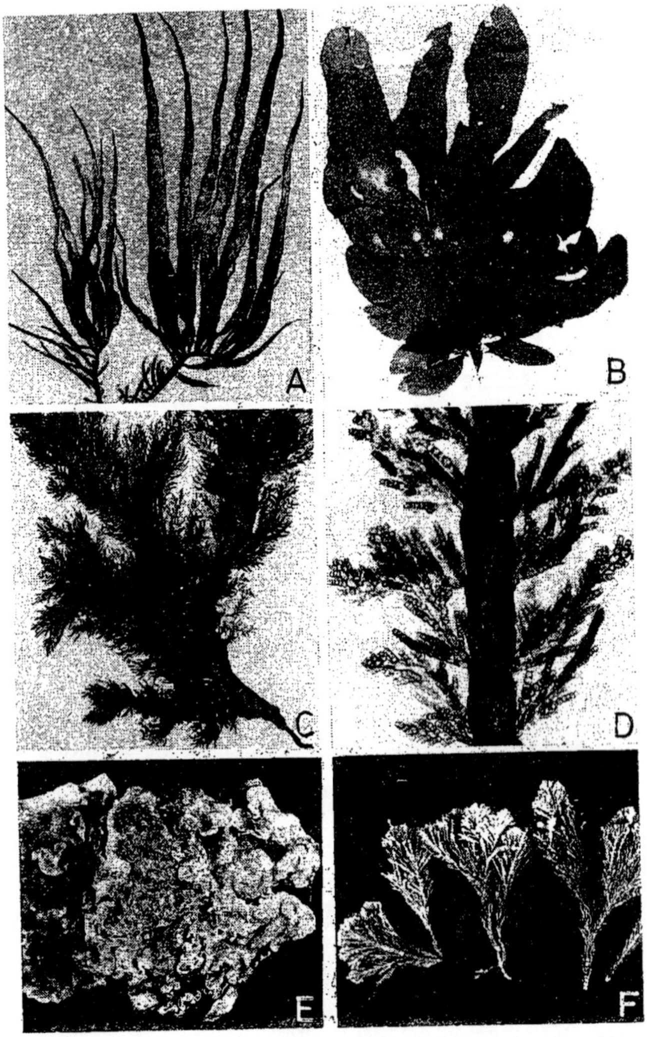

Fig. 42 A-F Morphology

A. *Grateloupia stipitata* showing stipate, foliose blades which are hollow and gelatinous (after South and Whittick, 1987).

B. *Cryptonemia borealis* Habit sketch (after Bold and Wynne, 1978).

C. *Halymenia venusta* Short stipe, cuneate, expanding abruptly into a much broadened flat thallus (after Srinivasan, 1969).

D. *Gloiosiphonia capillaris.* Showing uniaxial construction of the thallus. Main axis bears transverse whorls of short, simple branches (after Bold and Wynne, 1978).

E. *Lithophyllum lichenare :* Showing articulated coralline alga.

F. *Corallina vancouveriensis :* Showing nonarticulated Coralline alga that lacks genicula.

spermatia have been noted in a chain, apparently cut off from a single mother cell. The spermatia are beseiged in mucilaginous covering of the pustule and are shed *en masse*. Spermatangial mother cells in one pustule may be in various stages of mitosis. The metaphase period is long and the metaphase plate shows contracted and clumped chromosomes. The chromosomes counts are, thus, made only during the late prophase stages (Goff and Cole, 1973).

In *Choreocolax polysiphoniae*, spermatia are produced terminally from elongate spermatangial mother cell, which develop abundantly at the surface of a mature male gametophyte and cover the surface. The spermatangial mother cells divide by oblique successive cleavage cutting off alternatively spermatia. The spermatia mature acropetally and are being continuously released (Goff and Coleman, 1984) and immediately upon release their nuclei enter prophase. Final nuclear division is completed, most infrequently, upon the contact of the spermatium and the carpogonium. Either one or two male nuclei enters the trichogyne during the plasmogamy.

Pterocladiophilaceae : In *Pterocladiophila hemisphaerica*, spermatangia are cut off in a single chain at the distal end of the spermatangial mother cell (Delivopoulos and Kugrens 1985).

Corallinaceae The postition of origin of spermatangia within male conceptacles has been considered significant amongst the non-geniculate Corallinaceae. All taxa of the Melobesiodeae show spermatangia formation in the conceptacle cavity on both the chamber-floor and chamber-roof; whereas the non-geniculate Corallianaceae, with two exceptions, reveal spermatangia formation confined only to the chamber-floor. Male conceptacles of *Spongites fruticulosus* contain simple spermatangia restricted to the chamber-floor (Pendrose. 1991).

In *Pneophyllum fragile*, the male conceptacle roofs protrude 2-8 cells thick above the chamber or flush with surrounding thallus surface. The simple spermatangia are confined to the floor (Pendrose and Woelkerling, 1991). The male conceptacle roofs of *Melobesia membranacea* lack elongate cells and roof cells are quadrate and squat. In contrast, the male conceptacle roofs of *M. rosanoffii* (Fig. 43 A) have at least one layer of elongate cells that overarches layers of small, round to compressed cells (Wilks and Woelkerling, 1991). Spermatangia occur both on the floor and the roof of conceptacle chambers and both dendroid and simple spermatangial clusters are present.

In *Lithophyllum crouanii*, spermatangial conceptacles (Fig. 43 B) occur near the surface of the outgrowth; have relatively broad, shallow chambers bearing spermatangia across the conceptacle floor; the ostiole may be prolonged into a spout. Elongate spermatangia are born individually on spermatangial mother cells, and round off when released into conceptacle chamber (Chamberlain, 1988).

CARPOGONIAL BRANCH

Wecksiaceae : In *Constantinea rosa-marina*, carpogonial branch is 6-9 celled. The distal cells enlarge and cell 2-5, particularly 4, is lobed (Fig. 44 A, B). In *C. simplex* carpogonial branch 10-17 cells long (Fig. 44 C). Cells 2-7 enlarge and most to all of these cells are lobed.

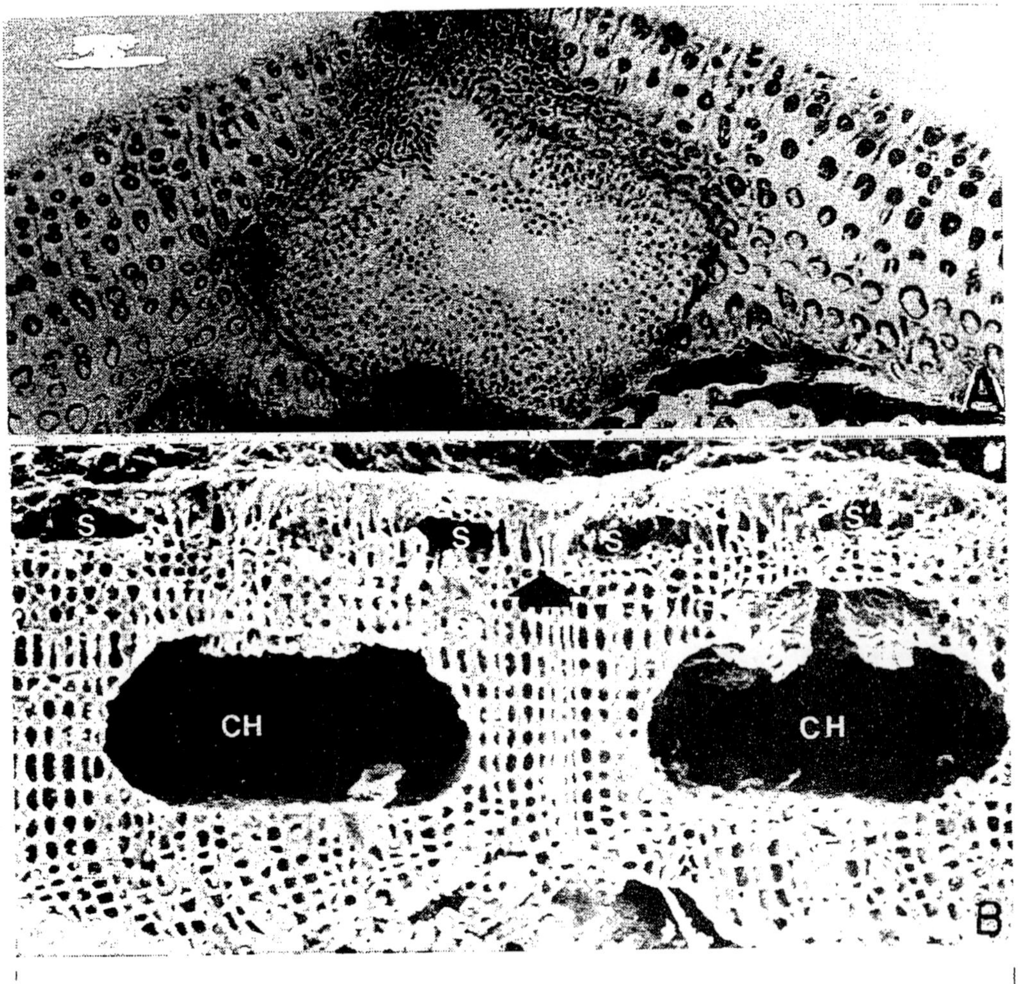

Figure 43 A,B. Reproductive Organs

(S. male conceptacle; CH, female conceptacle)

A. *Melobesia rosanoffii.* Section showing mature male concetacle (after Wilks and Woelkerling, 1991)

B. *Lithophyllum crouanii*: Vertical fracture of a carposporangial plant with two female conceptacles, surrounded by a spermatangial outgrowth with four male conceptacles; the outgrowth start at the level of arrows (after Chamberlain, 1988).

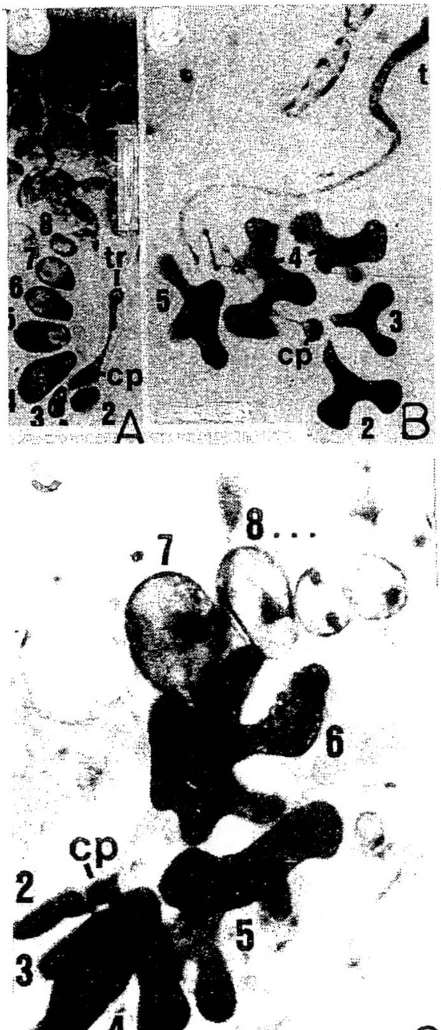

Figure 44 A-C. Reproductive organs : carpogonial branch

(after Lindstrom and Scagel, 1987)

(cp, carpogonium; tr, trichogyne; 2-8, cell number of carpogonial branch)

A. *Constantinea rosa-marina.* Immature carpogonial branch developing from inner cortical cells of upper surface of terminal blade.

B. *C. rosa-marina.* Distal humerus-like cells of mature carpogonial branch.

C. *C. simplex.* Mature carpogonial branch.

Dumontiaceae : The carpogonial and the auxiliary cell branches are born remotely from one another and the auxiliary cell branch is sparingly branched. The carpogonial branch in *Dasyphloea insignis* is 6 celled; in *Dumontia* species 9 celled; in *Cryptosiphonia* 15 celled; and in *Pikea* ranges from 12 to 20 cells long. The carpogonial branch cells of *Dasyphloea insignis, Dumontia* spp., and *Cryptosiphonia woodii* are a little modified from the ordinary vegetative cells. A few members even show wedge-shaped distal cells (Lindstrom, 1988).

Acrosymphyton is unique in having pinnate carpogonial branch and terminal, rather than intercalary, auxiliary cell. In *A. tenax* carpogonial branch is different from other species only in the bipinnation of some of the lower axial cells. In both *A. tenax* and *A. taylorii,* the lateral carpogonial branch pinnae have the potential to act as sites of gonimoblast initiation. In *A. tenax* the terminal 2 or 3 cells of the carpogonial branch are free of lateral appendages. The carpogonium is reflexed; the trichogyne is strongly recurved and twisted before entering the outer cortex. The auxiliary cell filament is long and considerably contorted, and tends to produce single rhizoidal filament from many of its proximal cells. In *A. caribaeum, A firmum* and *A. tenax* the auxiliary cell filaments are branched (Millar and Kraft, 1984).

In *Dudresnaya* species the carpogonial and auxiliary cell filaments are morphologically and spatially distinct and scattered throughout the thallus. The carpogonial filaments are simple or bear non-functional laterals. The fertilized carpgonium develops a primary connecting filament that fuses with one or more nutritive auxiliary cells in carpogonial filament. Secondary connecting filaments are produced from the fusion cell complex.

Choreocolacaceae : In *Choreocolax* and *Harveyella,* the carpogonial branches arise singly from the auxiliary cells which also bear one or two sterile branches (Fritsch, 1945). In *Choreocolax polysiphoniae,* the carpogonium is borne on a specialized branch system in the periphery of the female gametophyte. Nearly five hundred carpogonia may be found covering the surface of a single female gametophyte. The trichogyne of the carpogonium protrudes through the surface of the female gametophyte. The carpogonial branch is 4-celled (Goff and Coleman, 1984). In *Dawsoniocolax bostrychiae* a single gametophyte bears upto sixty carpogonial branches. The fully elongate trichogyne is either straight or curved at the tip. Carpogonial branches are constituted of four laterally displaced cells borne on a supporting cell, namely the intercalary cell, in the cortical filament (West and Calumpong, 1988).

Cryptonemiaceae : This family is characterized as being nonprocarpial, the carpogonia and auxiliary cells being formed in separate accessory branches which are much branched and show compact knotted appearance called as "ampullae". Carpogonial branch is generally 2-celled which may have several celled sterile laterals arising from the lower cell. In the ampulla, the supporting cell of the carpogonial branch is usually an interclary cell. After fertilization, one or more connecting filaments arise from the carpogonium and make direct contact with the auxiliary cell, from which the carposporophytes are produced toward the thallus surface (Bold and Wynne, 1981).

In *Halymenia asymmetrica,* the carpogonial branches and auxiliary cells are borne on subsidiary branch systems or ampullae and arise from the inner cortical cells. The carpogonial ampullae are more or less globular and more densely branched than the auxiliary cell ampullae. The supporting cell of the carpogonial branch is intercalary on the primary ampullar filament. The carpogonium tapers into a very long trichogyne that projects about 80-100 µm above the thallus surface. After fertilization, the carpogonium and the hypogynous cell form a fusion cell from which several thin, aseptate connecting filaments arise which fuse with auxiliary cells (Gargiulo and Tripodi, 1986). In *Halymenia acuminata,* both the carpogonial branches and the auxiliary cells are formed in separate flask-shaped ampullae. The carpogonial branch is 2-celled (Kawaguchi, 1991).

Corallinaceae : The carpogonial branches arise from the conceptacle floor and are always 2-celled, arising singly from a supporting cell. In *Synarthrophyton schielianum.* the carpogonial branch consists of a terminal carpogonium, a hypogynous cell and a supporting cell and on the same branch, a sterile cell is also borne distally on the hypogynous cell. Only those carpogonia that are beneath the conceptacle pore reach maturity. These carpogonia have long trichogyne which extend upwards into the pore canal (Woelkerling and Foster, 1989). In *Hydrolithon cymodoceae,* the carpogonial branch is 1 or 2 celled, and each terminate in a carpogonium. The fully mature carpogonia bear trichogynes (Penrose and Woelkerling 1992 a, b).

In *Pneophyllum fragile*, each female conceptacle produces many 2-celled carpogonial branches across the conceptacle-floor, The terminal portion of the centrally situated carpogonia becomes extended into elongate trichogyne (Penrose and Woelkerling, 1991). *In Lithophyllum crouanii,* the carpogonial conceptacle occurs near the thallus surface with supporting cell occuring on the conceptacle floor. In the centre of the conceptacle these may bear one or two carpogonial branches. The carpogonium extends into long trichogyne that protrudes well beyond the ostiole which is surrounded on the thallus surface by a ring of mucilaginous substance (Chamberlain, 1988).

CARPOSPOROPHYTE

Carposporophytes are second generation plants of red algae that produce carpospores and are parasite on the female gametophyte. They arise either directly or indirectly from the zygote.

Dumontiaceae : In *Acrosymphyton tenax,* after fertilization, diploidization of the auxiliary cell occurs when a connecting filament contacts the carpogonium from the top and fuses. A single gonimoblast initial arises from the connecting filament near its point of fusion to the auxiliary cell, and later many gonimolobes develop on this gonimoblast initial. Pit-connections between the auxiliary cell and the two immediate subtending cells enlarge and the contents of all the three cells atrophy thus supplying the energy to the developing gonimoblasts. The carposporophyte are subspherical and even cause a conspicuous bulge in the adjacent cortex (Fig. 45 A). The angular carpospores later escape through a distinct pore formed due to the lyses of a patch of cuticle above the spore mass (Millar and Kraft, 1984).

Dudresnaya shows two distinct cystocarpic forms. In the first group *D. hawaiiensis, D. japonica, D. minima, and D. patula* all have cystocarps that incompletely surround the auxiliary cell filament, leaving a narrow slit, that passes through one side of the carposporangial mass. In the second group namely *D. australis, D. capricornica* and *D. verticillata* have cystocarps that completely encircle the auxiliary cell filaments. Here two gonimoblast initial tend to recurve around opposite sides of the generative auxiliary cell whereas the notched cystocarp of the *D. hawaiiensis* group correlate with gonimoblast initials that do not recurve around the auxiliary cells. In all *Dudresnaya* species diploidization of generative auxiliary cells occurs in a fairly consistent manner. The secondary connecting filments fuse laterally with generative auxiliary cell after contact and later form pit-connections just beyond the point of fusion with the auxiliary cell. The zygote nucleus divides into two and one nucleus remains in the connecting filament next to the auxiliary cell, while the other continues in the rapidly extending apex of the connecting filament. The small section of connecting filament that is attached to the diploidized generative auxiliary cell inflates and gives rise to globular, gonimoblast initials (Robins and Kraft, 1985).

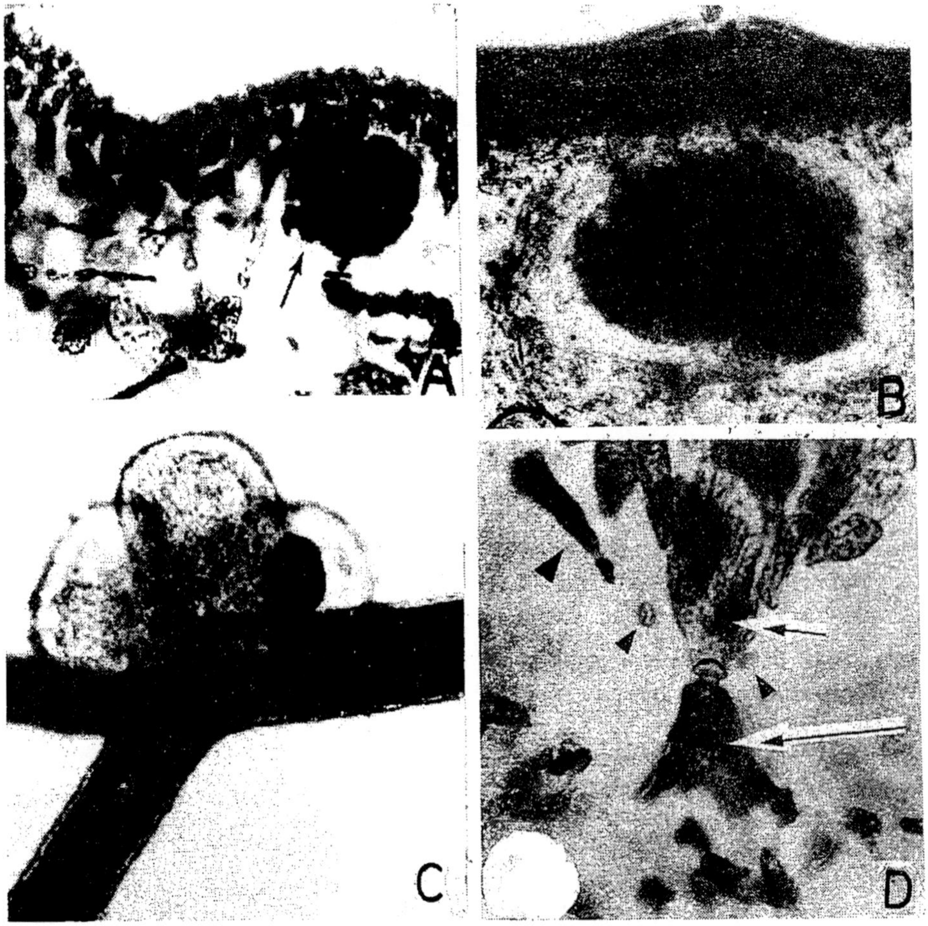

Fig. 45 A-D. Reproductive organs

A. *Acrosymphyton tenax.* Young cystocarp born on auxiliary cell filament and densely enclosed by the locally protuberant cortex (after Millar and Kraft, 1984).

B. *Halymenia acuminata.* Mature cystocarp (after Kawaguchi, 1991).

C. *Dawsoniocolax bostrychiae.* Female gametophyte with three developing cystocarps (after West and Calumpong, 1988).

D. *Halymenia acuminata,* Basal portion of mature cystocarp. The long arrow shows fusion complex while the short arrow indicates elongated gonimoblast initial (after Kawaguchi, 1991).

Cryptonemiaceae : In *Halymenia acuminata,* carposporophyte development starts with fertilization where connecting filaments are cut off from the enlarged zygote. The connecting filaments fuse with an auxiliary cell. After fusion, one gonimoblast initial is cut off from the auxiliary cell towards the surface, and the ampullary cells then begin to produce lateral branchlets. The initial cell cuts off successively serveral gonimoblast cells and these in turn divide to form the carposporangia. During further development of gonimoblast, the ampullary filament and the derivative branchlets elongate and form a thick pericarp. The neighbouring vegetative cells and their derivatives also contribute to the pericarp formation. The pericarp is distinct around the fully developed carposporophyte. The mature cystocarp is spherical and opens by a pore in the cortex (Fig. 45 B, D. Kawaguchi, 1991).

The cystocarps of *Halymenia porphyriaeformis* are large and are visible to the naked eye whereas in *H. asymmetrica* small, with ostioles, deeply immersed in the thalli and invisible to the eye. The involucre is formed by the ampullary cells that elongate as the gonimoblast grows (Gargiulo and Tripodi, 1986). In *Cryptonemia undulata* and *C. kallymenioides* the cystocarpic plants are rare (Wetherbee and Kraft, 1982).

Pterocladiophilaceae : In *Pterocladiophila hemisphaerica,* carpogonial filaments and early post-fertilization stages were not observed. Mature carposporophytes are found singly in a conceptacle and consist of a few gonimolobes on a large fusion cell. All cells develop into carposporangia (Stegenga and Vroman, 1986).

Chorcocolacaceae : In *Dawsoniocolax bostrychiae,* the mature cystocarps are hemispherical and each bear about 50-100 carposporangia. The pericarp is single layered and shows an ostiole (Fig. 45 C; West and Calumpong, 1988).

Corallinaceae : In *Hydrolithon cymodoceae,* the carposporophytes develop within the female conceptacles. Each mature carposporophyte consists of an ovoid fusion cell which gives rise either directly to a carposporangium or to a single marginal gonimoblast filament that forms a terminal carposporangium, The conceptacle roof is 2-4 cells thick and either protrudes slightly above or is flush with the surrounding thallus surface (Penrose and Wolekerling 1992b).

In *Pneophyllum fragile,* the carposporophyte develops within the female conceptacle after karyogamy; consists of a flat fusion cell; with marginal gonimoblast filaments that bear terminal carposporangia. The conceptacle roof of 2-10 cells thick either protrudes above the thallus surface or is flush with it (Penrose and Woelkerling, 1991).

TETRASPOROPHYTE

Weeksiaceae : In *Constantinea,* the tetrasporophyte are isomorphic, tetrasporangia are large, irregularly or regularly zonate and borne terminally on superficial cortical cells either on the upper/lower or rarely on both surfaces of the terminal blades. In *C. rosa-marina* and *C. sublifera,* the zonate tertrasporangia (Fig. 46 B) increase in size and are found only on the lower surface of the blade in discrete and confluent sori. In *Weeksia coccinea,* the tetrasporangia are cruciate to irregularly zonate; borne laterally on cortical cells; and 3 or 4 cells below the thallus surface (Lindstrom and Scagel. 1986).

Fig. 46 A-C. Reproductive organs

(*T.* tetrasporangium)

A. *Halymenia asymmetrica.* Cruciately divided tetrasporangium among cortical cells (after Gargiulo and Tripodi, 1986).

B. *Constantinea subulifera.* Zonate tetrasporangium maturing among paraphysis in sori on lower surface of terminal blades (after Lindstrom and Scagel, 1987).

C. *Pterocladiophila hemisphaerica.* Tetrasporangial conceptacles (after Stegenga and Vroman, 1986).

Dumontiaceae : In the less advanced taxa like *Dosyphloea insignis, Dumontia* sp., and *Cryptosiphonia woodii,* the tetrasporangia are attached basally or subbasally. In *Hyalosiphonia caespitosa* and more advanced taxa of Dumontiaceae, with the exception of *Dilsea carnosa* and *D. californica,* the tetrasporangial initials are also attached basally or subbasally when first cut off, but as they mature, with more enlargement occuring proximally than distally, they become attached laterally. In *Dilsea carnosa* and *D. californica* tetrasporangia are intercalary, a condition derived from lateral tetrasporangia. Intercalary tetrasporangia arose independently in *Dilsea carnosa* and *D. californica.* Both cruciate and zonate tetrasporangia are found among primitive Dumontiaceae. *Dosyphloea insignis* has zonate tetrasporangia whereas those of *Dumontia* sp. and *Cryptosiphonia woodii* are irregularly zonate to cruciate. The tetrasporangia of the Dumontiaceae become regularly cruciate although bisporangia have been reported in *Farlowia irregularis.* Only tetrasporangial initials are found in *Dilsea integra* (Lindstrom, 1988).

In culture, three species of *Farlowia* show a life history where upright gametophyte alternate with crustose tetrasporophytes (Fig. 48). The crustose stages have paraphyses and decussate-cruciate to irregular-cruciate tetrasporangia. The tetrasporangia of *F. compressa* and *F. mollis* do not differ substantially in size but those of *F. conferta* are small. In *F. mollis* and *F. conferta,* the production of tetrasporangia only at 15°C, 8:16 photoregime suggests a short-day response (DeCew and West, 1981).

In *Dudresnaya* tetrasporophytes are either isomorphic with erect gametophytes or show discoid polystromatic crust. Tetrasporangia are zonate or irregularly cruciate. *D. verticellata,* zonate tetrasporangia are borne laterally on outer cortical filaments. The tetrasporangia in *D. australis* are, however, borne terminally (rarely lateral) and show zonate pattern. If a cortical cell that bears hair becomes a sporangium, the hair is shed as soon as its subtending cell begins to differentiate. The tetrasporangia in *D. capricornica,* are produced both terminally and laterally; and in *D. hawaiiensis* are absent (Robins and Kraft, 1985).

The life history of *Acrosymphyton purpuriferum* is heteromorphic and probably diplohaplontic (Hoek and Cortel-Breeman, 1969; Cortel-Breeman and Hoek, 1970). In *A. purpuriferum,* the carpospore grows into a tetrasporophyte which resembles *Hymenoclonium serpens.* The Hymenoclonium phase produced tetrahedral tetrasporangia. The tetraspores grew into *A. purpuriferum* gametophyte. Short day conditions induced the formation of tetrasporangia. The Hymenoclonium-phase could reproduce itself vegetatively by 4-celled structures that resemble seriate tetrasporangia (Cortel-Breeman and Hoek, 1970 see also Cortel-Breeman, 1975).

Choreocolacaceae : In the *Dawsoniocolax bostrychiae,* the mature tetrasporophyte is similar in size to the gametophytes but has fewer lobes. The tetrasporangia are cruciate-decussate to nearly tetrahedral. Sporangial tetrads are liberated intact and individual spores separate within 2 or 3 minutes and immediately become spherical. Tetraspores germinate readily on the surface of both host and parasite but no hyperparasitism occurs, because the sporelings do not develop beyond the 2-celled stage when they are on the parasite.

Cryptonemiaceae : In *Halymenia acuminata,* the tetrasporangia are dispersed over the blade. Tetrasporangial initial are cut off laterally from the cortical cells. Mature sporangia are cruciate or decussate, broad, ellipsoidal, and submerged in the cortex (Kawaguchi, 1991). In *H. asymmetrica* (Fig. 46 B), the tetrasporangia are scattered over the thallus surface and are not grouped in sori. The tetrasporangial initials seem to arise as lateral outgrowths from the inner most cells of the outer cortex. Tetrasporangia are round and generally cruciate, although tetrasporangia with two periclinal cleavages have been observed (Gargiulo et al., 1986). In *Cryptonemia kallymenioides* tetrasporophytes bear many cruciate tetrasporangia that are submerged in the outer cortex (Scott *et al.*, 1982).

Gloiosiphoniaceae : *Gloiosiphonia capillaris,* in culture shows light red young tetrasporophytes that are smooth, orbicular with encrusting discs, composed of monostromatic basal layer, adherent to the substrate, and loosely packed vertical filaments. After several months, the tetrasporophytes become brittle, rough, dark and are covered with necrotic areas. The tetrasporangia are embedded and are either zonate, or cruciate. The tetrasporophytes, which persist even in the detached stage, form and release tetraspores at irregular intervals even upto one year. The tetraspores released later, although viable, failed to produce gametophytes (Edelstein and McLachlan, 1971).

Pterocladiophilaceae : In *Pterocladiophila hemisphaerica,* tetrasporangia develop terminally on short, one or two celled filaments that arise (Fig. 46 C) from the basal reticulum. The mature sporangia are zonate (Stegenga and Vroman, 1986).

Corallinaceae : Most taxa produce tetrasporangia in conceptacles. In *Epilithon, Lithothamnion* and *Phymatolithon,* the conceptacles are absent and the sporangia forming sori are embedded within the crusts.

In *Pneophyllum fragile,* tetrasporangial plants have uniporate conceptacles whose roofs protrude to 2-10 cells thick above chamber, or are flush with surrounding thallus surface (Penrose and Woelkerling, 1991).

In *Melobesia* sp., the tetrasporangial conceptacles are multiporate and the conceptacles roofs are formed by filaments interspersed amongst sporangial initials. The cells of adjacent roof filaments are sometimes joined by cell fusions. Columella is absent. Tetrasporangia are scattered across chamber-floor. The mature sporangia are zonate. In *Melobesia membranacea,* the conceptacles open to the external environment either through gametangial pore or via a number of tetrasporangial pores (Fig. 47B).Tetrasporangia are scattered across the concepatcles chamber-floor. Each sporangium contains 4, zonately arranged spores. Tetrasporangial initials develop only one cell layer below the thallus surface (Fig. 47A). Section through tetrasporangial conceptacles show variable number of tetrasporangia. A few conceptacles have numerous whereas others have a few tetrasporangia. The conceptacle roofs either protrude or are flush with the thallus surface. However, *M. rosanoffii,* the conceptacle roofs are flush with the thallus. Further, elongate cells, such as those found in male and female conceptacles roofs, usually occur in peripheral roof region of most, but not all, tetrasporangial conceptacles (Fig. 47C) Such elongate cells do not occur in the tetrasporangial conceptacle roofs of *M. membranecea* (Wilks and Woelkerling, 1991).

In *Hydrolithon cymodoceae,* the tetrasporangial conceptacles usually contain a single, zonate tetrasporangium lying horizontally within the chamber. Irregularly divided tetrasporangia, bisporangia and unidivided sporangia resembling monosporangia are also formed (Penrose and Woelkerling 1992b).

In *Lithophyllum crouanii,* the tetrasporangial conceptacles have globose chambers containing elongate tetrasporangia that are mostly tripartite with uninucleate upper and lower spores and a binucleate central one. Quadripartite spores are also seen. The tetrasporangia are borne on stalk cells and are interspersed with a few sterile filaments but lack pronounced columellae (Chamberlain, 1988).

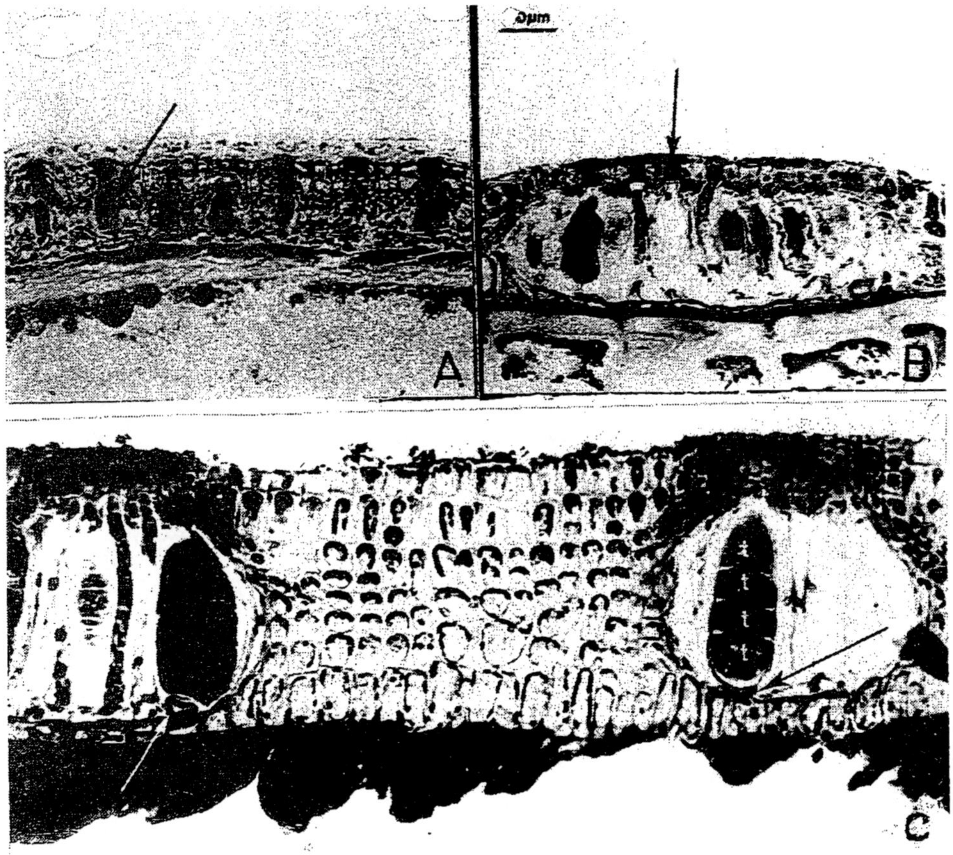

Fig. 47 A-C, Reproductive organs : Tetrasporangial conceptacle of *Melobesia* sp.
(after Wilks and Woelkerling, 1991).
(t., tetraspore.)

A. *Melobesia membranacea* Sporangial initial (arrow) with interspersed sterile filament only one cell layer overlies the initials.

B. *Melobesia membranacea.* Developing conceptacle showing chamber roof and undivided sporangia (arrow denotes pore-plug).

C. *Melobesia rosanoffii.* Section through mature tetrasporangial conceptacles. Note size difference between conceptables; tetraspores (arrow indicates stalk cells subtending tetrasporangia).

LIFE CYCLE STUDIES

DUMONTIACEAE

The members of this family exhibit alternation of gametophytes with an isomorphic or heteromorphic tetrasporophytes. The life cycle in *Cryptosiphonia woodii; Dilsea california and*

Neoditsea americana is isomorphic whereas in *Thurtellopsis pegiana, Pikea robusta* and *Dudresnaya columbiana* is heteromorphic. The genus *Farlowia* exist only as a gametophyte with the exception of a solitary report of zonate or cruciate tetrasporangia on the upright filament of *Farlowia conferta* (DeCew and West, 1981).

In *vitro* studies of *Farlowia mollis* and *F. conferta* (Fig. 48) indicate that both species have life history where the upright gametophyte alternates with a crustose tetrasporophyte. In *F. mollis and F. conferta* the tetrasporangia were produced only at 15° C, and 8-16 h photoregime. Interestingly, in *F. compressa,* tetrasporangia were lacking in culture, but the field collected material reveal crustose tetrasporophytes which produce upright gametophytes. Upright initiated only in long day condition for *F. compressa* and *F. conferta:* 15° C, 16 : 8 h and 10° C, 12 : 12 h for *F. mollis* (DeCew and West, 1981).

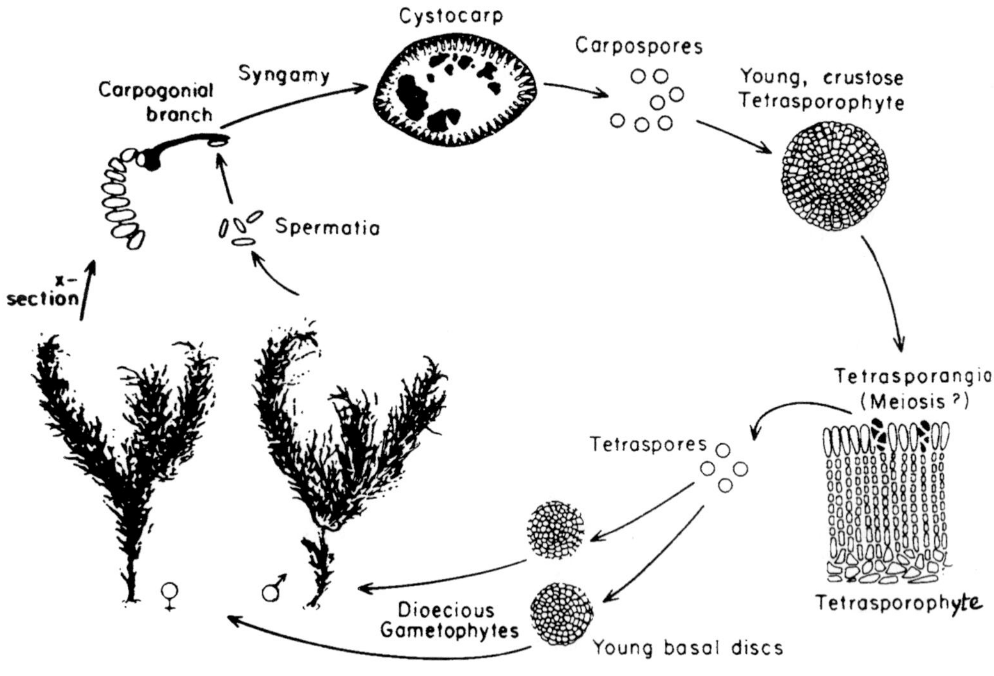

Fig. 48 *Farlowia conferta* : Diagrammatic representation of life-cycle.
(after DeCew and West. 1981).

The life history pattern of *Farlowia irregularis* (from east coast of Hokkaido in Japan) follows the Polysiphonia-type and differs from that of pacific north American species which possess upright gametophytes and crustose tetrasporophytes (Shimizu and Masuda, 1983).

Dudresnaya minima carpospores in culture give rise to compact crustose tetrasporophytes which produce tetraspores that germinate and form erect male and female gametophytes. Among the 13 species of *Dudresnaya* hitherto known, only 4 have isomorphic life cycle. The remaining 9 species show crustose tetrasporophytes which is heteromorphic with erect gametophytes (Robins and Kraft, 1985). *D. contorta* has an isomorphic life history with similarly branched erect gametophyte and tetrasporophyte (Wilce and Davis, 1984).

Acrosymphyton purpuriferum, from the mediterranean, is heteromorphic and diplobiontic and the carpospore grew into creeping plants that resemble *Hymenoclonium serpens*. This Hymenoclonium phase produced tetrahedral, tetrasporangia where the tetraspores give rise to *Acrosymphyton* gametophyte. The Hymenoclonium phase recycles vegetatively through seriate tetrasporangia. The *Acrosymphyton purpuriferum* gametophytes are summer annuals while tetrasporophytes are perennial that are independent of the gametophyte. The gametophyte is dependent for its existence on the tetrasporophyte, for it can only grow from tetraspores produced in tetrasporangia formed during the winter half-year, and it lacks means of vegetative propagation (Cortel - Breeman and Hoek, 1970).

CHOREOCOLACACEAE

Choreocolax has a triphasic life history (Fig 49) consisting of (i) haploid gametophyte generation which arises from the zygote and proliferates within the tissue of the gametophyte to

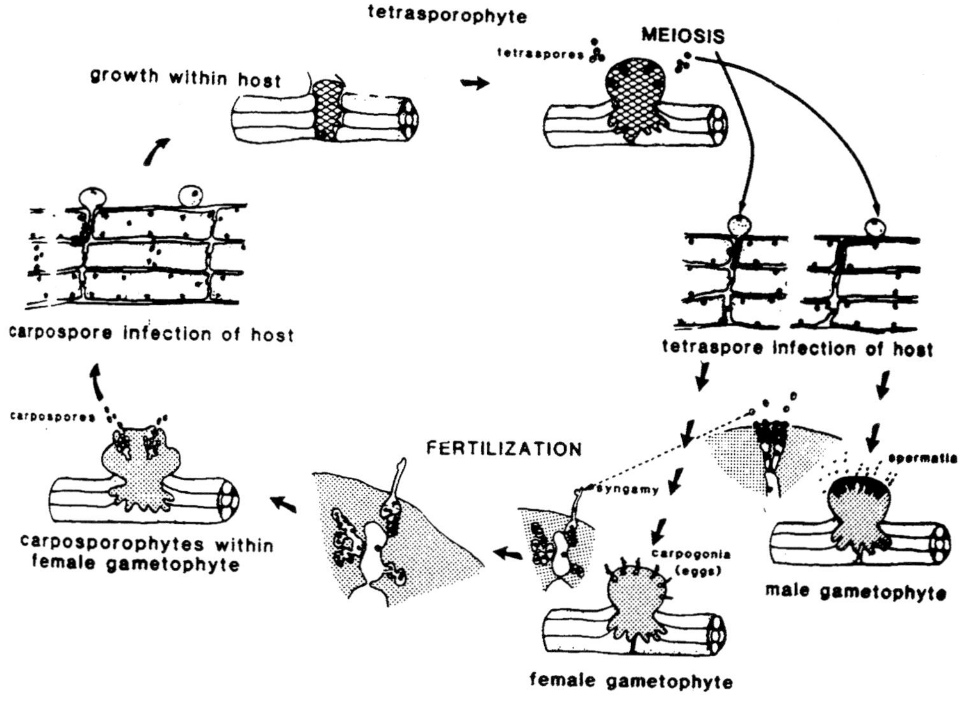

Fig. 49. ***Choreocolax polysiphoniae*** **: Diagrammatic representation of life-cycle.** (after Goff and Coleman, 1984).

produce diploid carpospores. (ii) diploid tetrasporophyte generation which arises from the germinating carpospores, (iii) meiosis results in tetraspores and fertilization involve the fusion of spermatium with the carpogonium. Subsequent fusion occurs within the connecting or the auxiliary cell (Goff and Coleman. 1984).

A study of chromosome number in the life cycle of *Harveyella mirabilis* (Fig 50) indicates that it has Polysiphonia - type of life history (Goff and Cole, 1976). The haploid chromosome number is found in male and female sexual stages while the diploid chromosome number is noted in carposporic and tetrasporic phases, exclusive of tetraspores. Meiosis occurs in the tetrasporangium and results in the formation of four haploid spores.

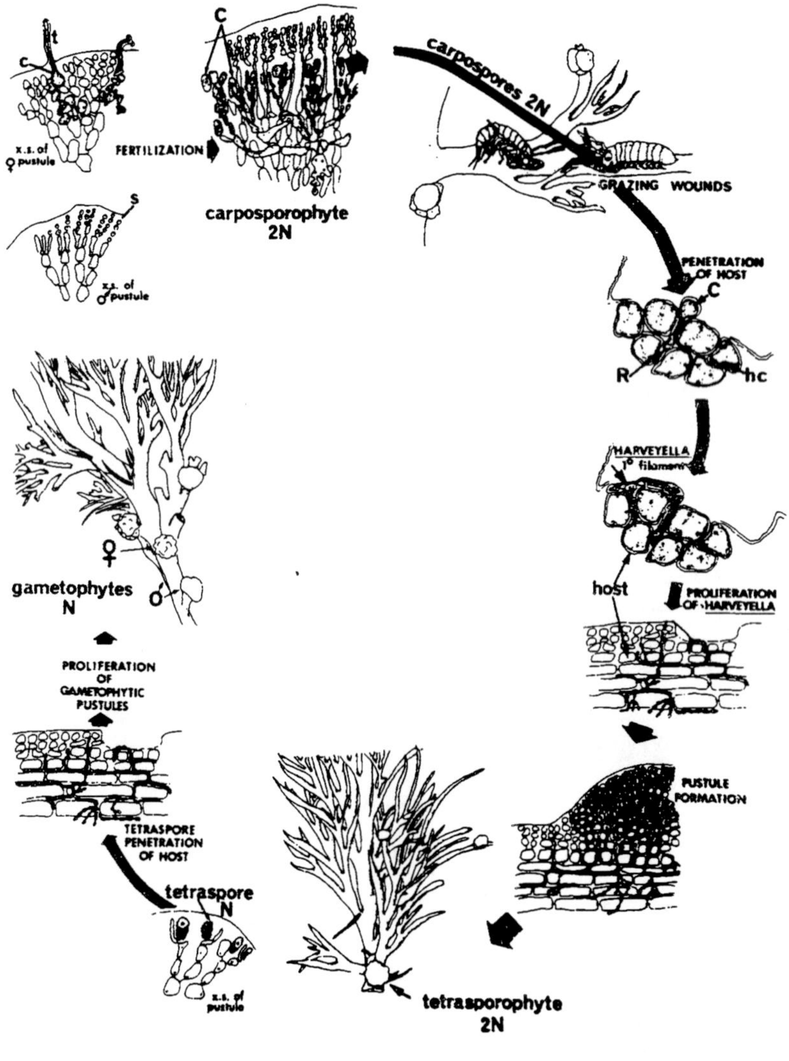

Fig. 50 ***Harveyella mirabilis*** **: Diagrammatic representation of life-cycle.** (after Goff and Cole. 1976)

CRYPTONEMIACEAE

In this family, the life history is generally diplobiontic with isomorphic tetrasporophytes and gametophytes. However, in *Halymenia floresia,* in culture, an anamolous pattern exists where the carpospores grew into Acrochaetium-like plants that eventually formed the monospores. Later, a few of Acrochaetium-like plants differentiated into *Halymenia* sporophytes which in monosporangial sori produce monospores that germinate into a new crop of gametophyte (Hoek and Cortel-Breeman, 1970).

GLOIOSIPHONIACEAE

In *Gloiosiphonia capillaris* and *G. verticillata,* the life history has been investigated (DeCew and West, 1981). In these taxa the carpospores give rise to small compact discs which are composed of erect filaments embedded in mucilaginous matrix but revealed monostromatic growth. These crusts were identified as *Cruoriopsis.* The cruciate, zonate or tetrahedral tetraspores develop in the crust, and when released germinate into new monoecious gametophytes. In culture, the basal system of the large gametophytes, that resembles the small crustose tetrasporophyte, persists for more than 18 months. The basal portion of the gametophyte and crustose sporophytes could both serve to maintain the alga during the season when the conspicuous gametophytic phase is lacking.

CORALLINACEAE

The life history is generally diplobiontic with isomorphic gemetophytic and tetrasporophytic stages. In *Pneophyllum lobescens* and *P. myriocarpum* a regular Polysiphonia type of life history occur. A self-perpetuating, bisporangial life history was confirmed in *P. lobescens, P. plurivalidum* and *P. zonale* (Chamberlain, 1987).

Fosliella zostericola, an epiphytic crustose coralline alga, in culture completed its life history, in the modified Grund medium at 15° C, 14 : 10 LD with 30 w/m^2. The carpospores gave rise to mature tetrasporophytes in 44-124 days. The liberated spores from the tetrasporophytes reached mature male or female gametophytes in 20 days. The carpospores were liberated from carposporophyte in 92 days. The germination-disc of both tetraspores and carpospores possess eight-celled central elements which produce intercalary trichocytes. *Fosliella* has 4-celled element and terminal trichocyte whereas *Pneophyllum* has 8-celled central element and intercalary trichocyte. *Fosliella zostericola,* in culture had 8-celled central element and intercalary trichocyte, and thus, is transferred to the genus *Pneophyllum* (Fujita, 1988).

ULTRASTRUCTURAL STUDIES

PIT-CONNECTION

Pit-connections in red algae are generally formed during the cell division. The primary pit-connections develop when the apical cells divide and, thus, occur only between the kindered cells and cells derived from the same apical-cell. Pit-connections between non-kindered cells are, however, common in many taxa. The origin of such secondary pit-connections are well known from the developmental studies either in the Ceramiales or studies from the red algal parasites. An unequal intercalary cell division produces a small conjuctor cell which eventually fuses with the adjacent cell resulting in non-kindered cells that share pit-connetion. The fused cell is, thus, binucleate. Cabioch (1971) reported that direct connections and perforations between non-kindered cells allow limited cellular continuity through a small aperture which is filled with pit-plug.

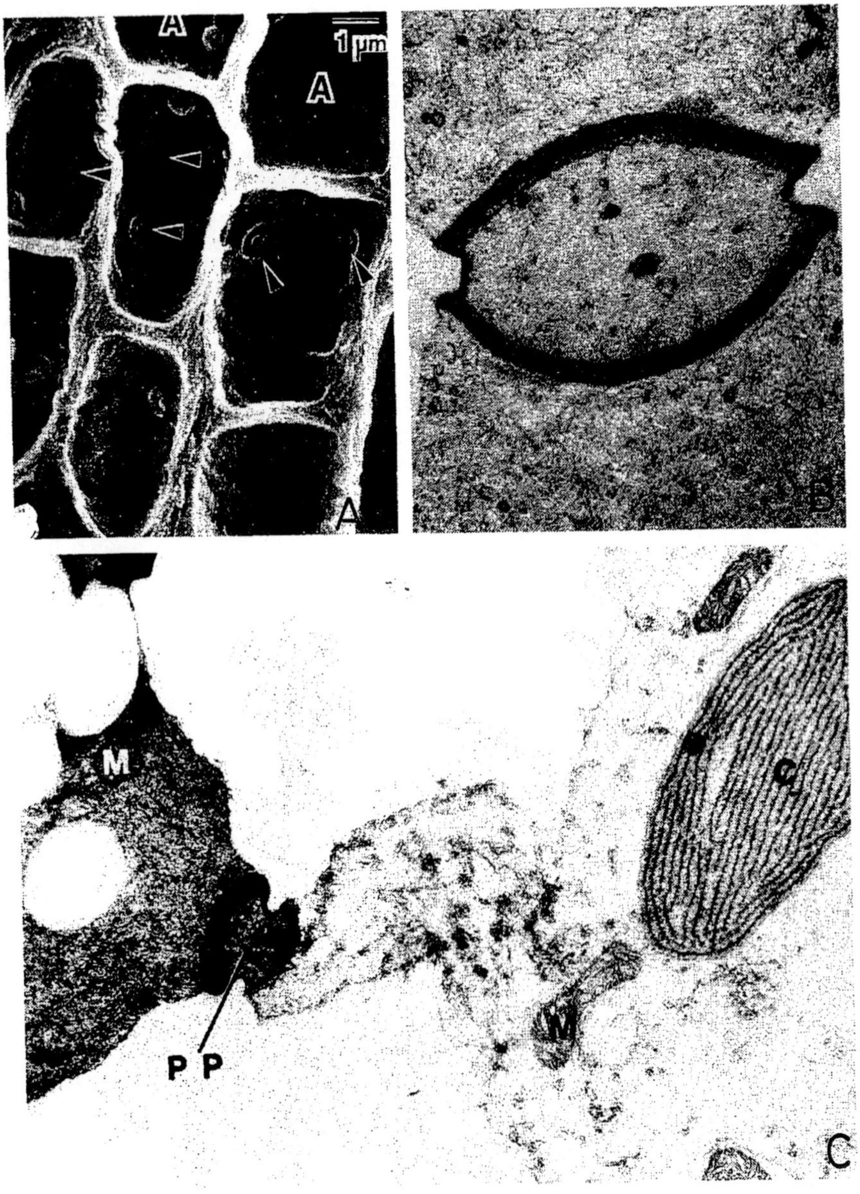

Fig. 51 A-C. Ultrastructure : Pit-connection

(A. empty wall of apical cell; c, chloroplast; fs, floridean starch; m, mitochondria; pp, pit-plug.

A. *Hildenbrandia*. Lateral walls of sub-apical cell have secondary pit-connections appearing as holes (arrowheads; after Pueshel, 1988b).

B. *Choreocolax polysiphoniae*. Pit-connection. (after Kugrens and West, 1973).

C. *C. polysiphoniae*. Host-parasite pit-connection (after Kugrens and West, 1973).

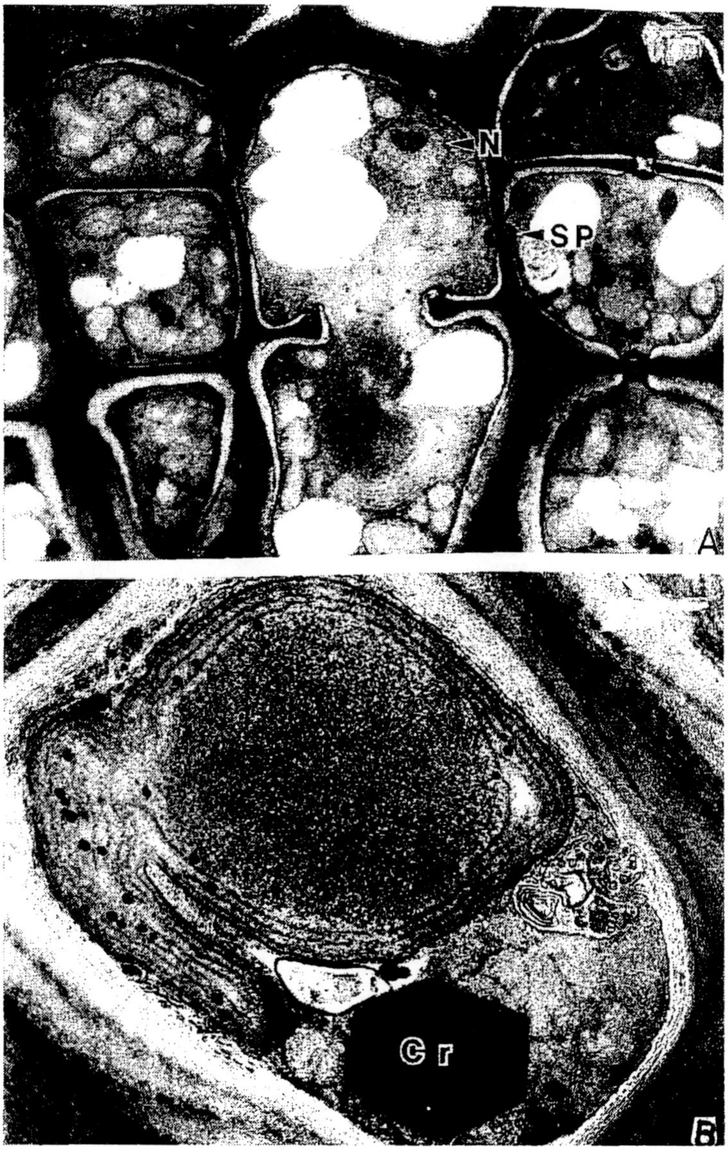

Fig. 52 A, B. Ultrastructure of *Hildenbrandia rubra*.

(after Pueschel 1988)

(N. nuclei; Sp, secondary pit-connection; Cr, crystal)

A. Ultrastructure of cell-sloughing showing response of surviving cells. One cell is in the process of cytokinesis, as indicated by the cleavage furrows. Nucleus is present in the cells (arrows). The apical daughter cell bears a secondary pit-conncection (SPC) to and adjacent sub-apical cell.

B. With the exception of the large granular inclusion the chloroplast appears normal. A protein crystal is present in the cytoplasm.

Cabioch and Giraud (1982) suggested that *Hildenbrandia* and many other non-calcified taxa possess secondary pit-connections formed by plug perforation method. Pueschel (1988) demonstrated abundant secondary pit-connections in anatomically homogenous crusts of three *Hildenbrandia* species. The connected cells are uninucleate confirming that secondary pit-connections in this group does not involve nuclei transfer as would be expected if conjuctor cells are involved. The systematic potentials of secondary pit-connections and related feature namely uninuclearity versus multinuclearity are important factors. The functions of primary/or secondary pit-connections are still a matter of debate. Scanning electron microscopic studies in *Hildenbrandia rubra*, *H. crouanii* and *H. occidentalis* reveal that each cylindrical cell in a vertical file is surrounded laterally by about six files of non-kindered cells. Secondary pit-connections are established between many of the laterally adjacent files. Even subapical cells have secondary pit-connections (Fig. 51 A). The cells of *Hildenbrandia* are uninucleate. This combination of secondary pit-connection and uninucleate cell amply demonstrates that secondary pit-connections in this taxon are produced directly without the formation of conjuctor cells (Cabioch and Giraud, 1981). In these simple crust forms such abundant pit-connections do provide a mechanical role and help in the adherence of vertical filaments (Pueschel, 1988).

A few members of Cryptonemiaceae show differentiation into blades and stalks. *Cryptonemia kallymenioides* is compared with *C. undulata* which lacks massive stalk. In both the taxa there is extensive periodic secondary cortication of the stalks that result in the formation of growth rings. The blades of *C. kallymenioides* appear to be seasonal and its stalk perennial, while that of *C. undulata* are apparently perennial but short lived than *C. kallymenioides*. The ultrastructure of pit-connections from cortex to medulla cells located progressively distant from the plant surface, show the following modifications.

(a) They increase dramatically in size, the largest being located near the stalk-centre where the diameter of the aperture can be over ten times greater than observed at the surface.

(b) The surface of the pit-connections become highly convoluted while the cap-membranes follow these contours.

(c) The surface area of the cap-membranes is often further increased by the formation of membranous tubules of uniform diameter which extend out into the cytoplasm of each cell.

(d) The core region becomes electron-transparent though still capped by electron-dense layers. Identical modifications of pit-connection structure occur in *C. undulata*, in which the width, internal volume and electron-transparency of the pit-connections also increase with distance from the thallus surface (Scott *et al.*, 1982).

Choreocolax polysiphoniae, an alloparasite, is one of the most evolved parasitic red alga. Host-parasite pit-connection is the only diagnostic feature to indicate that *C. polysiphoniae* is a valid member of Rhodophyceae. The pit-connections consists of an aperture that is traversed by plug (Fig. 51 B), which consists of an electron-dense border with finely granular-core material. Secondary pit-connections are formed between the host and parasite, the plug being that of the parasite (Fig. 51 C; Kugrens and West, 1973).

CELL-SLOUGHING AND CHLOROPLAST INCLUSION

Hildenbranda rubra (Hildenbrandeaceae) is unique in having two unusual characters (Pueschel, 1988).

(a) Large portion of the crust surface is necrotic; and the depth of the affected areas varies. The top layers of healthy cells that occur beneath the zone of dead cells showed regeneration through cytokinesis (Fig. 52 A).

(b) Chloroplasts of both living and dead cells have conspicuous spherical inclusions that are visible even under light microscope. These inclusions generally had a homogenously granular structure. Rarely, a dense, non-granular substance and phytoferritin are also present (Fig. 52 B). It is not known whether there is a casual connection between the presence of chloroplast inclusions and necrosis.

In *Constantinea simplex* and *C. sublifera* the chloroplasts of senescing blades contain electron-dense, core particles that were confirmed by X-ray microanalysis as phytoferritin, indicating the presence of iron (Pueschel and Parthasarathy, 1984).

MERISTEMATIC AND CORTICAL CELLS

Choreocolax polysiphoniae an alloparasite, which apparently represents one of the most evolved parasitic red alga, shows at ultrastructural level that the meristematic and cortical cells are narrow, spindle-shaped, rectangular or ovoid. The nucleus is usually found at the centre of the cell and possesses acentric nucleolus which occupies from one-third to one-half of the nucleoplasmic volume (Fig. 53A). Nuclear envelope shows many pores with complex ultrastructure. Numerous mitochondria are scattered within the cytoplasm. Chloroplasts are highly reduced and lack thylakoids development.

In the cortical cells dictyosome are active and release many vesicles that may be involved in wall formation. No difference between rhizoidal and vegetative cells exist. Both cells types are uninucleate, may or may not contain starch grains, and are approximately of the same size.

The medullary cells are large and contain large quantities of floridean starch which displaces the nucleus and other organelles to peripheral area of cytoplasm (Fig. 53 B.). In medullary cells the maturing faces of dictyosome vesicles are oriented away from cell wall and perhaps contribute to the formation of starch grains. The large cells form secondary pit-connections with one another (Fig. 53 B). Pockets of endoplasmic reticulum are located throughout the cells. Mitochondria and highly reduced chloroplast are abundant in the peripheral cytoplasm.

SPORE SIZE AND SPORE GERMINATION

Chihara (1973, 74) observed that spore size could be correlated with subfamily. Melobesioideae, with largest spores, are most advanced than families with small spores. Bressan (1980) summarised the pattern of segmentation and considered the possible relationship between them. The above authors, suggest that difference in germination pattern can be correlated, with subfamily grouping in the Corallinaceae.

Spore size and germination were investigated in four epilithic species namely *Pneophyllum myriocarpum*, *P. lobescens*, *P. plurivalidum* and *P. zonale* and four epiphytic species namely *P. lejolisii*, *P. limitatum*, *Fosliella farinosa* and *Melobesia membranacea* belonging to crustose Corallinaceae. All spores types examined, on release, were uncalcified and mature spores were surrounded by a clear halo which, in younger spores, were less distinct. These spores often changed their shape and upon anchorage secreted an adhesive substance (Fig. 54 A). Like *Corallina*, in the four *Pneophyllum* epilithic species, the spores become calcified even at an early stage of development.

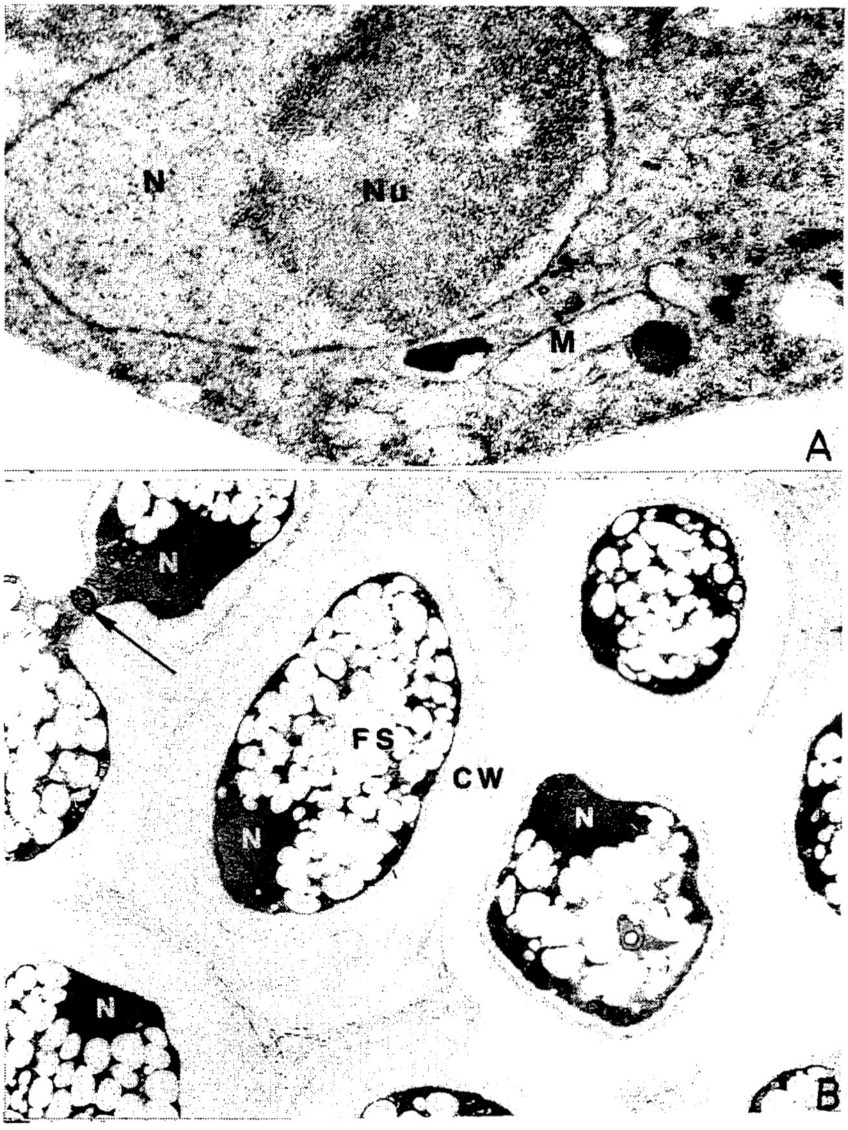

Fig. 53 A, B. Ultrastructure
(after Kugrens and West, 1973)
(N. nucleus; Nu, nucleolus; M, mitochondria;
Fs, Floridean Starch; CW, cell wall).

A. *Choreocolax polysiphoniae* : Cortical cell showing the acentric nucleolus and mitochondria.

B. Same, medullary cells are filled with starch. surrounded by very thick walls, and have nuclei that are displaced by starch grains; arrow indicates a pit-plug between 2 cells.

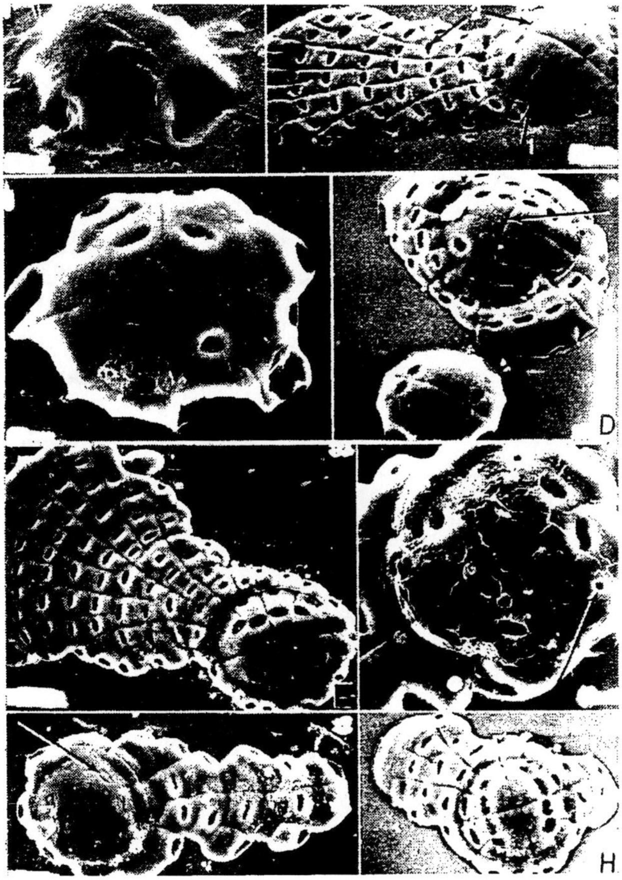

Fig. 54 A-H. Ultrastructure : Spore germination.

(after Chamberlain, 1984)

A. *Pneophyllum plurivalidum.* Bisporlings with adhesive substance.

B. *P. plurivalidum,* Bisporlings showing the growth of the young crust, the ring (1) indicating periphery of the original spore, and the epithallial concavities (2).

C. *P. lobescens.* Carposporelings stages grow peripherally, four of the eight central epithalia cells are covered with surface mucilage.

D. *P. myriocarpum.* Carposporelings at two stage of development, the direction of the initial cleavage (arrow), and eight central epithallial concavities indicating the position of the eight celled central element are clearly visible.

E. *P. lobescens.* Bisporelings.

F. *P. lobescens.* Tetrasporelings; two of eight central epithalial concavities are obscured by surface covering but hair (arrow) is present.

G. *P. lobescens.* Carposporeling; three of the eight central epithalial cells (arrow) are visible. The others are obscured by surface covering.

H. *P. myriocarpum.* Tetrasporelings.

In red algae, the initial stages of spore germination are significant and aid in taxonomic classification. The Corallinaceae are characterized by Dumontia-type of development, where the spores divides internally resulting in a germination disc of upto 32 cells before growth beyond the original spore periphery commences. Cabioch (1972) however, recorded Naccaria-type characterized by spores which upon germination, produce directly filaments. This was seen in parasitic and endophytic taxa namely *Choreonema thuretii; Amphiroa verruculosa: Amphiroa rigida* and *Schmitziella endophloea* (according to Woelkerling and Irvine, 1982, this last taxon is not a coralline alga). Cabioch (1972), also proposed, within the Corallinaceae, three modes of spore development and referred to them as *Lithothamnion, Negoniolithon* and *Lithophyllum* types. These types principally concern the morphogenesis of the young thalli. Chihara (1973, 74 a.b) observed Dumontia-type of development in a wide range of articulated Corallinaceae. He, however reported variation in disposition of cells within the germination disc with two principal disc patterns. On this basis two groups, were recognized, (a) *Amphiroa/Lithophyllum* group where spore germination pattern was bilaterally symmetrical, secondary pit-connections occurred, tetraspores were mainly less than 40 μm in diameter and reproduction occurred mainly in summer (Amphiroideae and Lithophylloideae), (b) *Corallina/Lithothamnion* group with a regularly grid-like spore germination pattern, cell fusion but no secondary pit-connections, tetraspores mainly more than 40 μm diameter and reproduction occur other than summer months (Corallinoideae and Melobesioideae).

Segmentation in sporelings of epilithic species of *Pneophyllum* proceeded by a series of cell divisions perpendicular to the substratum to form a germination disc of 16 cells. The initial division indicates spore polarity and the direction of the first wall often remained conspicuous throughout spore segmentation (Fig. 54 D-H). Each cell of the segmented spore cuts off an epithallial cell from distal end (Fig. 54 B) and occasionally, a hair base may be produced (Fig. 54 F). In scanning electron micrographs, the position of the epithallial concavity indicate the position of its underlying cells. Two rows of 4 central cells each are evident. Air-dried sporelings, right from early stages appeared robust and calcified. The sporeling growth is usually initiated from one or two points from the spore periphery (Figs. 54 B, C, D, E, G, H) producing radiating, hypothallial filaments, with each cell cutting-off one epithallial cell (Fig. 54 B). The periphery of the original spore often remained distinct even in older sporelings (Fig. 54 B). Sporelings from all spores types of the four epilithic *Pneophyllum* species were similar in appearance (Fig. 54 D-H) but differed mainly in size. The sporelings development pattern of the epiphytic species *P. limitatum* (Fig. 55 A) is identical to that of epilithic species.

The bispores of *Fosliella farinosa* (Fig. 55 B) exhibited a different pattern of development from *Pneophyllum*. The third division, as in *Pneophyllum*, resulted in 8-cells but, at this stage, each of the 4-central cells cuts of a small apical cell (Fig. 55 B) and later ceased to divide. All subsequent thallus growth is initiated by the four outer cells (Fig. 55 B). The 4-celled unit often remained conspicuous in the mature crusts. However, the four cells, occasionally became secondarily meristematic and produced filaments in older plants. A third pattern of development occured in *Melobesia membranacea* (Figs. 55 C, D) where spores segmented into 32-48 more or less isodiametric cells in a grid-like pattern (Chamberlain, 1984).

CARPOSPOROPHYTE

Postfertilization developments which are believed to be conservative, formed the basis of taxonomy of red algae (Kylin, 1930). The order Cryptonemiales is a notable exception in having

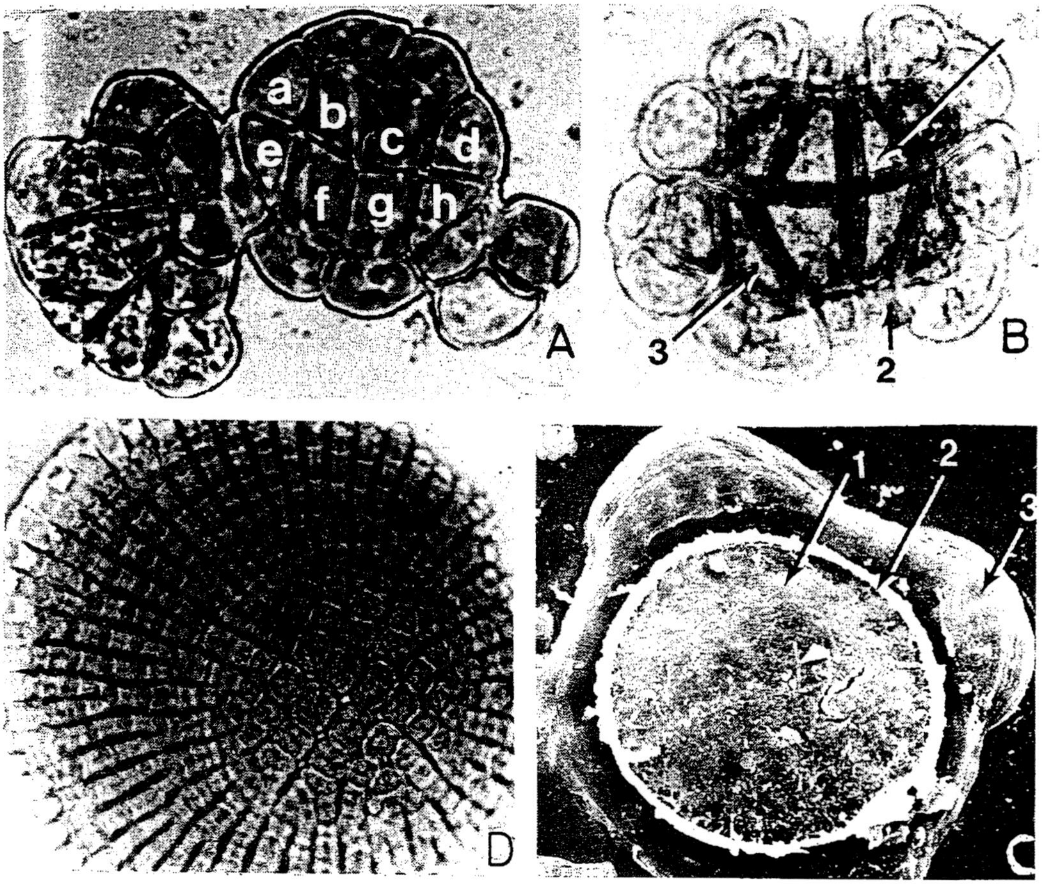

Fig. 55 A-D. Ultrastructure : Spore germination.

(after Chamberlain, 1984).

A. *Pneophyllum limitatum.* Bisporelings in which peripheral growth has started. (a-h cells of central element).

B. *Fosliella farinosa.* Bisporelings showing the four celled central element (1). small cells (2) and peripheral growths inhibited by the four outer cells (3).

C. *Melobesia membranacea.* Scanning electron micrograph of critical point dried sporelings. The many celled central element (1) is covered with a cuticle. The spore periphery (2) is prominent and thallus cells (3) develop around the entire periphery.

D. *Melobesia membranacea.* Sporelings showing the centre composed of more or less isodiametric cells.

a variety of post-fertilization stages ranging from procarpial to non-procarpial; and monocarpogonial to polycarpogonial. The complexity occurs primarily during early development of carposporophyte and consequently has received the most attention. The later events, however, are taken for granted.

The small size of reproductive structures necessiates the use of electron microscopy to unravel the cellular details, specially the cellular and cytological events, during carposporophyte formation and cell fusion to elucidate the relationship among the diverse taxa.

In *Gloiosiphonia verticillaris,* the auxiliary cell produces gonimoblast initials which again divide to produce two type of gonimoblast cells, (a) the nondividing vacuolate cells and (b) the terminal generative gonimoblast cells. The latter form clusters of carpospore initials which eventually differentiate into the carpospores. The earliest formed gonimoblast cells become eventually incorporated into the auxiliary cell to form the fusion cell. The gonimoblast cells between the fusion cell and carpospores, however, loose their storage materials, become highly vacuolate and transform into sterile gonimoblast cells. Thus mature carposporophyte shows three distinct components (i) The fusion cell, (ii) sterile gonimoblast cells, and (iii) peripheral groups of carpospores.

The auxiliary cell no longer participates in gonimoblast cell formation, instead, the first formed gonimoblast initial divides, to form more gonimoblast cells. As fusion begins, the auxiliary cell assumes a degenerate appearance followed by the enlargement of the septal-plug between the auxiliary cell and adjacent gonimoblast cells. This plug eventually ruptures and creates a cytoplasmic channel between these cells resulting in a temporary synchytium. Sites of cellular fusion are identified by both wall and setal-plug remnants. By the fusion, proplastid is the first organelle to be affected where their membranes become crenulated, or the plastids enlarge and eventually rupture. Nuclear membranes also become irregular, and intercell nuclear migration commences. Mitochondrial degeneration is characterized hypertrophied cristae. At this stage, proplastids have no structural identity. Nuclei are the last structures to be disrupted.

Generative gonimoblast cells contain large quantity of floridean starch (Fig. 56 A). In dividing generative gonimoblast cell (Fig. 56 A) the nuclei are situated at the poles and are separated by numerous starch gains, which are randomly distributed in cytoplasm of the presumptive daughter cells. Plastids occur in the peripheral cytoplasm and by this time form internal thylakoids. The carpospore differentiation begins only after generative gonimoblast cells cease division. Both the generative gonimoblast cells and carpospores, are indistinguishable during the early stages of carposporophyte formation (Fig. 56 B). The carpospore at maturity, occurs in terminal, hemispherical clusters at the end of uniseriate gonimoblast filament (Fig. 56 B).

Carposporogenesis is simple and involves only three developmental stages. (1) Initially granular vacuoles appear (Fig. 56 C). In proplastids internal thylakoids are formed by an invagination of the inner membrane of the peripheral thylakoids (Fig. 56D). Later on vacuolar contents appear to undergo degradation as carpospore differentiation continues (Fig. 56 B-E). Dictyosomes formed during this stage, produce vesicles having fibrils embedded in an electron-transparent and amorphous mucilage (Figs. 56E, 57A). The second stage is characterized by deposition of carpospore wall material Mature carpospore wall consists of two layers. The inner layer is the true carpospore wall since it is deposited during carpospore differentiation while the outermost layer represents a dark staining zone that results from the compression of fibrils during

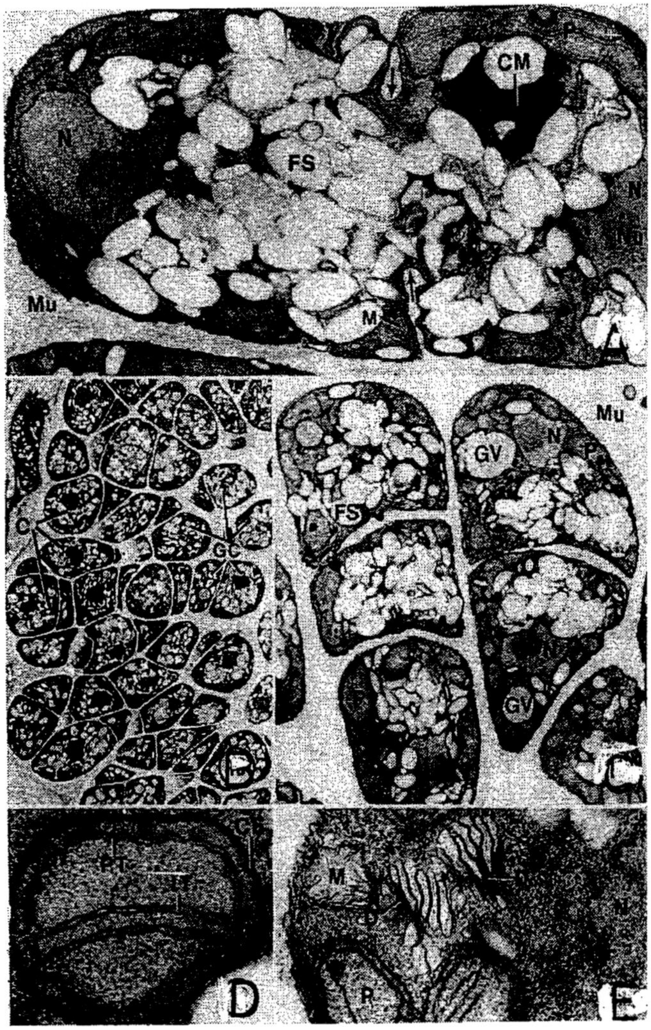

Fig. 56. A-E. Ultrastructure of carposporophyte of *Gloiosiphonia verticillaris.*
(after Delivopoulos and Kugrens, 1985).

(N, nucleus ; Nu, nucleolus; Fs, floridean starch; GC, gonimoblast cell; PT. peripheral thylakoids; IT, internal thylakoids; C, carpospore).

A. Dividing generative gonimoblast cell. The daughter nuclei are at the poles of the cell while the cleavage furrows develop by centripetal growth.

B. Part of the carposporophyte depicting groups of terminally situated carpospore.

C. Young carpospore showing cytological feature characteristic of these stage.

D. Formation of thylakoids from the invagination of inner membrane of peripheral thylakoids (arrow).

E. Characteristic straight profiled dictyosomes are present, Carpospores have a fibrous materials embedded in an amorphous matrix.

carpospore wall material deposition. The third and the final stage involves intense dictyosome activity; but instead of wall material deposition, cored vesicles are produced. A few vesicles expand slightly but are never released into existing wall. Mature carpospore (Fig. 57A) consists of a separating layer of compressed mucilage fibrils, a carpospore wall, peripheral chloroplast, cored vesicles, floridean starch and a centrally located nucleus. Tubular structures are formed (Fig. 57 B) and become associated with septal-plugs. A few tubules are continuous with plasma membrane (Fig. 57C). If the septal-plug is between two carpospores, the tubules are formed on both sides of the plug (Fig. 57 D) (Delivopoulos and Kugrens, (1985).

Mesophyllum lichenoides

The scanning electron microscope has not been extensively used for studies of marine macroalgal morphology. Interestingly the coralline algae with their well developed calcium carbonate cell walls are well studied. (see Woelkerling, 1988). *Mesophyllum lichenoides* showed epithallial (upper) and hypothallial (lower) surfaces which have different patterns of calcification. Besides, the conceptacles show greater calcification than the vegetative parts. The upper surface appeared as a series of round to oval irregular concavities showing the sites previously occupied by epithallial cells. The protoplasts of these cells disintegrate during speciman preparation, leaving only the cell walls and the surface depressions referred to as epithallial concavities (Garbary, 1978). These concavities were separated by calcium carbonate cell walls in which the crystals of calcite did not have a distinct orientation and thus result in a smooth surface (Fig. 58 B). Epithallial concavities on both tetrasporangial and carposporangial conceptacles were more calcified and tend to show small epithallial concavities and large intercavity matrix on conceptable surface. They had a tendency to split and thus, leave each concavity with its cell wall distinct from the neighbouring cells (Fig. 58B, C). This structural change might be caused by forces exerted during the uplifting of the epithallium during conceptacle formation. Tetrasporangial conceptacle pores are well delimited structures bounded by a ring of about eight cells that are slightly depressed into the conceptacle surface. The lower surface differed from the upper and consisted of a series of angled, overlapping and pointed arches (Fig. 58 A) each of which showed the apex of an hypothallial filament. In addition, the lower surface was not flat, but formed a series of irregular steps as successive rows of decumbent hypothallial cells reached the surface. A transition zone between the epithallium and hypothallium occurs at the margins of the crust, and in this region the plant is covered with thick cuticle (Fig. 58 D) (Garbary and Veltkamp, 1980).

BRANCH APEX

Bailey and Bisalputra (1970), through electron microscopy, studied the tissues in the branch apices of *Corallina officinalis* and other articulated coralline algae. They suggested that fibrillar materials in the cell walls is possibly cellulose and found that calcite appears to be secreted from the intercellular vesicles that empty into the walls. They also reported that the cuticle covering branch apices consists of lamellae which may be formed by secretion from vesicles in the meristematic cells (Matty and Johansen, 1981). According to Borowitzka and Vesk (1978), a dense fibrillar matrix packs the region between the adjacent genicular cells. This matrix has been found to be mucilage.

MODE OF ATTACHMENT OF EPILITHIC CORALLINE ALGAL SPORES

Many coralline algal spore in nature germinate and grow well on rough substrate which may be due to moulding and penetration of hypothallial filament into surface-interstities. In contrast, those spores grown in culture condition, using smooth glass surface, remain attached for many

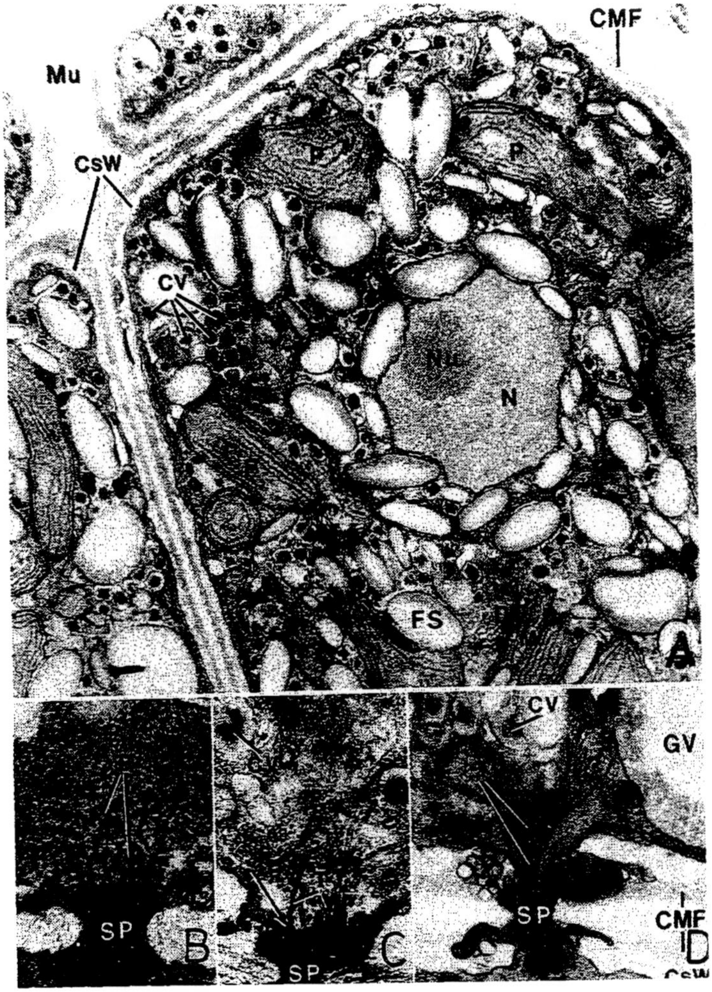

Fig. 57 A-D. Ultrastructure of carposporophyte of *Gloisiphonia verticillaris*
(after Delivopouls and Kugrens. 1985)
(N, Nucleus; Nu, Nucleolus; Fs, Floridean starch ;
Cv, cored vesicle; M, mitochondria).

A. Mature carpospore consists of separating layer of compressed mucilage fibrils, a carpospore wall, peripheral chloroplast, cored vesicle, floridean starch and centrally situated nucleolus.
B. Septal plug between young carpospores. Tubules are beginning to form and aggregate on one side of the plug.
C. Septal plug between a mature carpospore and a generative gonimoblast cell. Numerous tubules are directed towards the plug on the carpospore side. One of the tubule showing a continuity with the plasma membrane.
D. Septal plug between mature carpospores. Tubules are touching the septal plug on both sides.

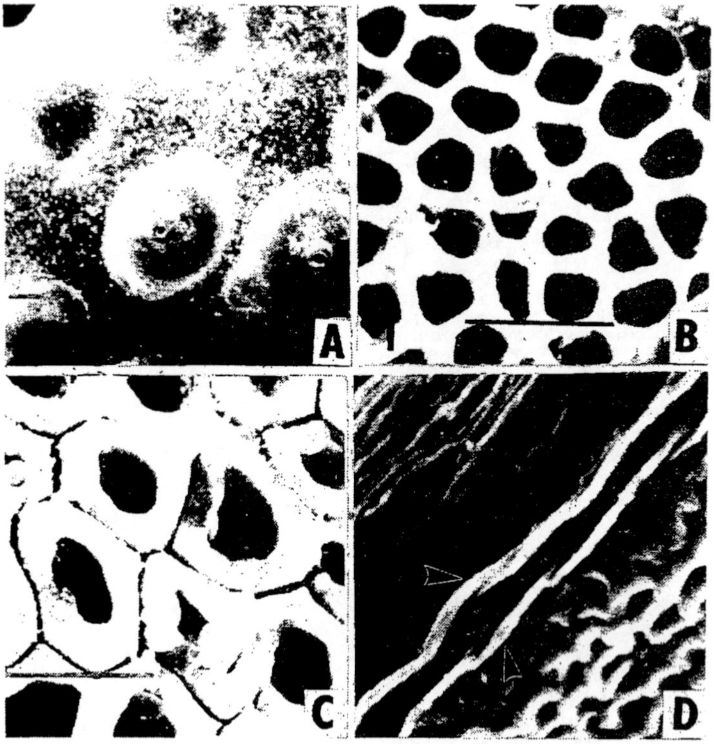

Fig. 58 A-D. Ultrastructure : *Mesophyllum lichenoides* (after Garbary and Veltkamp, 1980).

A. General view of cells on lower surface of plant.
B. Surface of non-reproductive area with numerous concavities and smooth cell walls.
C. Details of surface conceptacle wall, showing splitting of cell walls.
D. A view of lower surface at edge of crust with two prominent lines of thickening on cuticle (arrows)

months, which suggests a different mode of attachmen. Six epilithic, crustose coralline taxa belonging to genera Phymatolithon, Lithothamnion, and Lithophyllum which grow on glass slides revealed apart from development variations, several characteristics features in common. Both transmission and scanning electron microscopic studies reveal that the mucilage forms the adhesive for spore settlement and early development. As young crust shows progressive stages of clacification, the attachment involves two process. (i) Moulding of the hypothallial cell to surface irregularties and (ii) deposition of thick layers of crystals which fill the tiny gaps between the cell and substratum (Fig. 59 C) ensuring a precise fit. These crystalline deposits have an associated organics component. If a space is formed bebeath the crust much later the third method of attachment follows, where clusters of aragonite crystals bound by organic material are precipitated in the gap and resecure the attachment.

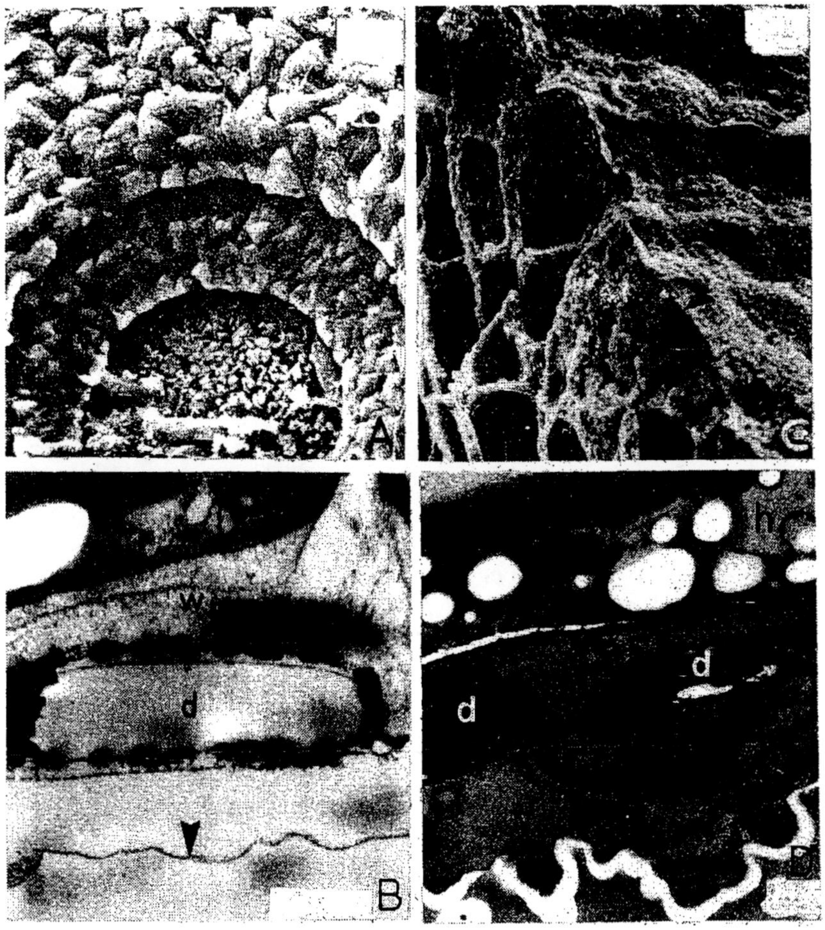

Fig. 59 A-D. Ultrastructure : Mode of attachment of six epilithic algae.

(after Walker and Moss, 1984).

(h. hypothallial cell; a, aragonite globule)

A. *Lithophyllum.* A crust detached from smooth glass shows successive layers of increasing crystal size that build up adjacent to the substratum (SEM).

B. *Phymatolithon lenormandii.* Showing thin layer of organic material (arrow) beneath the flat part of crust.

C. *Lithophyllum* sp. Fractured thallus on the under surface show thick layer of supplementary wall material (1) which has filled the gap between cells and substratum. Arrow denotes the lower surface in contact with the rock.

D. *Phymatolithon lenormandii.* Showing small aragonite globule (a) growing beneath hypothallial cells. Concentrically arranged organic material and diatoms remain in cell wall.

Plants grown in culture have a colourless halo of mucilage that surrounds the pigmented cells. After two weeks, the slides were transfered to flowing sea water where the older sporelings that had already been calcified, revealed marginal outgrowth bereft of mucilage. After a few weeks in culture, some young crust growing on the glass could be detached with a slight amount of pressure. The crust underside is flat and covered with deposit of granular materials.

In *Phymatolithon laevigatum*, the hypothallium filaments are seen, where granular deposits are absent. The fractured specimen of *Lithothamnion glaciale*, however, showed distinct granular material beneath the cell wall of hypothallium and is followed by an orderly arrangement of calcite crystals. The external coating was, however, interrupted at regular intervals by small circular areas. A small depression in centre presumably a scar left by the tearing away of some deposits stuck to the glass. In older specimens of *Lithophyllum sp.*, concentric rings within circular areas (Fig. 59A) suggest the sequential deposition of many different types of granular materials. In outer layers, the deposit is coarser and possesses tightly packed crystal that are embedded in a matrix.

In *Phymatolithon lenormandii,* many crusts are attached firmly to the glass surface. A section through a small aragonite globule (Fig. 59 D) shows that it is extracellular and the appearance of an associated organic component arranged in concentric layers suggest that it has grown by radiating outwards from a specific point. Larger globules showed the organic material to be dispersed throughout the mass, and appeared to bind the crystals. In parts of hypothallium, where crystal masses are absent, there is often a narrow layer of organic material beneath the cell wall (Fig. 59 B). Mechanically removed broken hypothallial cells still cemented to the substratum. A few crust margins curled underneath, and thus, raised the thalli portions away from slides. This weakens the adhesion. In *Phymatolithon polymorphum* and *Lithophyllum incrustans,* when the thalli have been raised from glass surface, the interphase regions are dotted with either colourless globular or crystalline formations. Some of these remain stuck to the glass even after removal of the crust. X-ray diffraction analysis revealed the presence of calcium carbonate in the form of aragonite.

ECOLOGY

DISTRIBUTION

The encrusting coralline algae are a major component of the rocky intertidal and shallow subtidal zones of the boreal north eastern pacific (the region from northern California to southeast Alaska) where they show vertical distribution (Steneck and Paine, 1986). The factors that govern zonation and spatial distribution are desiccation, wave action and shading from canopy of the brown algae. *Corallina vancouveriensis.* which often form nearly continuous strands covering 5.40 m^2 on shaded and slightly dampened surfaces and exposed to heavy wave action, is abundant in the intertidal region. *Pseudolithophyllum neofarlowii* is particularly abundant in cave mouths and on vertical-walls between tidal hights of + 1.52 to 0.61 m MLW. It can grow even lower, but at the - 0.30 m level is confined to rock prominences. *P whidbeyense,* occurs above the upper limit of a kelp *Hedophyllum sessile*, where grazing is intense. However, in three abundant crustose corallines, *Lithophyllum impressum, Lithothamnion phymatodeum* and *Pseudolithophyllum muricatum* are abundant only under the canopy of the kelp *Hedophyllum* or in the tide-pools. The relative abundance of these taxa are complex functions of their relative competitive abilities, growth rates and susceptibility to grazing. These five taxa along with 3 or 4 species of articulated corallines and fleshy crusts such as *Ralfsia, Hildenbrandia* and *Peyssonnelia* form the major components of the algal understory from about 0.61 to – 0.45 m/level. Below this point, *Laminaria setichelli* becomes dominant and replaces *Hedophyllum* (Goff and Cole, 1990).

Mesophyllum vancouveriense is an epiphyte, but unlike other coralline epiphytes, it grows off its primary host (*Laminaria setchelli)* onto surrounding substrata where it persists indefintely. According to Steneck and Paine (1986), this is the only alga which changes ontogenetically its epiphytic habit.

Most of the species are intimately associated with and dependent on the herbivores to eradicate potentially lethal fouling epiphytes. The above species also reveal herbivore induced injuries in the form of grazing trails on their surface and healed wounds within the perithallus, which even affect the morphology of some corallines. Most of the primary substratum in the low intertidal, exclusively occupied by Kelp holdfasts, is occupied by calcareous crusts. This ecological affinity is usual for the west coast of north America (Steneck and Paine, 1986).

Plants of *Synarthrophyton schielianum*, at depths greater than 10 m, were abundant, showed no evidence of grazing, and commonly harbour on the ventral surface of lamellae mixed communities of encrusting invertebrates. In contrast to nongeniculate corallines, this geniculate taxon appears adept to dominate and coexist, rather than compete and perish, with other encrusting organisms.

The ecography of *Synarthrophyton schielianum* appears to differ in many ways from these generalizations and thus provide insight into the diversity of characteristics of these plants which may have contributed to the success in environment where grazing, competition for space, and water motion are acute.

Synarthrophyton schielianum has a dorsal cuticle and an epithallial layer. The former may inhibit adhesion or penetration by recruits of other species, and sloughing-off the latter reduces overgrowth (see Woelkerling and Foster, 1989). Surface fouling may also be eliminated by new lamellae overgrowing older ones. That such anti-fouling mechanisms are more likely than grazing or chemical inhibition is further suggested by the presence of encrusting animals on the ventral surface of the lamellae. This surface has no evident cuticle and no epithallial layer, and is pitted, thus providing a better texture for adhesion. Many non-geniculate Corallines attach the whole thallus to substratum and thus competitive interactions with other species are high in environments where space is a constraint. These interactions may, however, be mediated by grazing. Steneck (1986) has argued that foliose Corallines with their raised margins are adapted to dominate other species, and that overgrowth probably results in the death of that portion of the organism which is overgrown. He suggests that foliose morphology is vulnerable than encrusting forms to damage from surge. *Synarthrophyton schielianum* appears to coexist rather than compete, with the plant it dominates, and the pressure provides a mechanical advantage over unsupported and free lamellae. These traits indirectly enhance growth by making more surface area available for nutrient uptake. In subtidal habitats animals such as sponges, tunicates, and bryozoans are very abundant on steep slopes, and with depth become usually abundant and are important competitors for space. Morphological characteristics of *Synarthrophyton schielianum* show alternative solution to some of the problems of life on the rocky subtidal benthos. These characteristics help in the coexistence and not competitive ability.

Horizontal and vertical growth rates and calcium carbonate deposition were determined. The results indicate that significant differences exist between the two growth rates and the plants with vertical growth showed formation of reproducive tissues. Thus, in an average, 2.25 yrs old *Lithophyllum* plants revealed initiation of reproductive phase and growth rates are identical to other arctic and temperate encrusting red algae. However, the magnitude of growth is less when compared to the tropical condition (Edyvean and Ford, 1987).

The distribution of encrusting *Pneophyllum-Fosliella* (Corallinaceae), from sublittoral and eulittoral environments on *Amphibolis antarctica* (seagrass) was examined both in terms of density and relative fertility on host plant. The distribution of individuals of *Pneophyllum* and *Fosliella* on *Amphibolis antarctica* leaves is influenced by a variety of factors. such as the nature of habitat, the position of particular leaves within an *Amphibolis* leaf cluster, the orientation of the leaf surface (adaxial or abaxial), and the segment of the leaf surface (distal, mid and proximal).

Plants of *Pneophyllum - Fosliella* are better adapted to grow on sub-littoral plants of *Amphibolis antartica* than on rock pool plants. The data on cover (Table 1) and density (Table 2) showed that the sublittoral environment is more conducive to the development of *Pneophyllum - Fosliella* plants than is the rock pool environment.

TABLE 1. Numerical data on cover of *Pneophyllum - Fosliella* plants on *Amphibolis antarctica* leaves.

(after Bramwell and Woelkerling, 1984)

Data reported as mean percentage of cover per leaf segment (n = 45)

Position	Leaf surface	Segment	Eulittoral pool		Sublittoral meadow	
			Mean	s.d.	Mean	s.d.
Outer	Abaxial	Distal	9.7	14.2	70.2	22.5
		Mid	8.5	11.8	55.5	26.3
		Proximal	7.2	18.0	40.8	26.8
	Adaxial	Distal	9.6	10.1	67.2	22.4
		Mid	12.2	17.1	61.0	27.1
		Proximal	3.5	8.7	31.4	29.5
Inner	Abaxial	Distal	4.1	9.1	30.1	25.5
		Mid	0.8	3.8	27.7	27.3
		Proximal	0.2	1.5	11.1	28.1
	Adaxial	Distal	2.0	4.2	16.8	12.0
		Mid	0.5	1.7	7.2	13.8
		Proximal	0.2	1.0	1.2	2.9

TABLE 2. Numerical data on density of *Pneophyllum - Fosliella* plants on *Amphibolis antarctica* leaves.

(after Bramwell and Woelkerling, 1984)

Data reported as mean number (n = 45) of individuals per 10 mm^2 host leaf surface

Position	Leaf surface	Segment	Eulittoral pool		Sublittoral meadow	
			Mean	s.d.	Mean	s.d.
		Distal	8.2	10.6	34.7	11.7
	Abaxial	Mid	3.7	6.1	32.5	10.4
		Proximal	1.9	6.8	26.6	11.2
Outer						
		Distal	6.2	6.8	32.3	10.9
	Adaxial	Mid	5.6	7.8	29.4	11.8
		Proximal	2.6	6.1	23.0	17.1
		Distal	3.5	3.7	27.4	14.2
	Abaxial	Mid	0.8	0.9	22.2	15.5
		Proximal	0.8	0.7	10.0	12.4
Inner						
		Distal	2.6	6.4	17.6	13.6
	Adaxial	Mid	1.0	0.3	11.0	14.2
		Proximal	0.3	1.7	1.4	4.5

TABLE 3. Numerical data on relative fertility of *Pneophyllum - Fosliella* plants on *Amphibolis antarctica* leaves.

(after Bramwell and Woelkerling, 1984)

Data reported as means of 45 measurements.

Position	Leaf surface	Segment	Eulittoral pool		Sublittoral meadow	
			Mean	s.d.	Mean	s.d.
		Distal	26.8	36.0	24.2	18.7
	Abaxial	Mid	16.1	32.0	20.1	20.4
		Proximal	12.6	30.0	12.0	15.9
Outer						
		Distal	22.4	32.3	25.2	19.1
	Adaxial	Mid	11.9	25.0	23.8	22.3
		Proximal	6.3	20.3	7.8	19.2
		Distal	8.4	22.8	9.9	19.1
	Abaxial	Mid	2.2	14.9	3.8	6.2
		Proximal	0.7	1.9	0.4	0.2

Inner						
		Distal	10.4	20.8	1.9	6.1
	Adaxial	Mid	0.0	-	0.4	14.9
		Proximal	0.0	-	0.0	

In both environments, similar pattern of epiphytic distribution on host leaves were evident (Tables 1 and 2). Thus, along all leaf surfaces whether adaxial or abaxial, significant decrease in epiphyte cover and density occurred from the distal to the proximal regions. This may be partly due to the relatively greater levels of illumination at the distal ends of leaf surfaces and partly because the distal parts of the leaves are the oldest.

There was no significant differences in epiphyte cover and density between adaxial and abaxial leaf surfaces except on the inner leaves of leaf clusters on sublittoral plants; here cover and density of *Pneophyllum - Fosliella* plants were greater on the abaxial surface. Same results were obtained when proximal, mid and distal segments of various adaxial and abaxial leaf segments were compared. The cover and density of *Pneophyllum - Fosliella* plants were significantly greater on the second outermost leaf than on the third innermost leaf. This is partly due to relatively greater level of illumination on the outer leaves and partly from relatively greater water and nutrient flow past the outer leaves. The data on relative fertility for entire sets of leaves (outer or inner), for abaxial or adaxial leaf surface showed no significant difference between populations of adaxial surface of inner leaves and no clear cut patterns emerge from particular segments of leaf surfaces of sublittoral and rock pool plant. There is a significantly greater level of relative fertility (Table 3) in sublittoral populations of *Pneophyllum - Fosliella* than in rock pool populations. Thus, along all leaf surface examined, significant decreases in relative fertility always occurred from the distal to the proximal leaf segments.

There are number of environmental factors which affect the distribution of seaweed, like substratum, light, temperature, salinity, water movement, nutrient and grazing. Each factor affects the ecology of the red algae in its own way.

SUBSTRATUM

Subtidal calcareous sand can support many species of red algae. *Corallina officinalis* prefers natural habitats and coarser substratum. *Callocolax neglectans* (found on *Callophyllis*) and *Choreocolax polysiphoniae* (found on *Polysiphonia*) are restricted to a single host. Most of them are poorly pigmented and some even lack photosynthetic pigments. The parasitic nature of several species has been confirmed by the translocation of C^{14} labelled compounds from host to parasite. Endophytic and parasitic red algae are highly host specific, although the reason is enigmatic (Kain and Norton, 1990).

LIGHT

Red algae tolerate a wider range of light levels than any other group of photosynthetic plants. The deepest known plant is an undescribed Rhodophyceaen coralline alga. Other members of this group abound on tropical shores. Crustose corallines have very low requirements, the lowest quoted being that of *Leptophytum laeve*. Fresh water genus *Hildenbrandia* show preference for low light regime (Kain and Norton, 1990).

COMPETITION

The development of *Porolithon* on coral reef is attributed to the increased insolation that results from the removal of frondose algae by wave action.

Many calcareous algae, both crustose and erect forms, also continuously slough their outer layers, and thus, shed the potential fouling organisms. Gastropods are major competitors of many intertidal seaweeds, and on some shores colonization by *Chondrus crispus* and *Endocladia muricata* may be reliant upon predator removing gastropods and making space available.

TEMPERATURE

Quantitative studies on field population of 5 species of *Farlowia* and on a cultured population of 2 species of *Pneophyllum* (Syn. *Heteroderma) - Fosliella* complex (Corallinaceae) have led to the conclusion that the trichocyte (a) presence or absence and (b) arrangement are influenced by light and temperature and/or vary within field populations (Jones and Woelkering, 1984).

The production of tetrasporangia only at 15° C, 8 : 16 h photoregime in *Farlowia mollis* and *F. conferta* suggests a short day response, but no critical day length tests were carried out. Tetrasporangia are formed in *Thurtellopsis* under short day condition ((9 : 16 h photoregime), but a short day response was not tested (Dixon and Richardson, 1969). A short day response, not controlled by a phytochrome system, was demostrated for tetrasporangial initiation in *Acrosymphyton purpuriferum* (Dumontiaceae), (Cortel - Breeman and Hoek, 1970).

SALINITY

Natural hypersalinity gradient (37 - 55%) in Shark Bay, Australia, affecting the Corallinaceae epiphytic on leaves of *Amphilbolis antarctica* has been examined. Four species of *Fosliella* and one species each of *Pneophyllum* and *Melobesia* were identified. The species composition of the epiphytic community did not change with the gradient and both the basic germination disc pattern and position of trichocytes within each taxon remained constant. With the exception of stations, at 47.2% and 49.3% salinity, the overall density of epiphytic coralline algae on *Amphibolis* leaves decreased markedly as salinity level increase. At all stations, most individuals on the oldest *Amphibolis* leaves lacked germination–discs. In *Pneophyllum* the proportion of plants sans germianton-disc was least 88% at salinities up to 49.3% but this significantly was reduced at highest salinities.

In *Fosliella,* the production ranged from 74 to 95% with no relationship to salinity level from 1 : 1 at 37.3% to 4 : 1 at 55%. Development of the epiphytic coralline community within leaf cluster occurred in a consistent sequence at all stations. On the young leaves, spores settled and divided to produce germination-discs and intact thalli. On older leaves, the plants became fragmented but remained sterile. Subsequently, conceptacles were produced, almost exclusively on portions lacking germination-discs. Increasing salinity, slows the growth rate and development of both the host and the coralline community. (Harlin *et al.*, 1985).

5 GIGARTINALES

CLASSIFICATION

Fritsch (1945) characterized the order Gigartinales as follows :

1. Uniaxial or multiaxial plant body.
2. Distinct or indistinct heterotrichy in the young stages.
3. Diplobiontic life-cycle.
4. Auxiliary cells sometimes remote from the carpogonia, both borne on accessory lateral which are sometimes aggregated in special fertile tracts (nemathecia, conceptacle).
5. Cruciate or Zonate arrangement of tetraspores.

He further classified this order into 16 families on the basis of presence or absence of procarps (when auxiliary cells and carpogonia form parts of the same branch-system the aggregate is procarp).

(A) WITHOUT PROCARPS

1.	Calosiphoniaceae	*Bertholdia, Calosiphonia*
2.	Nemastomaceae	*Nemastoma, Platoma, Schizymenia*
3.	Sebdeniaceae	*Sebdenia*
4.	Furcellariaceae	*Furcellaria, Halarachnion, Neurocaulon*
5.	Solieriaceae	*Agardhiella, Anatheca, Eucheuma, Flahaultia, Metistotheca, Opuntiella, Sarcodiotheca, Solieria, Thysanocladia, Turnerella*
6.	Rissoellaceae	*Rissoella*
7.	Rhabdoniaceae	*Areschongia, Catenella, Erythroclonium, Rhabdonia*

(B) WITH PROCARPS

8.	Rhodophyllidaceae	*Acanthococcus, Calliblepharis, Craspedocarpus, Cystoclonium, Rhodophyllis, Wurdemannia*
9.	Hypneaceae	*Hypnea*
10.	Plocamiaceae	*Plocamium*
11.	Sphaerococcaceae	*Heringia, Phacelocarpus, Sphaerococcus, Stenocladia*
12.	Gracilariaceae	*Ceratodictyon (Marchesettia), Curdiea, Gracilaria, Melanthalia, Tylotus*

13. Mychodeaceae — *Mychodea.*
14. Acrotylaceae — *Acrotylus, Hennedya*
15. Phyllophoraceae — *Ahnfeltia, Ceratocolax, Gymnogongrus, Phyllophora, Stenogramma.*
16. Gigartinaceae — *Chondrus, Gigartina, Iridaea*

According to Bold and Wynne (1981), this order is the largest in the Rhodophycota and contains 28 families. The characteristic features of this order are :

1. Large, heterogeneous and admittedly ill-defined assemblage.
2. Auxiliary cell not borne on a special, or accessory filament but an intercalary cell of an ordinary vegetative filament.
3. Separation of families with uniaxial or multiaxial constructions; cruciate or zonate arrangement of tetraspores; and the spatial relationship of carpogonium to auxiliary cell (Procarpial versus non-procarpial).
4. That most members are either moderate to large and fleshy or crustose, soft and gelatinous (*Calosiphonia*) or moderately calcified (*Titanophora* and *Rhodopeltis*).
5. Different familes are classified on the basis of following key :

1\. Tetrasporangia zonately divided .... **Solieriaceae**

1\. Tetrasporangia cruciately divided **2**

2\. The gonimoblast developing towards the interior of the thallus **3**

2\. The gonimoblast developing towards the exterior of the thallus ....**Gracilariaceae**

3\. Medulla composed of large cells,
pseudoparenchymatous in appearance .... **Phyllophoraceae**

3\. Medulla composed of narrow filaments,
nonparenchymatous in appearance ....**Gigartinaceae.**

Dixon (1982) characterized the order Gigartinales as follows :

1. The thalli may be crustose and discoid or erect and frondose, the degree of aggregation of the constituent filaments ranging from loose to compact, and may be of uniaxial or multiaxial construction.
2. Thalli bearing different types of reproductive structures may be similar, slightly dissimilar, or of totally different morphology,
3. Spermantangia are formed superficially, singly or in clusters.
4. The carpogonium arises from the apical cell of an undifferentiated filament of the cluster.
5. Fertilization is followed by the transfer of the zygote nucleus to an auxiliary cell, which is always an unspecialised vegetative cell of the thallus.

6. The carposporophyte is a compact mass of tissue producing terminal carposporangia which liberate either a single carpospore or four carpotetraspores.

The order contains 27 families and important members of each family are :

1.	Gymnophlaeceae	*Nemastoma, Schizymenia, Platoma*
2.	Gracilariaceae	*Gracilaria*
3.	Sebdeniaceae	*Sebdenia*
4.	Calosiphoniaceae	*Bertholdia, Calosiphonia*
5.	Petrocelidaceae	*Petrocelidea*
6.	Phyllophoraceae	*Ahnfeltia, Phyllophora*
7.	Gigartinaceae	*Chondrus, Gigartina, Rhodoglossum, Iridaea.*
8.	Chondriellaceae	*Chondriella*
9.	Polyideaceae	*Polyides, Rhodopeltis*
10.	Nizymeniaceae	*Nizymenia, Stenocladia*
11.	Rhizophyllidaceae	*Rhizophyllis, Desmia, Ochtodes*
12.	Acrotylaceae	*Reinboldia, Flennedya, Acrotylus, Ranavalona, Amphiplexia.*
13.	Plocamiaceae	*Plocamium, Plocamiocolax*
14.	Sphaerococcaceae	*Sphaerococcus*
15.	Phacelocarpaceae	*Phacelocarpus*
16.	Sarcodiaceae	*Sarcodia*
17.	Furcellariaceae	*Furcellaria, Halarachnion, Neurocaulon*
18.	Solieriaceae	*Turnerella, Gardneriella, Neoagardhiella*
19.	Hypneaceae	*Rhododactylis, Hypneocolax, Hypnea*
20.	Rissoellaceae	*Rissoella*
21.	Rhabdoniaceae	*Catenellocolax*
22.	Cubiculosporaceae	*Cubiculosporum*
23.	Rhodophyllidaceae	*Cystoclonium*
24.	Mychodeaceae	*Mychodea*
25.	Dicranemaceae	*Dicranema, Peltasta, Reptaxin, Tylotus*
26.	Cruoriaceae	*Cruoria*
27.	Blinksiaceae	*Blinksia, Californica*

MORPHOLOGY

Gigartinales is one of the most diverse orders of the Florideae and according to Bold and Wynne (1981) is the largest and heterogeneous. The thalli may be crustose and discoid or erect and frondose, the degree of aggregation of the constituent filaments range from loose to compact and are of either uniaxial or multiaxial construction (Dixon, 1982). Fritsch (1945) classified the various taxa on the basis of uniaxial or multiaxial thalli with or without procarps.

The uniaxial or multiaxial types of thalli are controlled by apical cell/cells and are related to the manner of growth patterns. In the former, the single apical cell is conspicuous, dome-shaped structure and cuts off proximally segments which constitute the uniaxial type of growth. In the latter, many filaments are produced and each terminates in an apical initial which divides constituting the multiaxial type of growth.

UNIAXIAL FORMS

Uniaxial thallus is encountered in *Gracilaria,* which grows as tangled clumps of irregularly, dichotomously or trichotomously branched, cylindrical, terete or flattened thalli with a cartilaginous texture. *Gracilaria chilensis* exhibits slender thallus. On the contrary, *G. confervoides* is a widely distributed, northern lithophyte, found at low-tide levels, in the Mediterranean. Fronds are dark-red, terete, up to 50 μm long and bear more or less numerous tapering branches. A few branches arise from a large, fleshy-disc, composed of compact and vertical threads. *G. multipartita* and *G. dendroides* are described from the Bay of Naples. The latter species has a typical branching pattern simulating the dendroid habit, lanceolate branchlets which become long and taper in the distal part of the branchlets. *G. dendroides,* thus, differs from *G. verrucosa* in branching pattern whose thalli branch in the proximal part and the second order branches being longer than the frequently obscure main axis arising from the basal disc. (Garguilo *et al.,* 1985).

The thallus of *G. corticata* is pale red, flat, erect and shows dichotomous branching (Fig. 60 A) (Manoj Kumar, 1996).

Table 1 reveals the thallus morphology of *Gracilaria* that occur in Mediterranean and N.E. Atlantic (after Garguilo *et al.*, 1985).

TABLE I

Comparison between Mediterranean and N.E. Atlantic taxa.

SPECIES	THALLUS MORPHOLOGY
G. foliifers	Partially flat
G. bursa-pastoris	"
G. cervicornis	"
G. heteroclada	"
G. corallicola	"
G. arcuata	"
G. armata	Richly, spinous branchlets.
G. dura	Cartilaginous
G. conferta	Distally branched
G. verrucosa	Proximally branched
G. divergens	Proximally branched
G. vieirae	Rhizoidal base

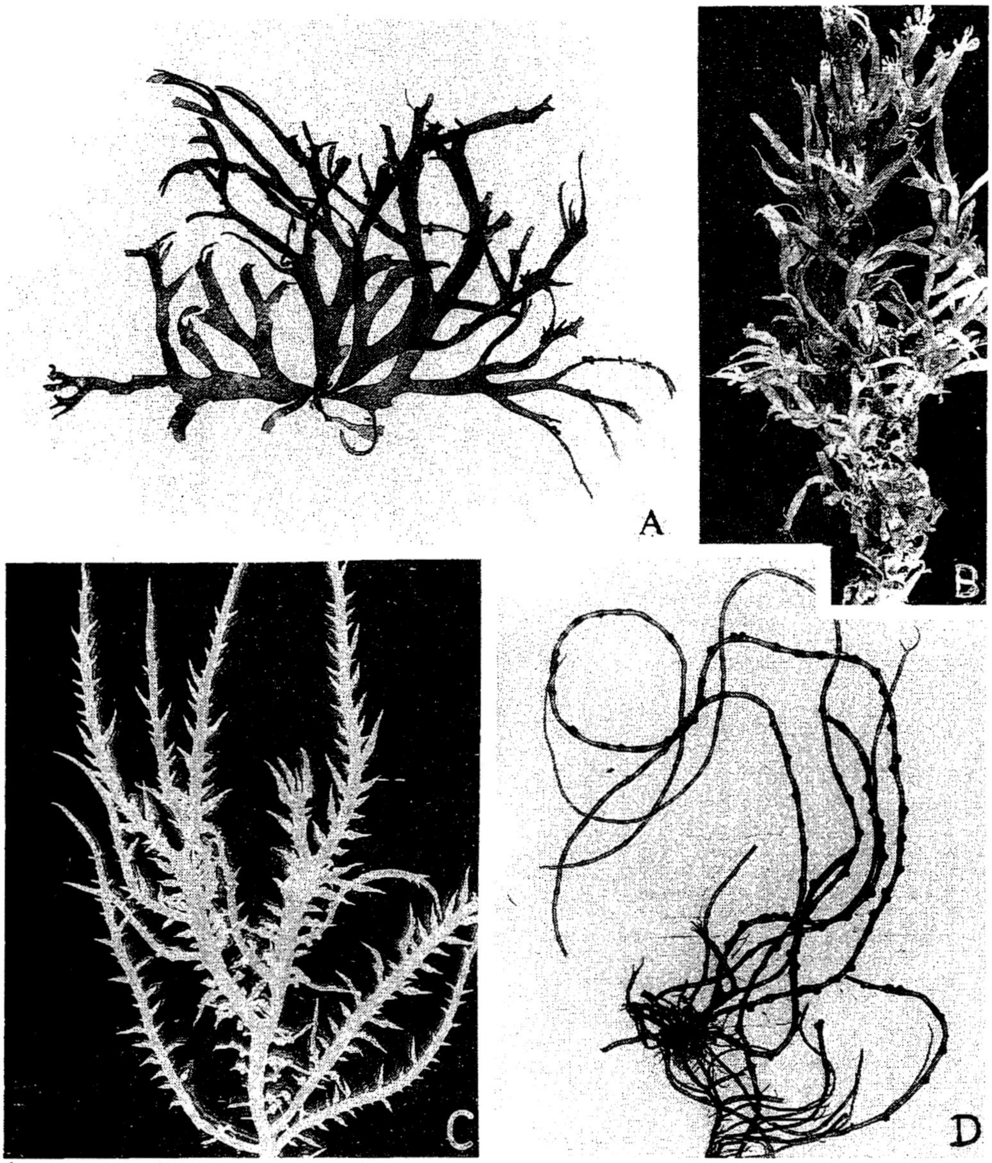

Fig. 60 A-D. Morphology

A. *Gracilaria corticata* showing slender thallus. (after Manoj Kumar, 1996)

B. *Nizymenia australis.* A mature plant showing profusely and irregularly branched thallus. (after Vijayaraghavan and Bhatia, 1991)

C. *Gigartina* sp. The dichotomously branched thallus has basal, stalk-like portions.

D. *Gracilaria verrucosa* A portion of mature plant with cystocarps. (after Manoj Kumar, 1996).

In *Gracilaria,* branching pattern varies greatly according to the environment as seen in Table 2 (after Bird and Oliveira, 1986).

TABLE 2

Comparison of three terete species from America.

Taxa	Branching
G. costaricensis	Sparse; secondary branches remote or absent.
G. lemaneiformis	Primary branches flagellar; subsequent order (upto three) markedly small.
G. tenuifrons	Freely branched, upto two to four orders.

Cystoclonium purpureum is a common lithophyte that occurs between tide-marks in Britain but elsewhere grows submerged. The cartilaginous thallus is terete, dull purple; reaches upto 6 cm in length, has a prominent main axis that bears numerous and richly branched laterals. A few branches may end in tendrils.

Erythroclonium, Catenella and *Hypnea* show identical mode of segmentation and apical structures. *Catenella opuntia,* a perennial alga, is widely distributed near high water-mark and many species are attached by the holdfasts to the roots of mangroove plants. *Hypnea musciformis* has fleshy terete thallus that grows upto 45 cm, and shows numerous, short and long branches. The latter curl at the tips and function as tendrils. In *Cystoclonium,* there is an uniform enlargement of the inner cells of the laterals so that the axial cells are evident. Hairs are often abundant in the plants growing in well illuminated situations.

Plocamium coccineum has delicate thallus, often slightly flattened, richly branched in one plane, in a sympodial manner. The successive sympodia bear adaxially and on their flanks curved branches; the uppermost ones are strongly developed; the axes deflected to one side and appear as an unbranched laterals. The growth, however, continues by the adjacent branches, which, in turn are displaced by their uppermost laterals. The secondary branches, thus, occupy alternate flanks of the successive segments of the sympodium. The tips of branches, on contact with firm substratum, develop broad, sucker-like holdfasts.

Nizymenia australis is uniaxial in construction and the axial filments are distinguished close to the apex. The thallus is flat, profusely and irregularly branched (Fig. 60 B). (Vijayaraghavan and Bhatia, 1991; Bhatia *et al,* 1990).

MULTIAXIAL FORMS

Ahnfeltia plicata, a perennial lithophyte, has dark purple or black thallus; the plant is attached to the substratum by a broad disc composed of vertical rows of almost cubical cells with firm walls. In the deep water, robust forms of *Ahnfeltia plicata* are found. The mature thallus shows irregular to dichotomous, ramification with promiscus adventitious branches on the older parts.

Foliose members belong to Phyllophoraceae and Gigartinaceae. Many species of *Phyllophora* are wide-spread lithophytes with small thalli that branch dichotomously in one plane. The mature thallus of *P. membranifolia* bears numerous forked fronds with the convex surface facing the light source. In the major dichotomous, one fork develops into a fan, while the other forms a continuation of the stalk.

The Gigartinaceae include important, commercially exploitable seaweeds like *Chondrus crispus* and *Gigartina stellata.* Both produce carrageenan and are dichotomously branched with basal stalk-like portion, *C. crispus* has a multiaxial construction, both in the apices and in sections of the mature parts. The fronds of *Gigartina teedii* usually found on rocks near low-water mark, often possess incurved edges. The fertile plants has numerous outgrowths which harbour the cystocarps; a distinctive feature of all *Gigartina* species. In *G. acicularis,* the thalli are almost terete, whilst in G. spp., they are broad and unbranched (Fig. 60 C).

REPRODUCTIVE STRUCTURES

SPERMATANGIA

Yamamoto (1975) recognised 3-basic types of spermatangia (i) Verrucosa (ii) Textorii and (iii) Chorda. Later studies led to the addition of two more types namely Symmetrica and Henriguesiana types leading to five spermatangial developmental patterns.

(i) **Verrucosa type** : This type of spermatangium is found in deep crypts penetrating the subcortex.

(ii) **Textorii type** : Here the spe.matangia are seen in the cortical regions in shallow concep. acles.

(iii) **Chorda type** : The spermatangi. are superficial and occur in continuous layers, and are forn. ed from outer cortical cells.

(iv **Symmetrica type** : In this type spermatangia are superficial like Chorda type but the are discontinuously formed.

(v) **Henriguesiana type** : This type of spermatangia resemble Verrucosa type but are aggregate.

Species	Locality	Spermatangia type	Source
Gracilaria	N. California	Textorii	Schreider & Searles, 1973
blodgettii	Japan	Verrucosa	Ohmi, 1958
	Japan	Textorii	Yamamoto, 1978
G. mammillaris	N. Carolina	Textorri	Schreider, 1975
	Cuba	Verrucosa	Dawson, 1962
G. foliifera	England	Textorii	McLachlan & Edelstein, 1977
	India	Verrucosa	Umamaheswara Rao, 1972

Gracilariaceae: *Gracilaria tenuifrons* described from Alagoas, in north eastern Brazil, has superficial spermatangia, which are of Chorda type and are produced over nearly the whole thallus surface (Bird and Oliveira, 1986). The sorus is, however, interrupted by sterile and slightly protruding cortical cells that occur either singly or in small groups. The spermatangial mother cells, spermatangia and the overlying cuticle is about 10-12 μm in thickness.

In *G. lemaneiformis,* the spermatangial mother cells at maturity are long and slender; the thickness of fertile layer is approximately upto 17-30 μm. The spermatangial sorus is continuous and sterile cortical cells are occasionally present. Interestingly, *G. panamensis* from eastern Pacific ocean, lacks spermatangia. *G. megaspora,* from India show Textorii type of spermatangia (Bird and Oliveira, 1986).

Spermatangial conceptacles reported in *Gracilaria verrucosa* and *G. dendroides* are ovoid and immersed in the subcortical layer (Garguilo *et al.*, 1985; Fig. 67 E). The outermost layer of the cortex is not modified (as seen in tetrasporangia), whereas the subcortical cells that limit the conceptacles are elongate. *G. verrucosa* and *G. arcuata* have Verrucosa type of spermatangial conceptacles whereas *G. foliifera, G. bursa-pastoris* and *G. cervicornis* reveal Textorii type (see Garguilo *et al.*, 1985).

Gracilaria chorda, an ecological form of *G. salicornia* and *G. eucheumoidea,* lacks gametophytes. In *G. salicornia,* spermatangial pits of Verrucosa type appeared in culture (Yamamoto, 1991).

A very unique kind of spermatangial arrangement is reported in *Gracilaria mixta* (Abbott *et al.*, 1991) where a mixture of spermatangial types occur that recently have ben used to separate genera in the Gracilariaceae; the simple, pi-like, Verrcosa type and the multicavited, Polycarvenosa type arrangements coexist. The first type is seen in the youngest branch portions, gradually dedifferentiating in the older parts upto the second type.

Sarcodiaceae: *Sarcodia dentata* is dioecious where male sori and spermatangia are formed directly by divisions of outer cortical cells (Fig 62 A). These cortical cells divide to form two spermatangial cells that elongate and divide anticlinally to form spermatangium that eventually releases a single spermatium. The process seems to be continuous. New spermatangia are segmented from the same spermatangial mother cell and spermatangia before release, are often arranged in a short uniseriate series (Fig. 62 A). On the thallus, spermatangial sori are seen as light patches (Fig. 61 B; Norris, 1987). Spermatangial mother cells are elongate and raised above the surface of the thallus. The development is different from many Solieriaceae and Gracilariaceae where, in the outer cortex, small clusters of spermatangia are developed in shallow cavities.

Gigartinaceae : In *Chondrus crispus,* the gametophytes are dioecious and spermatangial sori are superficial (Brodie *et al.*, 1991). Spermatangia are long, wide and arise singly or in pairs from the outermost cells of the cortex. Each spermatium is released through the thick, mucilaginous cuticle.

Nizymeniacea: In *Nizymenia australis* the development of spermatangia takes place from outer cortical cells of the thallus. The spermatangial sori are covered by a thick mucilage (Fig 63 C, D; Vijayaraghavan and Bhatia, 1991). A few outer cortical cells, adjacent to the epidermis become fertile and give rise to one or two cells. Each of these cells divides by centripetal transverse walls to form two cells out of these cells, the terminal one divides and thus, a 3-celled filament result. The uppermost of these cells act as the spermatangial mother cell which divides by centripetal furrow to form the upper spermatangium and lower basal cell (Fig. 63 A, B). Each spermatangium releases one spermatium. Wommersley (1971) reported in Nizymeniaceae, that spermatangia occur in clusters on the upper cells of uniseriate branched filaments that occur in tufts. But in *Nizymenia australis,* spermatangial filaments are unbranched (Vijayaraghavan and Bhatia, 1991).

Fig. 61. A-C Reproductive structures of *Sarcodia dentata.*

(after Norris, 1987)

(C, cystocarp)

A. Female gametophyte showing the cystocarps that are greatly elevated from the surface of the thallus and sometimes stallked.

B. Male gametophyte showing spermatangial sori that appear as light blotches on the thallus surface. The marginally proliferous habit of the thallus is typical for this taxon.

C. Tetrasporangiate plant whose thallus shows two or three orders of pseudodichotomous branching pattern.

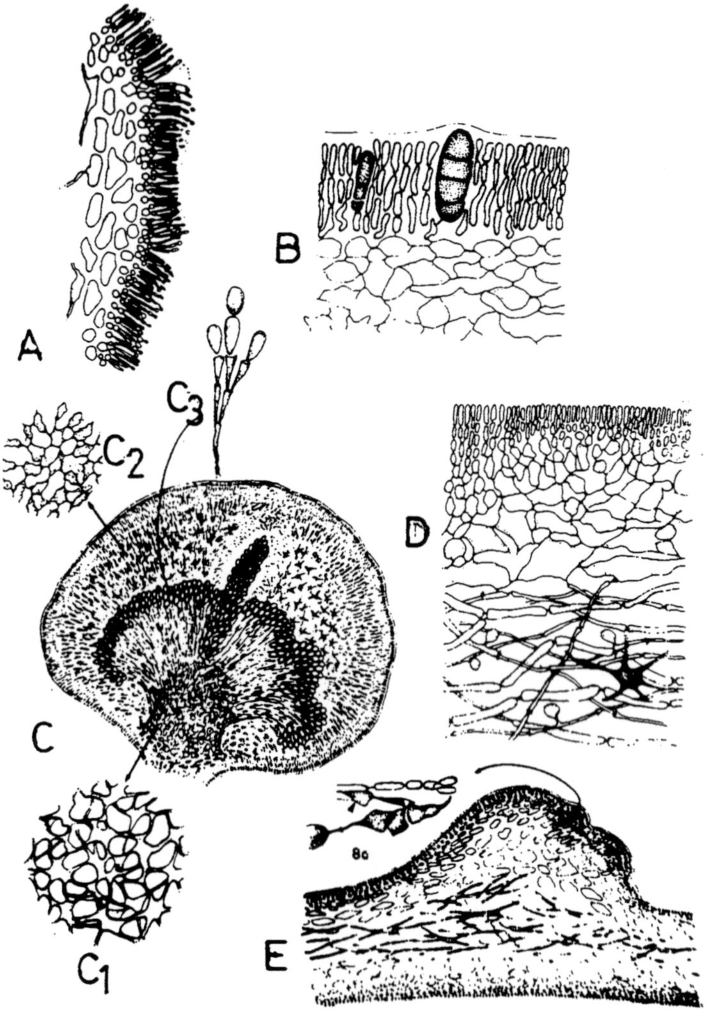

Fig. 62 A-E Reproductive structures of *Sarcodia dentata*

(after Norris, 1987)

A. Section of a male sorus showing elongate spermatangial mother cells.

B. Cortical region to show zonate sporangia.

C. Section of a cystocarp showing carposporangia. The highly magnified tissues showing details, are the part of the reticulate fusion cell (C_1), part of interior tissue of the pericarp (C_2), and gonimoblast filament (C_3), with terminal carposporangia (C_3).

D. Transverse section of thallus showing cortex and part of the medulla. The stellate cell and medullary filament are attached to a cell of the inner cortex. Globular cells on medullary filaments probably represent developmental stages in secondary pit-connections.

E. A carpogonial branch has produced a connecting filament which is attached to the auxiliary cell (arrowhead).

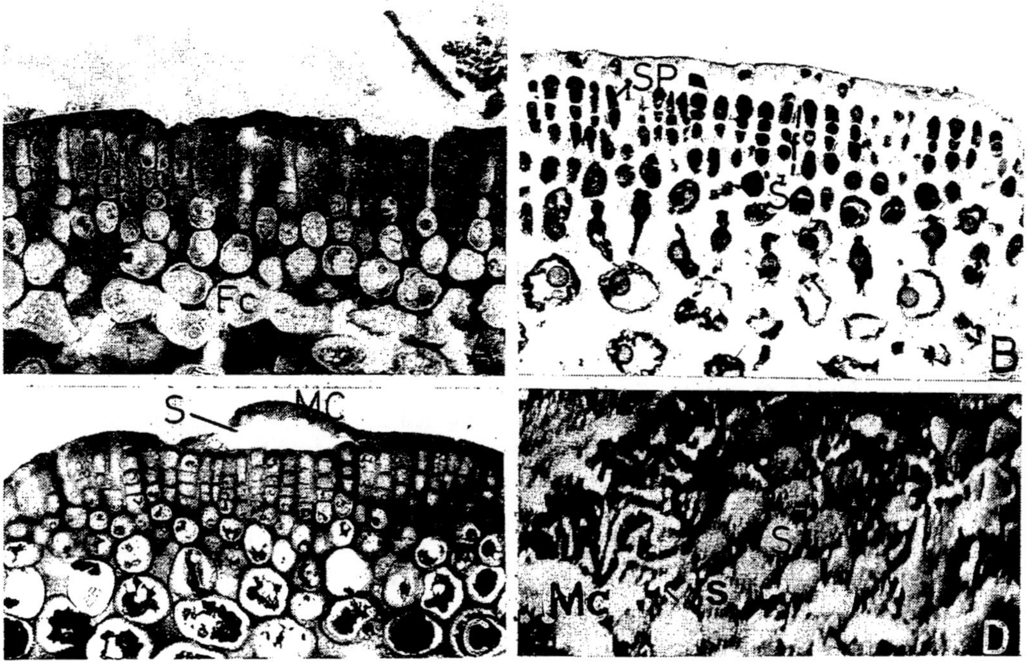

Fig. 63. A-D Reproductive structure and histochemistry in *Nizymenia australis*.

(after Vijayaraghavan and Bhatia, 1991)

(Fc, fertile cortical cells; Mc, mucilage coverings; S, spermatium; Sf, spermatangial filament; Sm, spermatangial mother cell; Sp, spermatangium).

A. Longitudinal section of spermatangial sorus showing spermatangial mother cell and fertile cortex differentiating into spermatangium. (TBO)

B. Spermatangial sorus showing 4-celled spermatangial filament with protein-rich spermatangiun.

C. Spermatangial sorus showing gelatinization of mucilage covering for the release of spermatium. (PAS)

D. Spermatangial sorus showing gelatinization of extracellular mucilage that forms a passage for the release of the spermatia. (SEM)

CARPOGONIAL BRANCH

Sarcodiaceae : In *Sarcodia dentata,* the carpogonial branches consist of 2 or 3 cells attached to a supporting cell which is an intercalary cell of the cortex (Fig. 62 E) and extends directly towards the surface of the thallus (Norris, 1987). This is in contrast to the carpogonial branch of *Anatheca montagnei,* which is attached to the interior side of the cortical cell, the branches extending towards the medulla and only the trichogyne reflexed towards the thallus surface. The form and position of this type of carpogonial branch is typical for the Solieriaceae. Female reproductive system of *Sarcodia montagneana* differs from *Anatheca dentata* in the following aspects : (i) the carpogonial branches of *S. montagneana* are not borne on thallus protuberance but, instead, at random in ordinary cortical regions of the thallus; (2) carpogonial branch system of *S. montagneana* may often be polycarpogonial, whereas in *A. dentata* it is monocarpogonial; (3) the connecting filament is formed directly by the carpogonium in *A. dentata* whereas in *S. montagneana,* it is the hypogynous cell that forms a connection with the auxiliary cell (Norris, 1987).

Nizymeniaceae : In *Nizymenia australis,* the carpogonial branch projects at various angles in the inner cortex and consists of carpogonium, hypogynous and supporting cells (Fig. 64 A) The carpogonia are formed from the apical cells of the 2-celled filament which arise in the clusters from the same supporting cell. The carpogonium has an elongate trichogyne which at maturity becomes reflexed. The post-fertilization carpogonium shows lysis of trichogyne (Fig. 81 A) The cells of carpogonial branch are interconnected through pit-connections (Vijayaraghavan and Bhatia, 1991) Similar situation is found in *Rissoella verruculosa* (Rissoellaceae) where a single large auxiliary cell is associated with many carpogonial branches (Fritsch, 1945).

Gracilariaceae : The carpogonial branch in *Gracilaria epihippisora* is 2-celled structure whereas in *Chondrus crispus,* is 3-celled and develops within the cortex. The multinucleate supporting cell is connected both to the filament in which it develops and to the nearby vegetative cells through pit-connections. It also, serves as auxiliary cell after fertilization. Neither the supporting cell nor the carpogonial branch bears sterile side branches (Brodie *et al.*, 1991). The carpogonium is connected laterally to the second cell of the carpogonial branch and lies in close proximity to the supporting cell. It is uninucleate and bears an elongate and often slightly swollen trichogyne whose tip either protrudes beyond thallus surface or is continuous.

The female gametophytes bear generally cystocarp on the main axis, but in *Gracilaria dendroides* no cystocarps occur either on the main axis or on the second order branches. The female gametophyte bears hemispherical cystocarps which are located on lanceolated branchlet (Garguilo *et al.*, 1985).

Solieriaceae : A very typical kind of carpogonial branch is found in *Agardhiella, Opuntiella* and *Eucheuma.* The female plants directly produce 3-celled, carpogonial branches with reflexed trichogynes (Bold and Wynne, 1981). Following fertilization, a single connecting filament emenates from each carpogonium and fuses with a remote auxiliary cell from which carposporophyte develops.

In *Gardneriella tuberifora,* the 3-celled carpogonial branch develops from a multinucleate and vacuolate intercalary cell that is 3-8 layers from the thallus surface. This intercalary cell serves as the support cell for one to three inwardly directed carpogonial branches. The carpogonium is conical in shape, with a reflexed apical trichogyne that sinuously ascends to the surface and protrudes through the outer wall (Goff and Hommersand, 1982).

Upon fertilization, one or two connecting filaments develop through protuberance from the carpogonium and initially grow towards the thallus surface. Prior to fertilization, remiform and large auxiliary cells are evident. Diploidization of the auxiliary cell occurs when the connecting filament fuses laterally or basally with the auxiliary cell. Later, small, densely staining, uninucleate cells are cut off from the surrounding vegetative cells. The former divide to form short (2-4 cells) filaments of uninucleate cells which grow in the vicinity of the diploidized auxiliary cell, but exhibit no directional growth towards it.

CARPOSPOROPHYTE

Gracilariaceae : The carposporophyte development is very characteristic and develops outside the thallus. In *Gracilaria verrucosa,* cystocarps are usually prominent, hemispherical structures, projecting from the thallus surface (Fig. 60 D) ; a pericarp is developed over the gonimoblast filaments, with an ostiole through which the carpospores are released. A section through the cystocarp, reveals nutritive filaments connecting the carposporophyte to the overlying pericarp and the presence of a relatively large celled central region (Bold and Wynne, 1981). The presence of these nutritive filaments is a variable character even within a single species. In *Gracilaria corticata* cystocarp shows an ostiole and striated outer surface (Fig 67 A) (after Manoj Kumar, 1996)

In *Gracilaria dendroides,* from the Bay of Naples, no cystocarps occur along the main axis or on the second order branches; but were located on lanceolate branchlets. The cystocarps are hemispherical in shape (Garguilo *et al.,* 1985).In *G. tenuifrons,* cystocarps are produced on the axes and the main branches (Bird and Oliveira, 1986). The carposporophyte of *G. dendroides* is lobed (Fig. 67 C) and is connected with the pericarp by many filaments (Fig. 67 D). The cells forming the pericarp are round, and are densely arranged both in anticlinal and radial rows (Garguilo *et al.,* 1985). A lobed fusion cell is present in the core of gonimoblast (Fig. 67 C). The proximal part of the gonimoblast filaments is associated in definitely oriented groups of irregularly shaped and thick walled cells, from which chains of elongate and progressively small cells arise the carpospore being terminal. The structure and distribution of carposporophytes helps in the taxonomy of many *Gracilaria* species. *G. arcuata* is characterized by curved, slightly compressed main branches and scattered carposporophytes. *G. dura* has cystocarps all along the frond, which is cartilaginous when dry and does not adhere to herbarium sheets. *G. verrucosa* bears tetrasporangia and spermatangial conceptacles with cystocarps scattered over the most of the thallus surface. The character also distinguishes *G. dendroides* from *G. verrucosa.* The former species has gonimoblast tissue formed by a number of small irregular cells gathered in precisely oriented filaments whereas in the latter gonimoblast tissue is composed of a few irregular and elongate cells.

Gracilaria tenuifrons reported from the tropical western Atlantic, was initially misidentified as *G. sjoestedtii.* It was further distinguished from serveral other species of subgenus *Gracilariella* by vegetative and reproductive structures (Bird and Oliveira, 1986). This species has got distinct cystocarp and pericarp. The cystocarps are borne on main branches, and are rostrate. A slight ostiolar bulge is, however, often present. The pericarp is very thick, 30-75 µm and constructed of 6-8 cell layers (Fig. 68 A) It increases slightly in the ostiolar region to 80-95 µm in thickness and 8-11 cell layers. Its outermost cells are subquadrate or slightly radially elongate. Inner cells of the pericarps are periclinally compressed to various degrees and generally maintain radial files despite becoming progressively large inwardly, to an innermost, discontinuous layer of subglobose

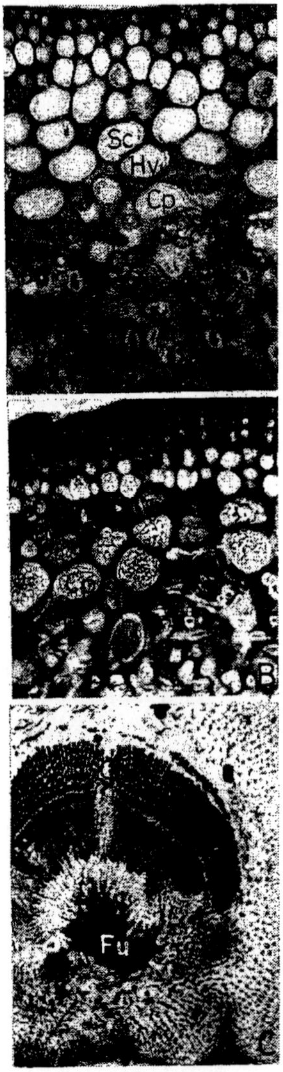

Fig. 64. A-C Carpogonium and carposporophyte of *Nizymenia australis.*

(after Vijayaraghavan and Bhatia, 1991)

(Ca, carposporangia; Cp carpogonium; Fu, fusion cell; Gf, gonimoblast filament; Hy, hypogynous cell; Sc supporting cell; Tr, trichogyne).

A. The carpogonial branch consists of carpogonium, hypogynous and supporting cells. The carpogonia are formed from the apical cells of the 2-celled filament which arise in clusters from the same supporting cell.

B. Carpogonial branch showing carpogonial cells that possess conspicuous cytoplasm and gorged with polysaccharides.

C. Section of cystocarp showing distribution of proteins in fusion-cell, which has many finger-like projections. Radiating gonimoblast filaments bear terminal chains of carposporangia.

cells 12-28 μm broad. The gonimoblast is broad based with flaring mass of ovate cells, and is occasionally shallowly lobed. Carposporangia are seriate on the gonimoblast filaments, the terminal mature sprorangia measuring 24-34 μm in diameter or less when laterally compressed and obovoid. In *G. costaricensis,* carposporangia are shorter than *G. tenuifrons,* In the cystocarp of *G. lamaneiformis,* greater cortical development is reflected in its thick pericarp with a prominent palisade layer of surface cells (Bird and Oliveira, 1986).

Melanthalia obtusata, a distinct genus within Gracilariaceae, has cystocarp cavity with a minimal space between the gonimoblast and pericarp and an inner pericarp composed of small, darkly stained multinucleate cells (Fredericq *et al.*, 1990). There is formation of small fusion centres in the floor of the cystocarp. Thus, the terminal gonimoblast cells fuse directly into inner pericarp cells and the fusion products incorporate still additional cells. This formation occurs in sterile gonimoblast cells. While studying interspecific, intraspecific crosses as well as sterility barriers among species of *Gracilaria,* from the Sao Paulo littoral, Brazil, a very different kind of cystocarp has been reported (Estela *et al.*, 1988) which was produced by female plants of *Gracilaria mammillaris* in the presence of male branches of *G. cervicornis* and called as pseudocystocarp. These cystocarpic structures did not produce carpospores and a section showed only an empty cavity surrounded by a pericarp. The production of pseudocystocarps is described for *G. follifera* and *G. tikvahiae,* (McLachlan *et al.*, 1977) where syngamy regulates the production of the pericarp (see also Estela *et al.*, 1988). Bird and McLanchlan (1982) believe that such structures are not an indication of greater taxonomic affinity, because they were also found in the absence of males. This was not the case with *G. mammillaris* and *G. cervicornis* for the pericarp was produced only in the presence of males indicating the need for some sort of male stimulus. These studies are mainly aimed to solve taxonomic problems and search for more productive strains, should continue.

Sarcodiaceae: In this family, peculiar type of carposporophyte development occurs with distinct fusion cells and gonimoblast filaments. In *Sarcodia dentata,* the cystocarp strongly protrudes from the surface of the thallus on which the carpogonial branch is borne (Fig. 61 A). This occurs due to the increase in thickness of cortical tissue surrounding the gonimoblast cells, which finally results in a thick pericarp (Figs. 62 D; C_1, C_2). After fertilization, the formation of secondary gonimoblast tissue is, thus, rapid and is closely coordinated with the proliferation of adjacent gametophytic tissue resulting in deeply embedded fertilized carpogonium and the juxtaposed auxiliary cell. Eventually, the thick pericarp develops short stalk that supports the spheroidal cystocarp. Proximal gonimoblast filaments fuse among adjacent cells and eventually a large, finely reticulate, fusion cell results (Fig. 62 C_1). This forms the central region of the cystocarp from which carposporangia bearing gonimoblast filaments radiate (Fig 62 C_3), terminally into short chains of carposporangia (Norris, 1987). This entire region, the major part of the gonimoblast, is free from connections with adjacent vegetative tissue except through the large-celled, nutrient-rich primary gonimoblast which is situated mainly at the base of the gonimoblast-stalk. Usually, only one cystocarp is produced but upto three cystocarps have been observed in the same papillum which indicates that similar number of fertilization have occurred.

The presence of finely reticulate fusion cell in the cystocarp and the occurrence of carposporangia in chains are distinguishing characters of Sarcodiaceae. The relationship of Sarcodiaceae and Dicranemaceae seems to be very close due to strong similarities of gonimoblasts and structure of cystocarps. In all the taxa studied, there is formation of either a reticulate fusion cell or a group of gonimoblast initials forming a 'placenta' consisting of large, nutrient-filled cells that may be partially fused with one another or with adjacent vegetative cell (Norris, 1987).

Gigartinaceae : Carposporophyte development in *Chondrus nipponicus* has great taxonomic importance not only from *C. crispus,* but also throws light on the relationship among three families Gigartinaceae, Petrocelidaceae and Phyllophoraceae.

In *Chondrus nipponicus,* after fertilization. the trichogyne contracts and the carpogonium fuses directly with the supporting cell which enlarge and functions as the auxiliary cell. Intercalary divisions of vegetative cells are occasionally found in the vicinity of the young auxiliary cell. Several gonimoblast initials are cut off inwardly from the auxiliary cell and these produce gonimoblast filaments that grow into the medulla. Swollen uninucleate gonimoblast cells become linked by secondary pit-connections to multinucleate vegetative medullary cells, during differentiation into carposporangia (Brodie *et al.,* 1991). It is found that prior to fertilization, the structure of the procarp of *C. nipponicus* resembles that of *C. crispus* and other members of the Gigartinaceae. However, post-fertilization events appear of differ in the two species. In *C. nipponicus,* a layer of cells is sometimes formed around the fertilized procarp. This phenomenon is probably associated with a need to provide nutrition for the developing gonimoblast. Such an arrangement of pericarpial cells does not appear to occur in Atlantic plants of *C. crispus* (Brodie *et al,* 1991).

Both *Chondrus nipponicus* and *C. crispus* lack an involucre in the mature cystocarp. The presence or absence of an involucre (also known as enveloping tissue, pericarp, pericarpial filaments, pericarpium, proprium and special medullary filaments) is recognized from an early stage as an important generic feature of the Gigartinaceae. In *Gigartina teedii,* the involucre has a nutritive role and, thus, *Chondrus* is distinguished from other members of the Gigartinaceae.

Similar intercalary divisions of the medullary cells in the area of the gonimoblast formation are also reported in *Mastocarpus.* The occurrence of these cells and absence of an involucre in the mature cystocarp in *Chondrus* (Gigartinaceae) and *Mastocarpus* (Petrocelidaceae) suggest a relationship that merits further study at the familial level.

Carposporophyte nutrition in the Gigartinaceae, Petrocelidaceae and Phyllophoraceae is of fundamental importance for an understanding of the phylogeny of these closely related procarpial families, but in our present understanding no definite relationship is possible either at the familial or generic level.

Solieriaceae : In *Gardneriella tuberifera,* the salient features of carposporophyte development are : (a) vegetative cells to which the gonimoblast fuses, initiate numerous multicellular filaments which resemble connecting filaments and lack pit-connections at the cross-walls; (b) a distinct involucre is absent. The vegetative cells surrounding the gonimoblast may, however, become greatly stretched giving the appearance of involucre; (c) the filaments that extend from the peripheral vegetative cells separate the developing cystocarp into distinct lobes, each lobe radiates from the several densely staining cells of central placenta; (d) the mature cystocarp is placentate as in Sarcodiaceae and numerous short chain of uninucleate carposporangia radiate from a common centre; (e) ostiole is absent. The unpigmented carpospores are released upon the dissolution of outer vegetative cells. *Gardneriella,* a parasitic red alga, appears to be most closely related to placentate solieriacean genera *Agardhiella, Sarcodiotheca* and *Meristiella* and should be included in the Gigartinales (Goff and Hommersand, 1982).

In *Solieria robusta* mature cystocarp is embedded in the medullary region (Fig. 82 A). The medullary filaments are elongate and show lax arrangement around the cystocarp. The mature carposporangium has multilayered cell wall that shows parallel arrangement of microfibrils and the carpospores are packed with pleomorphic floridean starch grains (Fig. 82 C).

In *Solieria robusta* the mature cystocarp shows a large fusion cell in the centre from which both branched and unbranched gonimoblast filaments emanate. The upper two or three cells of branched gonimoblast filament mature into carposporangia. The carpospores are gorged with floridean starch grains (Fig. 82 B).

Nizymeniaceae: *Nizymenia australis* resembles members of Gracilariaceae and Gigartinaceae in its carposporophyte development. Here the auxiliary cell is formed from the same supporting cell, develops secondary pit-connections with the carpogonium and eventually forms a large fusion cell which has many finger-like projections. The carposporophyte consists of gonimoblast filaments that radiate from the fusion cell and bear terminal chains of carposporangia (Fig. 64 C). The carposporangia are produced in basipetal fashion, are rectangular to ovoid, whereas the carpospores are globose to spherical (Fig. 65). The cortical cells proliferate to form a massive pericarp (distinctive feature of Gracilariaceae) which enclose carposporophyte (Vijayaraghavan and Bhatia, 1991). The cystocarp, at maturity, opens through ostiole and the carpospores beseiged with mucilage are released en masse (Fig. 66 C).

Caulacanthaceae: In *Heringia*, the cystocarps are stalked and have thick, multicellular pericarps with transverse ostioles and a large central fusion cell (Hommersand and Fredericq, 1983).

TETRASPOROPHYTE

The members of Gigartinales have been broadly classified on the basis of two types of tetrasporangia namely zonate and cruciate (Bold and Wynne, 1981). The former type distinguishes the family Solieriaceae where it is scattered over the thallus surface among the cortical cells as seen in *Agardhiella* and *Eucheuma* In *Opuntiella,* the tetrasporophytes are unkonwn. In *Gardneriella tuberifora,* zonate tetrasporangia are reported which helped the inclusion of this genus in the Solieriaceae (Goff and Hommersand, 1982).

Gigartinaceae: the tetrasporangia are always cruciate; and are differentiated from the subcortical cells or arise from special medullary filaments, as in *Iridaea* and *Chondrus* and, thus, are accessory structures.

In *Chondrus crispus,* the immature tetrasporophytes are morphologically similar to young gametophytes. At maturity, plants bear tetrasporangia in distinct sori visible as red, round or oval patches on the lower part of the main axis or on marginal proliferation. Tetrasporangia are transformed either from uninucleate cells; or terminal-intercalary chains of uninucleate cells that are produced secondarily from the medullary cells. A tetrasporangial initial enlarges, undergoes transverse division and is followed by longitudinal division resulting in four cruciately arranged spores. The pit-connections linking tetrasporangia lyse. The tetraspores release are identical to carpospores release and occur through breakdown of the cortex. Each sorus, initially, occupies a distinct region of the medulla. If more than one sorus develops in the same area, then the thallus may become almost completely filled with tetrasporangia. New tetrasporangia continue to develop in a sorus, at the same time, the early formed ones mature and are released. The sori, thus, appear to be productive for a long time (Brodie *et. al.*, 1991).

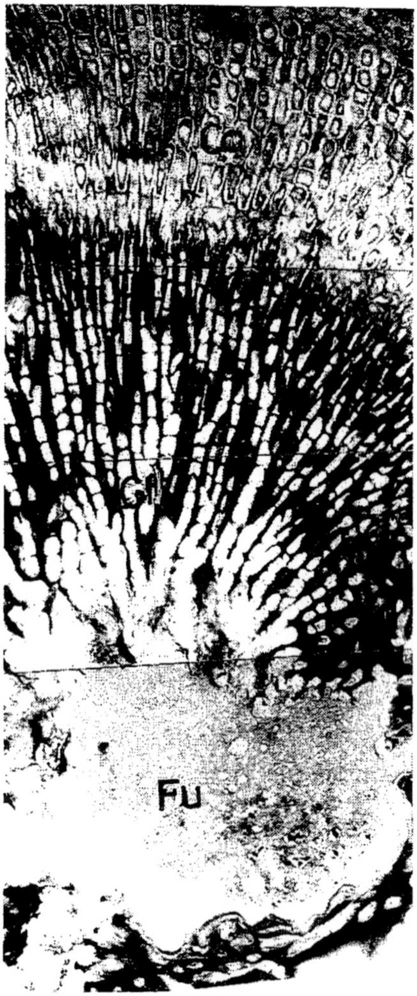

Fig. 65. Carposporophyte of *Nizymenia australis*

(after Vijayaraghavan and Bhatia, 1991)

(Ca, carposporangia; Fu fusion cell; Gf, gonimoblast filaments).

Longitudinal section of cystocarp showing gonimoblast filaments radiating from the fusion cell. The gonimoblast filaments bear terminal chains of carposporangia which are produced in basipetal fashion, and are rectangular to ovoid. (TBO)

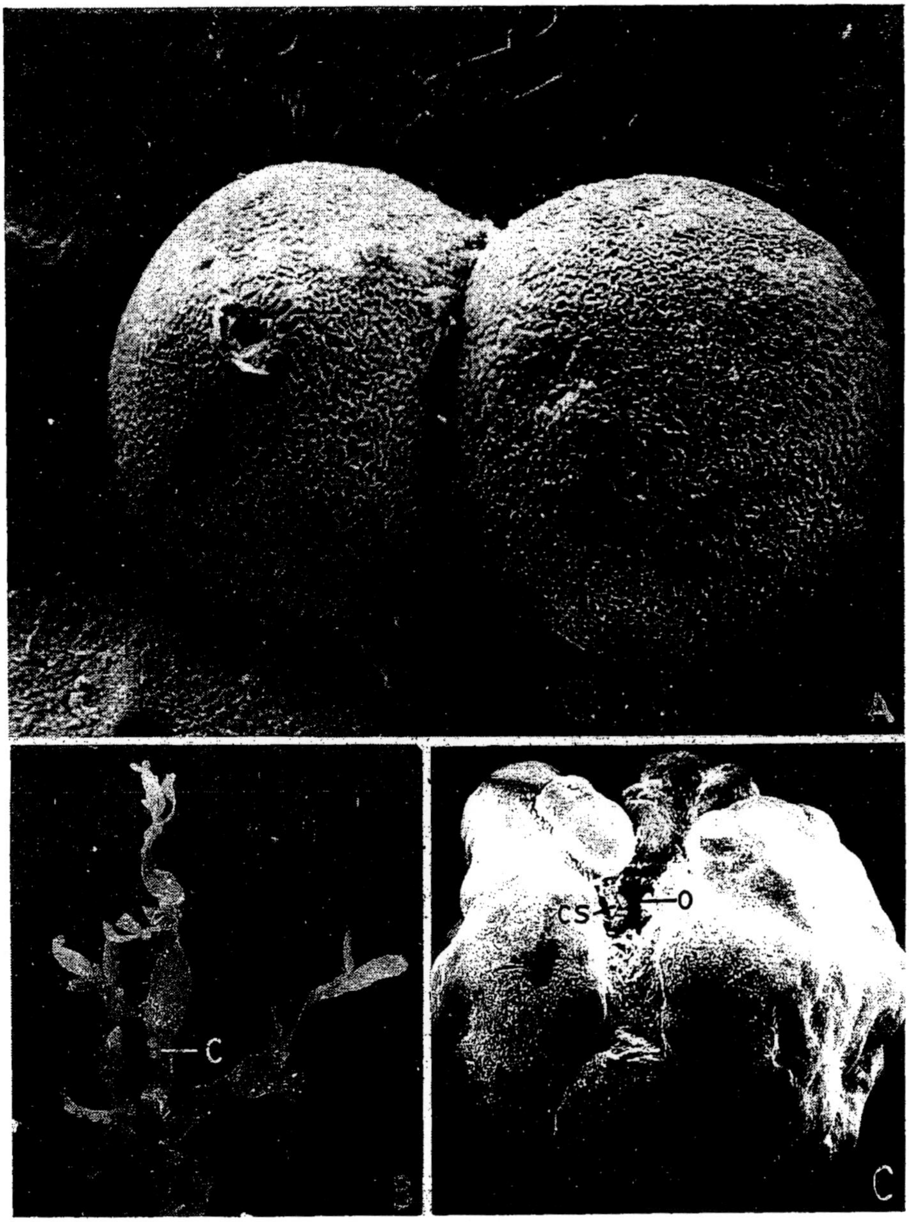

Fig. 66 A-C Cystocarps

A *Gracilaria corticata* SEM of cystocarps showing prominent ostiole. (after Manoj Kumar 1996).

B *Nizymenia australis.* A mature plant showing prominent cystocarps. (after Vijayaraghavan, and Bhatia 1991).

C *Nizymenia australis.* SEM of mature cystocarp. The carpospores are released en masse through the ostiole. (after Vijayaragavan, and Bhatia 1991).

The tetrasporangial development in both *Chondrus nipponicus* and *C. crispus* takes place by the transformation of chains of cells that are formed secondarily from medullary cells. The mature tetrasporangia in former are large and wider than those reported for *C. crispus* from the British Isles (Brodie *et al.,* 1991).

Phyllophoraceae: The tetrasporangia occur in nemathecia which are raised areas on the surface, rather than immersed in the thallus. In *Phyllophora tuncata,* culture and cytological observations demonstrated an abbreviated life-history, with a diploid tetrasporangial phase on a haploid, monoecious, gametophyte, seemingly as a result of fertilization. Meiosis occurred in the tetrasporangia, and the tetraspores developed into apparent gametophytes. In *Gymnogongrus leptophyllus* the crustose tetrasporophyte bore catenate series of tetrasporangia (DeCrew and West, 1981). In *Ahnfeltia gigartina,* an intercalary catenate series of tetraspores and in *A. plicata,* a crust bearing terminal tetrasporangia are reported (DeCrew and West, 1981).

Gracilariaceae: The tetrasporangia are cruciately divided like Gigartinaceae and Phyllophoraceae (Bold and Wynne, 1981). In *G. epihippisora,* the ovoid tetrasporangia divide cruciately, and are borne in sori. *G. blodgetti* bears tetrasporangia in nemathecial swelling above the constricted base of the ultimate branchlets (Kumar, 1990).

In *Gracilaria tenuifrons,* the ovoid tetrasporangia are decussate, cruciate or occasionally pseudotetrahedral in arrangement (Fig. 68 B). They are scattered throughout the cortex, compress the adjacent sterile cells but do not modify the tissue (Bird and Oliveira, 1986). *G. verrucosa* bears tetrasporangia with cystocarps which are found to be scattered over most of the thallus surface (Garguilo *et al.,* 1985) In *G. dendroides,* tetrasporangia are localized in the upper cortex of lanceolate branchlets (Fig. 67 A). Tetrasporangia are cruciately divided, although this is somewhat variable (Fig 67 B); the tetrasporangia are elongate and are 40-60 x 18-25 µm. They are immersed in a modified cortex, the cells of which are irregularly shaped, less vacuolate and possess thick cell walls.

Sarcodiaceae : In *Sarcodia dentata* tetrasporangiate thallus shows 2 or 3 orders of pseudodichotomous branching pattern (Fog. 61 C) ; tetrasporangia are zonately divided like members of Solieriaceae. Tetrasporangia are scattered in the outer cortex (Fig. 62 B) and are produced terminally on cortical filaments (Norris, 1987).

LIFE CYCLE

The life cycle of order Gigartinales is diplobiontic and includes two phases :

(i) The sexual phase

(a) Taxa with procarps,

(b) Taxa without procarps

(ii) The tetrasporangiate phase

(i) THE SEXUAL PHASE

Gigartinales, as defined by Kylin (1916), include all those Florideae in which the auxiliary cell is formed prior to fertilization and is an intercalary cell of one of the ordinary vegetative laterals. The sexual reproductive apparatus includes forms where the carpogonial branches and the auxiliary cells are spatially separated and others in which they are combined to form a procarp.

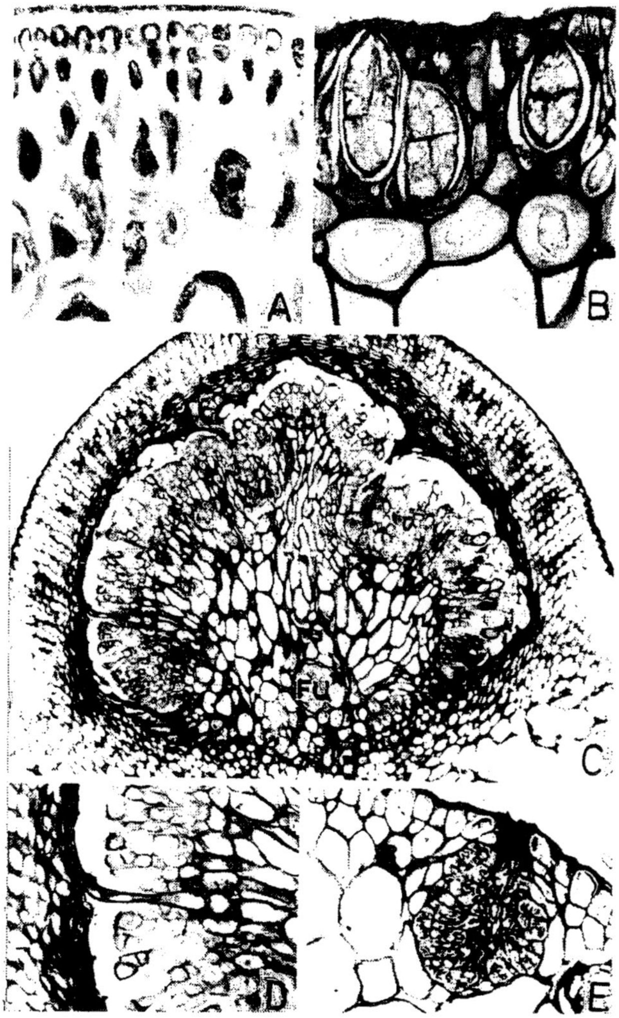

Fig. 67. A-E ***Gracilaria dendroides***

(after Garguilo *et al.*, 1985)

(Fu, fusion cell)

A. A part of thallus showing 4 or 5 layered cortex.

B. Cruciate tetrasporangia.

C. Cystocarp showing lobed carposporophyte and pericarp with radial rows of cells and lobed fusion cell.

D. Detail of a filament connecting the gonimoblast tissue and pericarp. The cells forming the pericarp are round and are arranged in anticlinal and radial rows.

E. Spermatangial conceptacle immersed in the subcortical layer whose cells limit the conceptacles and are slightly elongate.

(a) Taxa with procarps

In diverse Gigartinales with prominent procarps (*Cystoclonium, Rhodophyllis, Calliblepharis, Hypnea)* auxiliary cell is the basal cell of a lateral borne on the supporting cell of the carpogonial branch. Besides, there is elaboration in the mode of nourishment of the gonimoblast. The gonimoblast initial develops into an elaborate, well-branched, thread that penetrates into the central cavity of cystocarp. The cells fuse with one another, as well as with the auxiliary cell, to form an irregularly branched, tubular fusion-cell that bears numerous tufts of carposporangia. There is abundant production of hyphae near the developing gonimoblast. The antheridia form extensive sori on the younger branches of the male plants.

(b) Taxa without procarps

The important members which show this type of life cycle are *Platoma, Schizymenia, Calosiphonia,* and *Nemastoma.* In *Platoma bairdii,* the colourless 3-celled carpogonial branches arise from the inner cells of the laterals while the more numerous auxiliary cells are located near the base of other laterals. A carpogonial branch may be exceptionally borne directly upon an auxiliary cell. The auxiliary cells are considerably enlarged and protrude on one side. Several connecting filaments may fuse with many auxiliary cells, contact being made with the protruded part of the latter. The gonimoblast initial is cut off on the outer side of auxiliary cell and gives rise to a mass or short ill-defined pore in the cortex, through which carpospores escape.

(ii) THE TETRASPORANGIATE PHASE

The tetrasporangia are either cruciate or zonate, the latter type is found in all genera, except *Platoma, Calosiphonia, Nemastoma, Sebdenia, Gracilaria,* and *Melanthalia.* The sporangia are often terminal on the laterals and embedded at the surface of the thallus. The large sporangia which develop from the hypodermal cells, form a zigzag line along the axis of the stichidium and are covered by a single layer of cells.

GRACILARIACEAE

The life-cycle of a *Polycavernosa debilis* is isomorphic (Polysiphonia type life history, sensu Dixon, 1982); reveals three reproductive phases i.e. haploid male, haploid female (including cystocarps) and diploid tetrasporic. It was reported that morphologically identical but different phases in life-history of this alga showed no conspicuous differences in physiological or ecological performances. The questions were specially focussed on whether or not there are adaptive advantages and concomitant costs associated with ploidy levels and whether these are measurable.

The costs and benefits of ploidy level, as measured by net photosynthesis, calorific content, structure make up and resistance to predation are not sufficient to result in statistically significant differences for the various reproductive phases (Littler and Littler, 1987)

The lack of significant functional responses at the scale of those parameters investigated suggests that genetic/reproductive constraints may be predominant factors involved in maintaining isomorphic reproductive phases. However, if there is adaptive value in both gametic reproduction (eg. resource exploitation, genetic recombination) and sporic reproduction (e.g. environment resistance, dispersal) and a single morphological form represents the "best" compromise in fitness, then isomorphic, diploid and haploid phases could be maintained.

The life-history of the *Gracilaria* species follows the Polysiphonia type (Yamamoto and Sasaki, 1988). However, there are instances where only a few or no gametophytes exist in the

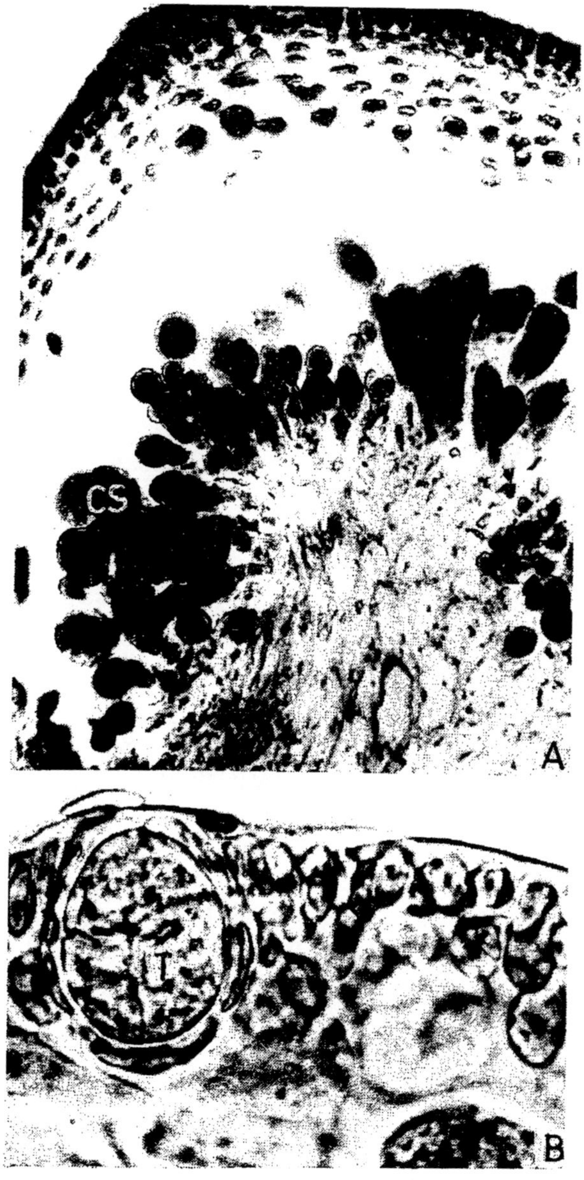

Figure 68. A, B Reproductive structures of *Gracilaria tenuifrons*

(after Bird and Oliveira, 1986)

(CS; cystocarp; T, tetrasporangium)

A. Portion of a cystocarp showing thick pericarp of six to eight cell layers. The pericarp inner cells are periclinally compressed to various degrees and generally maintain radial files. A slight ostiolar bulge is often present.

B. Section of a thallus showing slightly protruding tetrasporangium and compressed adjacent cortical cells.

natural populations as seen in *G. chorda* from Hokkaido, one of the ecological forms of *G. salicornia* and *G. euchemoides* and is also considered to be tetrasporophyte – tetrasporophyte type.

G. salicornia has two forms which can be distinguished both morphologically and ecologically. In the former type the nodes of segments are markedly articulated and fronds grow solitarily and the three reproductive phases are common. In the other type form, a mat consisting almost only of entangled and less constricted tetrasporangial or sterile fronds are seen. Yamamoto (1990) carried out culture studies to reveal the life history and the morphological modifications of this mat-type, which showed the Polysiphonia type and the morphology of fronds is similar to that of the articulated rather than of the mat-type.

The life history of *Gracilaria acruata* from Philippines was studied in culture (Yamamoto, 1990). This species also follows the Polysiphonia type of the history.

PHYLLOPHORACEAE

The histories in this family are extremely diverse. The tetrasporophytes and gametophytes are isomorphic in *Phyllophora, Stenogramme, Ozophora, Petroglossum,* and *Schottera.*

The genus *Ahnfeltia* exhibits two types of life histories. In *A. olicata,* an upright thallus bearing monosporangial nemathecia alternates with a crust bearing, terminal tetrasporangia, previously referred to as *Porphyrodiscus simulans.* The structure previously described as "hairs" on the nemathecia, are actually carpogonia, and thus the upright portion is actually a gametophyte (DeCrew and West, 1981).

In *Ahnfeltia gigartinoids* and *A. concinna,* an upright gametophyte with internal gonimoblasts alternate with a crust bearing and intercalary catenate series of tetrasporangia (*Petrocelis - Erythrodermis* stage). On the basis of these fundamental differences, it was suggested that these two species be merged into a single taxon. *Ahnfeltiopsis.*

A few species of *Gymnogongrus* also have a heteromorphic life history. *G. furcellatus, G. martinensis* and *G. flabelliformis* have a Petrocelis-type tetrasporophyte. A few other species possess a tetrasporoblastic life history in which the tetrasporophyte develops directly on the gametophyte after fertilization (DeCrew and West, 1981). No isomorphic life history is known for this genus. For most of the nearly thirty species of *Gymnogongrus,* however, the complete life history is either unknown or uncertain. In *G. linearis* and *G. leptophyllus,* from the Pacific coast of North America only gametophytes are reported. In culture, it was found that in the life histories of both taxa, upright gametophytes alternate with crustose tetrasporophytes, but some isolates of *G. leptophyllus* appear to be apogamous and bypass the tetrasporophyte generation. The tetrasporophyte of *G. lincaris* superficially resembles *Petrocelis middendorffii,* but differs in several aspects like size, colour, number of tetrasporangia per filament, and distal dichotomous branching of perithalial filaments. *P. middendorffii* generally occurs higher in the intertidal zone than the tetrasporophyte of *G. lincaris,* but there is some overlap. The two entities are difficult (though not impossible) to distinguish without microscopic examination. The cultured tetrasporophyte is identical to *Petrocelis- (=Erythrodermis) haematis,* California. It is, however, important to note that a Petrocelis-stage is found in life histories of both Gigartinaceae and Phyllophoraceae.

GIGARTINACEAE

The life history of *Chondrus nipponicus* collected from Japan show Polysiphonia type and identical events as seen in *C. crispus,* popularly called "Irish moss", with sequence of isomorphic plants consisting of gametophytes, carposporophytes and tetrasporophytes. The carposporophyte is parasitic on the female gametophyte. The tetraspores germinate into gametophytes, whereas the carpospores develop into tetrasporophytes (Brodie *et al.,* 1991).

In *C. nipponicus,* the plants are dioecious. Spermatangia develop single or in pairs from apical cells. It is likely that there is a photoperiodic response in the formation of carpogonial branches and carposporogenesis which occur both in LD and SD conditions - although development is more rapid (as reproductive maturity is reached) in tetrasporophyte, probably due to induction of higher photon exposures, in LD conditions. In culture, tetraspores, from plants collected from Oshoro Bay, Hokkaido, and Iwamuro, Niigita Prefecture, give rise to dioecious male and female gametophytes, but do so only under SD condition. In the presence of male plants, female plants formed carposporophytes that released carpospores, which germinated in turn to form tetrasporangial plants at 15° C, under both LD and SD conditions. All phases, except spermatangial development, of life cycle is under photoperiodic control.

All other members of Gigartinaceae have generally isomorphic life cycle. It is also reported that a Petrocelis-stage is found in life histories of both Gigartinaceae and Phyllophoraceae. *Petrocelis middendorffii* is the tetrasporophyte of *Gigartina papillata* while *P. cruenta* is the tetrasporophyte of *G. stellata.* Apogamy, to a higher extent, is reported in *G. stellata* and approximately 100 isolates from the Atlantic, had a higher degree of gametophytic recycling, and a lesser degree of alternation with *P. crucita.* The *Gigartina* sp. in culture revealed the greatest amount of direct recyling at higher latitudes (DeCrew and West, 1981).

SOLIERIACEAE

Gardneriella tuberifera is a parasitic, multiaxial and nonprocarpic alga where the life cycle pattern is unique in showing the following criteria with other genera of this family. (i) inwardly directed carpogonial branches (ii) auxiliary cells which upon diploidization, cut off single gonimoblast initial, (iii) gonimoblasts develop inwardly (toward the medulla), (iv) carposporangia are borne in chains or terminally as see in *Anatheca, Solieria* and some *Euch*euma species and (v) zonate tetraspores. It also resembles other placentate solieriacean genera (*Placentophora, Agardhiella, Sarcodiotheca* and *Meristiella)* in lacking a central fusion cell and in having a cellular cystocarp. It is most closely related to the predominantly north American taxa *Agardhiella* and *Sacodiotheca.* It is suggested that these should be included in newly erected tribe Agardhielleae due to following reasons : (i) carpogonial branch morphology (ii) production of one or two unbranched connecting filaments from a diploidized carpogonium (iii) characteristic reniform, uninucleate auxiliary cell which bears a striking resemblance to that of the host *Agardhiella gaudichandii* and (iv) early stage of gonimoblast formation which results in a placentate cystocarp with carposporangia borne in short radiating chains.

Gardneriella differs from other members of the tribe Agardhielleae in having multiple carpogonial branches and auxiliary cells borne on the same filament system and lack an involucre, compact central placenta, auxiliary cell complex and ostiole. A few or all of these differences may be the result of reduction due to the parasitic habit.

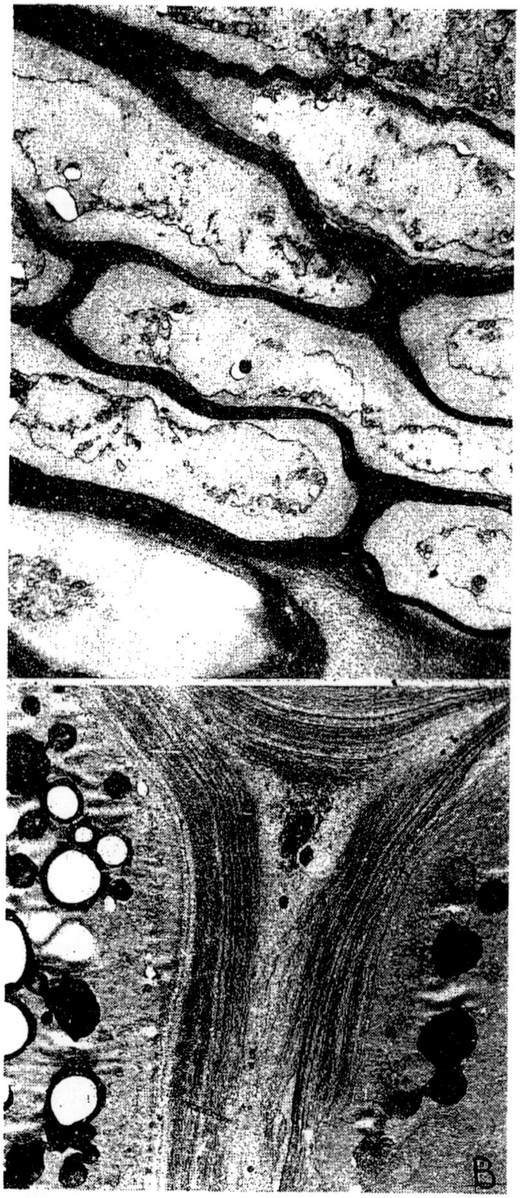

Fig. 69 A, B. Ultrastructure

A. *Solieria robusta* showing thick walled and vacuolate pericarp cells. Degenerating chloroplast lamellae are seen in the cell cytoplsm. The intercellular mucilage shows reticulate arrangements of fibrils (after Nashier, 1996)

B. *Gracilaria corticata* hold-fast cell showing multifibrous cell wall (after Manoj Kumar, 1996).

ULTRASTRUCTURAL STUDIES

CUTICLE

In *Chondrus crispus* the fine structures of cuticle varies according to the region of the thallus and in the different phases of the life cycle. Cuticle varies from a multilamellar structure in the apical region to a single amorphous layer in the stipe and holdfast regions. In the central region of the frond, however, alteration of multilamellar and amorphous cuticle coexist. Besides, cuticular variation is also found both in the gametophyte and the sporophyte stages. In the former, by contrast, lamellae are fibrillar, well-defined, regular, lack fusion and have an interlamellar microfibrillar network. In the sporophyte, the lamellae are, however, less numerous especially in the apical region; irregular, frequently are connected with one another and the fibrillar materials in the interlamellar space is sparse. Further, the older lamellar cuticle seems to peel off as a single layer, thus, exposing the amorphous material. In the gametophyte, by contrast, several lamellae are together peeled-off, but many remain in place, and prevent direct contact between the underlying tissue and the environment.

The protective role of cuticle in algae is neglected, despite its constructive role in higher plants. The variation in cuticle organization may be important in determining differential susceptibility to infection by algal endophytes and other pathogens (Corea and MacLachan, 1988).

CELL WALL

In *Gracilaria verrucosa* the ultrastructure of cell wall upon steam-explosion* revealed the architecture of the skeletal components. The ultrastructural (SEM, TEM) and cytochemical organization of the skeletal and matrix polysaccharides of the fresh and steam-exploded thalli were compared to chemical analyses of the extracted agar. Upon steam-explosion, surface microfractures appeared in lemented areas of the cortex. SEM preparations indicated that apical and medullary regions are the most affected, resulting in detachment of adjacent cells. In TEM sections, the cuticle of the surface layer exhibited deeply altered zones with a spongy appearance. In the cytoplasm, starch grains are disrupted. The skeletal components of the immediate cell wall is maintained with a different organization of the microfibrils and is positive to periodic-acid thiosemicarbazide silver proteinate test. The intercellular matrix contain granules and fibrils and lack a ruthenium red reaction. These results are consistent with lower sulfate-ester and higher glucose contents detected in agars extracted using steam-explosion.

It is concluded that steam-explosion procedure, besides improving the extraction of agar as demonstrated by the low sulfate content and molecular weight of the obtained agar that are comparable with commercial ones, might provide additional information on the ultrastructure of the cell wall. It also might be a useful tool to identify structures organized to resist mechanical stress at various levels (Talarico *et al.*, 1990).

APICAL MERISTEM

Gracilaria verrucosa was studied for the structural indications of the cellular differentiation that are produced in the apex of the frond. Frond is a branching system that consists of short cladomes called ''microcladomes'' which fasten on to primary Cladome (Kling *et al.*, 1990).

* The steam-explosion generally applied to the higher plants, can be destructive when applied to algal material. However, certain structural integrity can be preserved by performing steam-explosion under milder conditions such as 140 kpa and for 6-15 seconds.

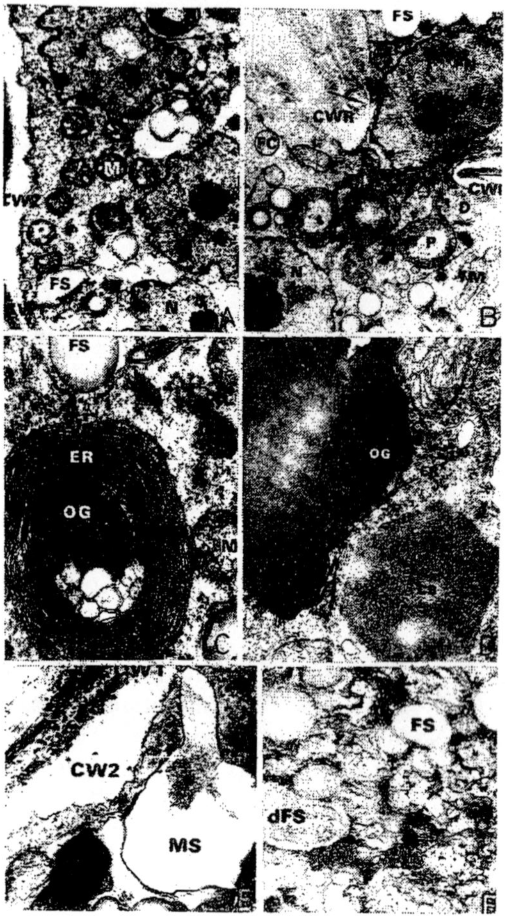

Fig. 70. A-F Ultrastructure of fusion cell in *Gracilaria verrucosa*

(after Delivopoulos and Tsekos, 1985)

(CB, crystalline body; CW, cell wall; D, dictyosome; dFS, disorganizing floridean starch; ER, endoplasmic reticulum; FC, fusion cell; FS, floridean starch; M, mitochondrion; MS, mucilaginous sacs; N, nucleus; OG, Osmiophilic granules; P, Plastid).

A. Part of the multinucleate fusion cell, showing mitochondria, proplastids and starch grains that are present in the cytoplasm. A two layered cell wall is also seen.

B. Fusion cell coalesing with an adjacent vegetative cell. Cell wall remnants (arrows) delineate sites of cytoplasmic fusion.

C. Fusion cell to show osmiophilic granules surrounded by concentric stacked aggregation of ER.

D. Same, to show large osmiophillic granules and crystalline body present in the cytoplasm.

E. Muclilage sacs discharging mucilage exterior to the fusion cell.

F. Degenerating fusion cell showing floridean starch grains and degenerating cytoplasm.

A longitudinal axial section in the apical region of the frond, shows 3 areas :

(1) The Apical dome	:	Consists of the primary and secondary cladomes.
(2) The Cortical layer/Cortex	:	Arise from the coalescence of the apical cells of the short secondary and tertiary cladomes.
(3) The Medulla	:	Formed in the inner part of the frond, with the proximal and vacuolised cells of the cladomes.

The apical cells of the dome (apical cells of the aged secondary, tertiary and quaternary axes) are conspicuous by heavy nucleoplasmic ratio (= 0.6), bulky nucleus and more active mitochondria and dictyosomes whereas those of the cortex (apical cells of the young secondary and tertiary axes) are noticeable by little nucleoplasmic ratio (= 0.2) and more differentiated plastids and mitochondria. The proximal cells of the axes, irrespective of their position in the microcladome, are more vacuolised and make the medulla.

FUSION CELL

The fusion cell of *Gracilaria verrucosa* is a large, multinucleate and irregularly shaped syncytium originating from cytoplasmic fusion of haploid and diploid cells (Fig. 70 A). Inspite of its importance, little is known of fusion cell structure and function because relatively few studies have been conducted. Most of the available information of this cell has been presented in papers dealing with carposporogenesis. To date, only a few studies exist describing the fine structure of fusion cells and it is apparent that there are fundamental differences in interpretation and actual structure of the fusion cell including the process of the fusion cell formation.

According to Delivopoulos and Tsekos (1985), the formation of fusion cell results from cytoplasmic fusions of fertilized carpogonium and surrounding haploid vegetative cells. Cell fusion occurs at the site of pit-connections (Fig. 70 B), but it is not clear whether the fusion involves either rupture or degeneration of the septal-plug. The sites of cellular fusions can, however, be identified by wall remnants in areas where narrow cytoplasmic channels have formed between fusion cells. No evidence of cytoplasmic degeneration is noted in any of the fusing cells and it would appear that nuclei can pass freely between fusion cells through these cytoplasmic channels.

In fusion cells which have both haploid and diploid nuclei, only diploid nuclei participate in the formation of gonimoblast cells and carpospores. Thus, the haploid, nuclei must somehow be inactivated or destroyed, so that they do not interfere with carposporophyte development. In the fusion cell of *Asterocolax gardneri*, a cytoplasmic barrier functions as an isolating mechanism. On the other hand, putative haploid nuclei in multinucleate gonimoblast cells of *Fucheocolax attenuata* appear as dark staining, spherical, naked masses and are inactivated through condensation and partial degeneration (Kugrens and Delivopoulos, 1985). The structure similar to those of *Feucheocolax* commonly occur in the fusion cells of *Gracilaria*. Such interpretations are primarily based on ultrastructural studies and biochemical tests must be carried out to determine whether these dark staining masses are, in fact, haploid chromatins.

Plastids, in the *Gracilaria* fusion cell, are arrested in their development and remain as proplastids; lack an internal thylakoid system and have only portions or a complete peripheral thylakoid. Starch granules are, however, present which most likely originate from the incorporated adjacent vegetative cells.

Mitochondria, in young cells, contain coiled lamellar bodies. These bodies may represent mesosome-like structures acting in mitochondrial-DNA replication. Alternatively, mitochondrial cristae might perhaps be derived from the coiled lamellae (Ramm-Anderson and Wetherbee, 1982). In contrast, during the last development stages of the fusion cell the mitochondria are characterized by hypertrophied cristae.

The most obvious features of the *Gracilaria* fusion cell is the dramatic increase of ER. During the young stage of fusion cell formation, concentric, stacked, aggregations of ER usually surround starch granules or large osmiophilic bodies (Fig. 70 C). It has been suggested that the close association of ER with starch grains and lipids is indicative of a possible role of the ER in starch (Borowitzka, 1978); Pueshchel, 1979; Tsekos, 1981) and lipid biosynthesis (Gunning and Steer, 1975; Alberts *et al.*, 1983). Besides, one of the additional functions of ER aggregations may be storage of materials which are utilized at a later stage of development (Fig. 70 D).

The production of large amounts of mucilage is an important feature of red algal carposporophyte development and carposporogenesis. The *Gracilaria* fusion cell produces mucilage sacs which secrete mucilage within the ER intracisternal space (Fig. 70 E). These mucilage sacs apparently discharge their contents into the extracellular space, either form the bulk of the fusion cell wall or surround the various carposporophyte components.

Meristematic activity (Tripodi and De Mari, 1977; Duckett and Peel, 1978; Kugrens, 1982; Delivopoulos and Tsekos, 1983) and secretory function (Kugrens and Arit, 1981) are ascribed to the fusion cells. These are due to presence of large quantities of floridean starch, osmiophilic bodies and crystalline inclusions which tend to support for the fusion cell of *Gracilaria* a nutritive-storage hypothesis (Fig. 70 F). The fusion cell in this taxon, thus, appears to be an active component of the carposporophyte and as this cell ages, the number of the starch granules increases considerably and are being continuously augmented as more and more vegetative cells are incorporated into the fusion cell. At maturity, many of the starch grains lyse and the fusion cell assumes a degenerate appearance, indicating a cessation of activity. These facts indicate an involvement of fusion cell only during early carposporophyte development stages.

More ultrastructural studies are needed on a broader spectrum of red algae before it can be determined whether or not these differences in fusion cell structure and function have any taxonomic or phylogenetic significance.

In *Gracilaria corticata* the hold-fast cells show multifibrillar cell wall (Fig. 69 B). In *Solieria robusta* ultrastructure of pericarp cells reveals thick-walked and vacuolate cells and degenerating chloroplast lamellae in the cell cytoplasm. The intercellular mucilage shows reticulate arrangement of fibrils (Fig 69 A).

CARPOSPOROPHYTE AND CARPOSPOROGENESIS

In *Plocamiocolax pulvinata*, a parasitic red alga, carposporophyte is simple and consists of 4 components :

(i) carpospores

(ii) generative gonimoblast cells (GGC)

(iii) storage gonimoblast cells (SGC)

(iv) a basal auxiliary cell.

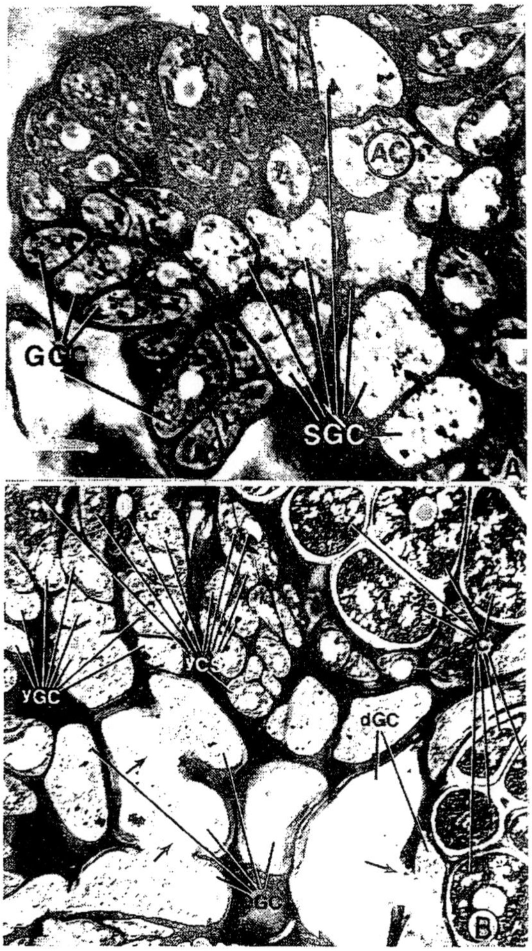

Fig. 71. A, B Ultrastructure of the carposporophyte and carposporogenesis in *Plocamiocolax pulvinata* (after Kugrens and Delivopoulos, 1986)

(AC, auxiliary cell; DGC, degenerating gonimoblast cells: GGC, generative gonimoblast cells; OGC, old gonimoblast cells; SGC, storage gonimoblast cells; YCS, young carpospores; YGC, young gonimoblast cells).

A. Young carposporophyte consisting of auxiliary cell, storage and generative gonimoblast cells. Numerous dark staining bodies occur in the cytoplasms of generative gonimoblast cells.

B. Portion of an older carposporophyte having a mixture of old and young gonimoblast cells and carpospores. Old carpospores are large and contain dark-staining cytoplasmic granules, whereas younger gonimoblast cells, which also have lighter staining cytoplasms, have produced smaller, incipient carpospores. Note the enlarged and distended septal-plug between storage gonimoblast cells (arrow).

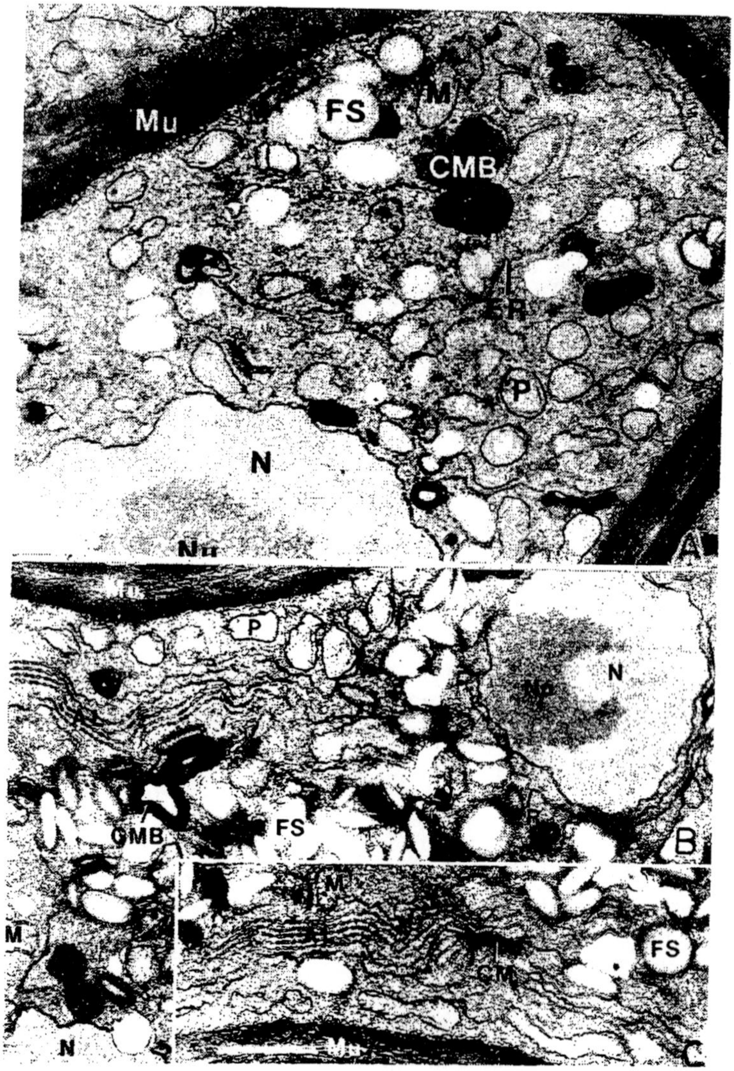

Fig. 72. A-C Ultrastructure of carposporophyte of *Plocamiocolax pulvinata*

(after Kugrens and Delivopoulos, 1986)

(CMB, compressed membrane bodies; ER, endoplasmic reticulum; FS, floridean starch; M, mitochondria; Mu, mucilage of future carposporangial wall; N, nucleus; Nu, nucleoulus; P, plastid)

A. Young auxiliary cell with a few starch grains, small vacuoles, compressed membrane bodies, numerous mitochondria, reduced plastids and a large nucleus.

B. Portion of a multinucleate storage gonimoblast cell with two nuclei and floridean starch. Annulate lamellae occur in the peripheral cytoplasm along with reduced plastids.

C. Annulate lamellae, in a SGC, attached to compressed membranes.

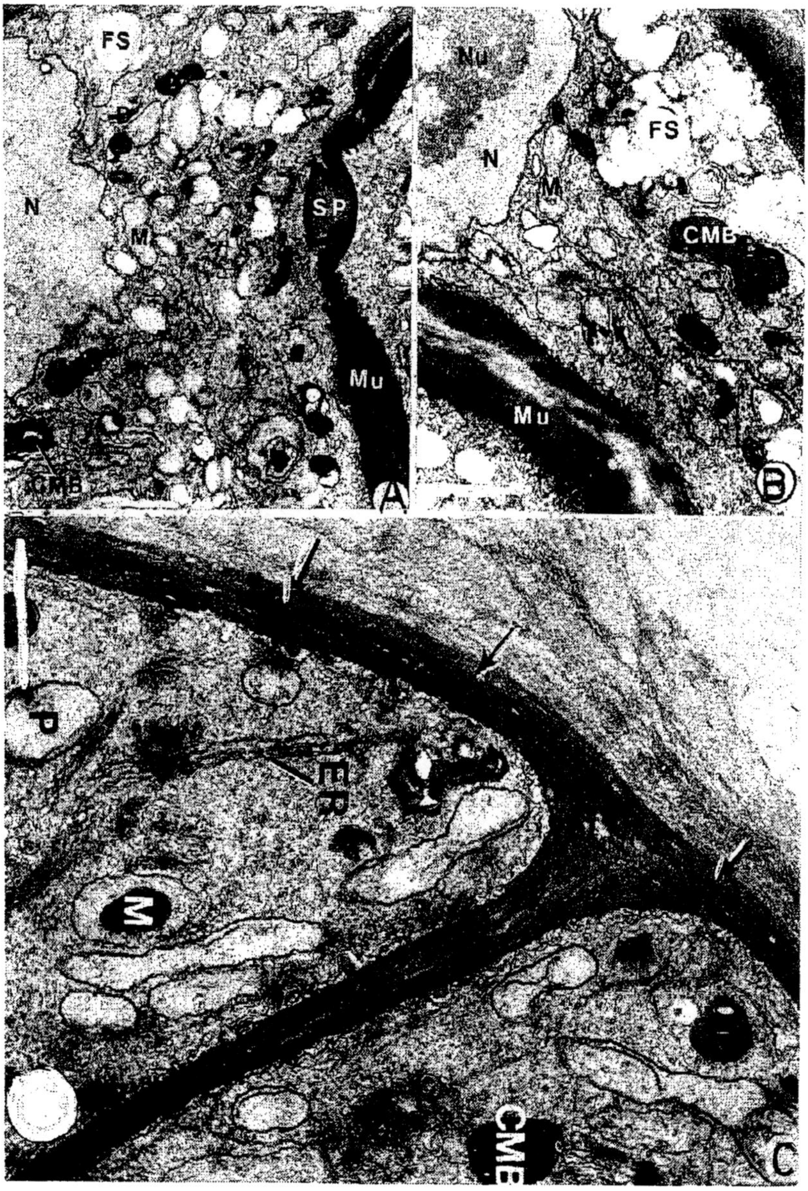

Fig. 73. A-C Ultrastructure of carposporophyte of *Plocamiocolax pulvinata.*

(after Kugrens and Delivopoulos, 1986)

(CMB, compressed membrane bodies; ER, endoplasmic reticulum; FS, floridean starch; M, mitochondria; Mu, mucilage; N, nucleus; P, plastid; SP, septal plug)

A. Degenerative auxiliary cell showing irregular nuclear envelope.

B. Mature storage gonimoblast cell showing reduction in starch grain size and incipient organelle layers.

C. Edge of a young carposporophyte lobe showing compression of mucilage fibrils (arrows), forming a more distinct layer around the young carpospores.

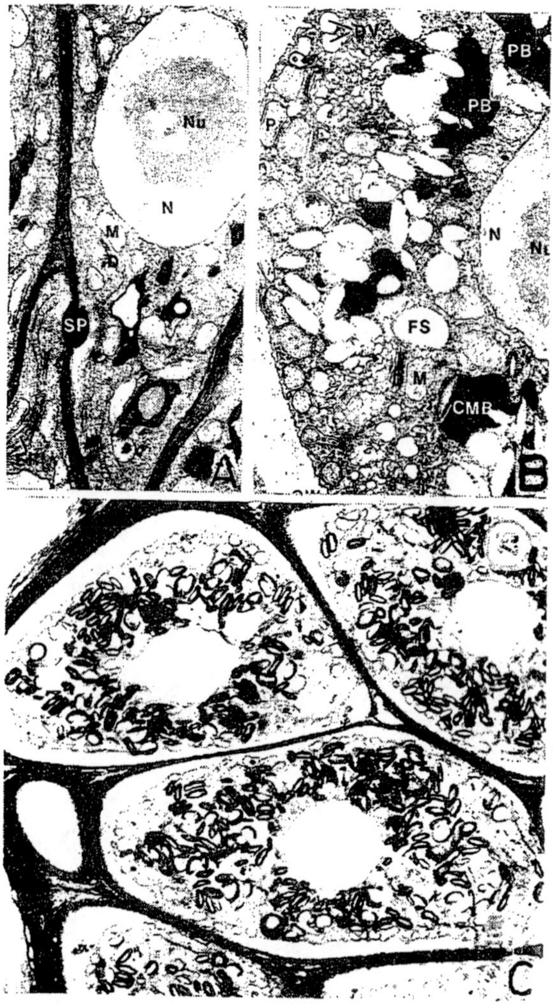

Fig. 74 A-C Ultrastructure of carpospores of *Polcamiocolax pulvinata*.

(after Kugrens and Delivopoulos, 1986).

(FS, floridan starch; M, mitochondria; Mu, mucliage; N, Nucleus; Nu, nucleolus; P, plastids; PB, polyphosphate body; Sp, septal plug).

A. Yound elongate carpospores connected by small, dark staining, septal plugs, and surrounded by a wall of dark staining, parallel oriented fibrils, Floridean starch is absent.

B. Carpospores with extensive dictyosome activity forming the first layer of the carpospore wall that is electron transparent. Wall fibrils are spaced further apart than those of carposporangial mucilage.

C. Carpospores showing synchronous differentiation.

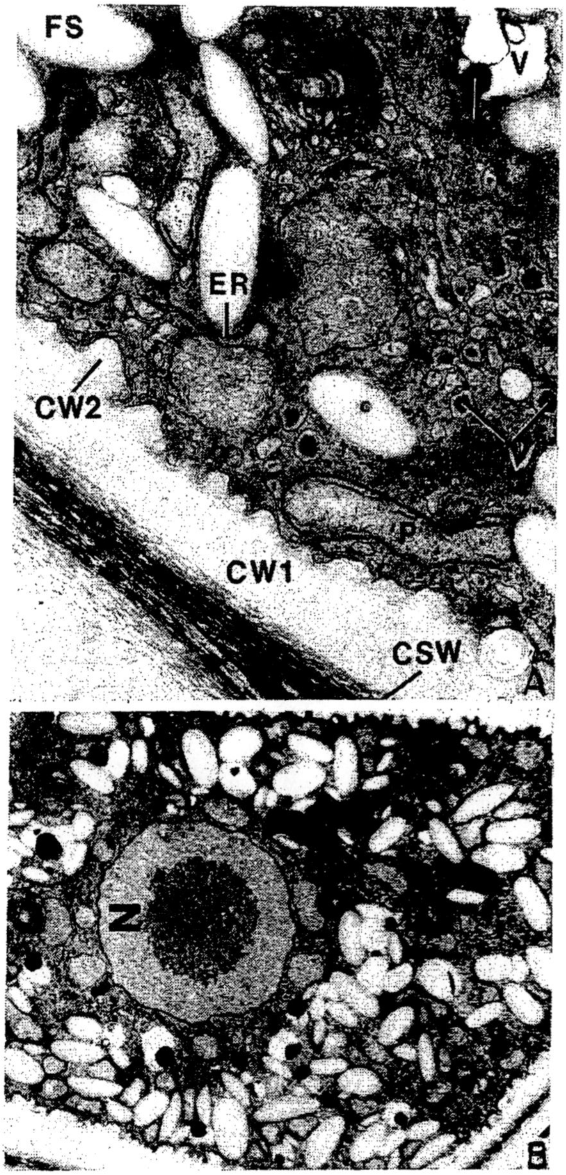

Fig. 75. A, B Ultrastructure of carpospores of *Plocamiocolax pulvinata.*

(after Kugrens and Delivopoulos, 1986).

(CW_1, first carpospore wall; CW_2, second carpospore wall; ER endoplasmic reticulum; FS, floridean starch; GC, gonimoblast cell; P, plastids; V, vacuole)

A.B. Slightly mature carpospores having curved dictyosomes, numerous granular cored vesicles presumably derived from protein body degradation. The plasmalemma profile is highly irregular, indicating intense deposition of the second carpospore wall layer.

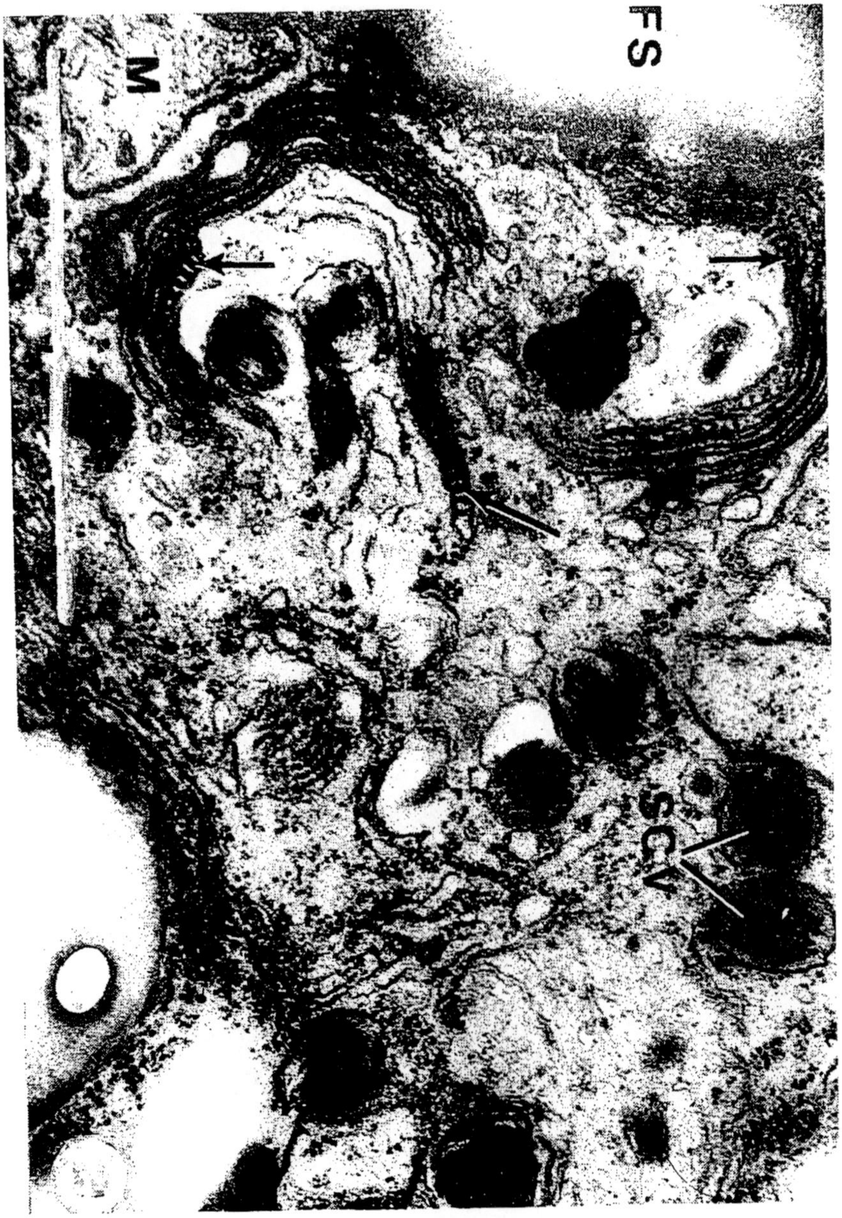

Fig. 76. Ultrastructure of carpospores of *Plocamiocolax pulvinata*.

(after Kugrens and Delivopoulos, 1986).

(Fs, floridean starch; M, mitochondria; SCV, striated cored vesicles)

Carpospores showing production of striated vesicles. Striations are visible in the cisternae of the maturing faces (arrow).

The latter two components, however, do not remain intact throughout carposporophyte differentiation, but begin to deteriorate (Fig. 71 A,B).

To begin with the process of carposporogenesis, the zygote nucleus is apparently transferred to the auxiliary cell which initiate gonimoblast cell production. The young auxiliary cell has a few starch grains, small vacuoles, compressed membrane bodies and numerous mitochondria (Fig. 72 A). These gonimoblast cells differentiate into storage or generative cell. Storage gonimoblast cells (SGC) are large and multinucleate, contain large quantities of starch and are located nearest to the auxiliary cell; whereas generative gonimoblast cells (GGC) are small, uninucleate, devoid of starch and form terminal lobes of carpospores. Besides, compressed membrane bodies and annulate lamellae are common in GGC cells (Fig. 72 B,C).

During the carposporophyte maturation, which takes place in 4 stages, the amount of starch in the SGC's decreases and eventually the auxiliary cell, as well as SGC's degenerate (Fig. 73 A,B). Generative gonimoblast cells cleave repeatedly to form carpospores which are interconnected through small pit-connections.

Stage I : Carpospores are recognized by their elongate shape, the formation of small dictyosomes, which begin to produce mucilaginous carpospore wall material, and the incipient formation of floridean starch (Figs. 73 C, 74 A.)

Stage II : Starch grains are abundant: numerous straight profiled dictyosome are present; a thick, electron-translucent carpospore wall layer and tightly compressed dark stained membrane bodies occur (Fig. 74 B). Carpospores generally show synchronous differentiation (Fig. 74 C). The function of numerous black bodies is unknown.

Stage III : A second mucilaginous carpospore wall layer is deposited either by curved, dictyosomes derived vesicles or from the endoplasmic reticulum (Fig. 75 A, B).

Satge IV : Dictyosomes produce striated and cored vesicles which condense to form dark staining adhesive vesicles (Fig. 76).

Mature carpospores consist of adhesive vesicles, centrally placed striated nuclei surrounded by endoplasmic reticulum, large quantities of floridean starch and a two layered mucilaginous carpospore wall. The function of adhesive vesicles was postulated to be spore adhesion. It is also proposed that the adhesive vesicles represent compact and stored wall material, to be utilized in the rapid formation of cell-wall during carpospore germination.

An important structure found between two carpospores is septal-plugs which remain intact throughout development and there is no apparent evidence to show that these structures either degenerate or begin to rupture while the carposporophyte is encased in the conceptacle.

Such studies which focus at ultrastructual details of carposporophyte and carposporogenesis of *Plocamiocolax*, an adelphoparasite, indicate post fertilization development changes are different from *Gigartina teedii*. In *Plocamiocolax,* most gonimoblast cells differentiate into carpospores, unlike *Gigartina teedii*, where only the terminal gonimoblast cells differentiate into carpospores. Although the carposporophyte structure is simple, carposporogenesis in *Plocamiocolax* is complex as compared to lower red algae like *Nemalion* and *Batrachospermum*. The results indicate that carposporogenesis is more similar among the Gigartinales, Rhodymeniales and Ceramiales than it is in the Nemalionales and Batrachospermales.

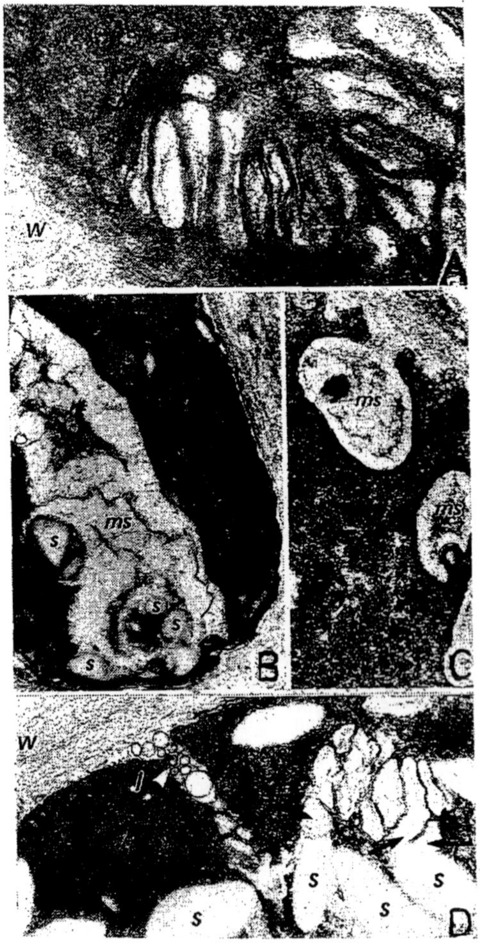

Fig. 77 A-D. Ultrastructure of carposporophyte of *Gigartina teedii*.

(after Tsekos, 1981)

(I, lomasome; ms, mucilaginous sacs; s, starch grains; w, wall)

A. Hypertrophied dictyosome producing considerable, inflated cisternae of a nearly mature carpospore.

B. Part of an immature carpospore with a central mucilage sac inside which many starch grains are present.

C. Mucilage sac of an immature carpospore, one of which is discharging its contents into the wall.

D. Part of a nearly mature carpospore with a lomasomelike body, which has released its contents into the cell wall. Hypertrophied golgi cisternae and vesicles (arrows) are continuous with starch grains.

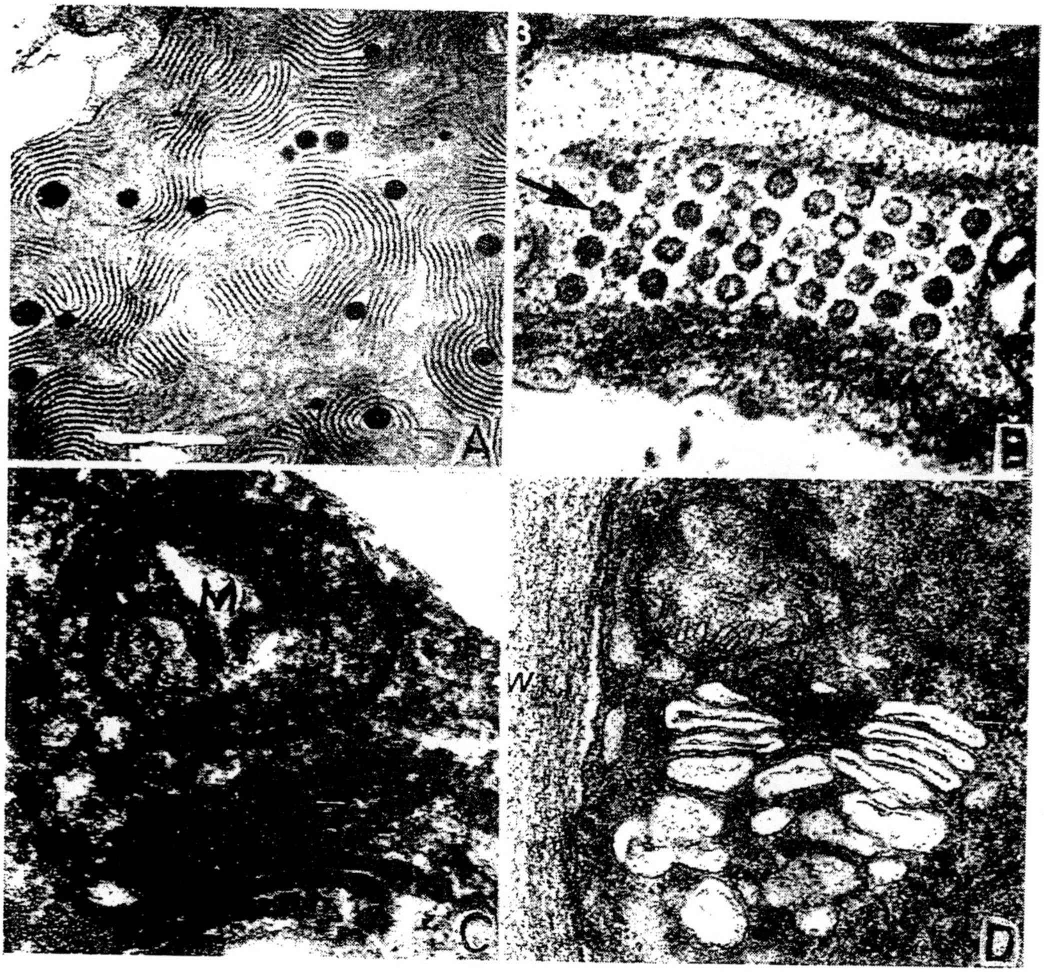

Fig. 78. A-D Ultrastructure of gall cells and carposporophyte

A. *Gracilaria epihippisora.* Ultrastructure of gall cell showing highly convoluted thylakoids. (after Apt and Gibor, 1991).

B. *Gracilaria epihippisora.* Ultrastructure of gall cells showing virus-like particles (arrow) arranged in a regular lattice.
(after Apt and Gibor, 1991).

C. *Gigartina teedii.* Ultrastructure of a carposporophyte showing typical form of dictyosome and its associated mitochondria in an auxiliary cell. (after Tsekos, 1981)

D. *Gigartina teedii.* Ultrastructure of carpospores showing part of an immature carpospore with hypertrophied dictyosome producing vesicles with fibrillar contents (after Tsekos, 1981).

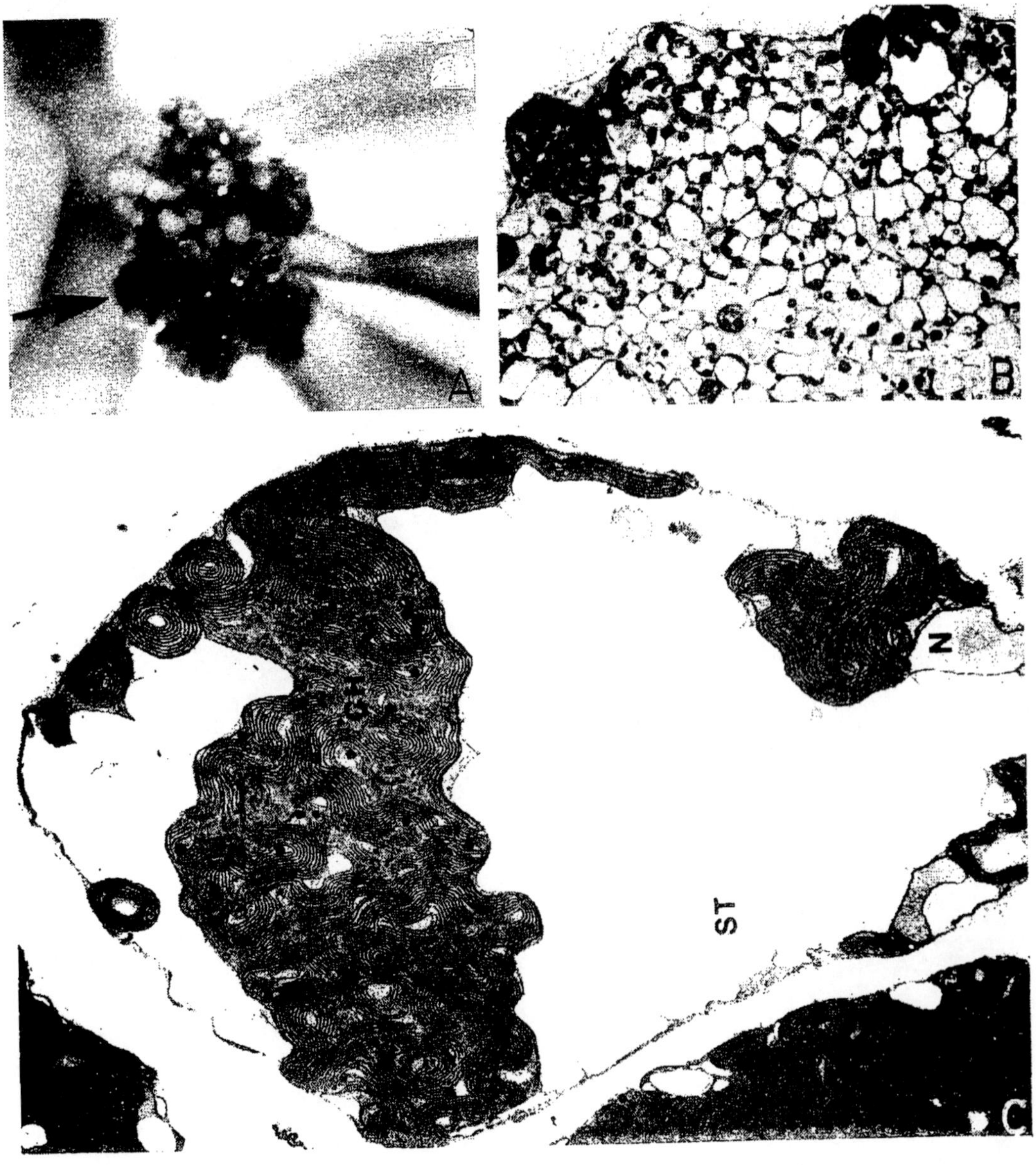

Fig. 79. A-C. Gall formation in *Gracilaria epihippisora*.

(after Apt and Gibor, 1991)

(CH, irregular shaped plastid; ST, starch grains; N, nucleus).

A. *In vitro* culture of gall to show highly convoluted gall growth attached to the thallus surface.

B. Cross section of gall-growth showing homogeneous nature of the tissue.

C. Cytoplasm of a gall cell showing large, irregularly shaped, plastid and starch grains scattered throughout the cytoplasm. Two multilobed nuclei are present.

In *Gigartina teedii,* the dictyosomes in differentiating carposporangia undergo extraordinary changes both in size and morphology. Dictyosomes of gonimoblast cells and auxiliary cell are composed of a stack of 5-7, straight cisternae and are approximately 0.18 um thick. From the structural and functional viewpoint, these are similar to dictyosomes in most other plants (Fig. 78 C). Hypertrophied golgi cisternae with fibrillar contents occur in immature and nearly mature carposporangia. (Figs. 77 A, 78 D). Secretory vesicles originate from the ends of cisternae. During the wall development of immature and nearly mature carposporangia, multivesicular, lomasome-like bodies are released from and peripheral mucilage sacs throughout plasmalemma (Fig. 77 B,C). Apart from other possible activities, these bodies seem to contribute to the plasticity of the expanding cell wall (Fig. 77 D). The dictyosome of older carposporangia are composed of a polarized stack usually with 7 - 11 cisternae; are very active with a semi-circular profile producing vesicles with an electron-opaque centre (Cored vesicle) (Tsekos, 1981).

GALL FORMATION

Various types of gall growths occur on a variety of marine algal taxa. These abnormal growth are characterized by host cell divisions and cell enlargements producing an abnormal and unorganized cell proliferation that projects outward from the thallus surface (Fig 79A). The resulting growth is morphologically similar to galls and tumours found in higher plants. The ultrastructure of this association is examined, in culture, in a strain of *Gracilaria epihippisora* (Apt and Gibor, 1991).

The characteristic gall ultrastructure features are :

(i) These growth can be excised and grown separately, where they retain an undifferentiated morphology between surface and internal cells, characteristics of normal thallus tissue (Fig. 79 B).

(ii) The gall tissues consists of a single morphological cell type without any differentiation between surface and internal cells as is characteristic of normal thallus tissue (Fig. 79 B).

(iii) Gall cells contain the usual complement of organelles and a prominent vacuole, although there are several other distinct features.

(iv) The large multilobed plastids have an extensive proliferation of thylakoid membranes, which form an arrangement of loops and spirals (Fig. 79 C). The thylakoids are highly convoluted (Fig. 78 A)

(v) The cell wall on the exterior surface of the gall growth is significantly thin than that of a thallus with normal morphology.

(vi) Presence of virus-like particles (Fig. 78 B).

The gall-growths of *Gracilaria epihippisora* possess several features that are previously unreported in red algae. An obvious feature is the undifferentiated and unordered cell enlargement (intercalary cell division) unknown in typical members of the order Gracilariales. The gall-growths are autonomous when excised from the parent thallus. This is unlike other marine algae but identical to higher plants. The gall-growths do not revert to a normal thallus morphology and maintain only gall morphology. This indicates that the biochemical and physiological constraints of the gall-growth is stable.

Gall originates from outer cortical layer. However, in gall formation, cell growth and cell division are not coordinated, and the cells have apparently lost their ability to differentiate as in normal development. As a result, general development is pronounced departure from the normal morphogenetic restraints of the thallus.

HISTOCHEMISTRY

Various histochemical studies have been done on various members of this order which are potent source of polysaccharides namely agar-agar and carrageenan. These studies can be broadly classified under the following:

(1) Polysaccharides

(2) Proteins and Polyphenols

(3) Extracellular secretions in *Nizymenia australis*

(4) Biomineralization

POLYSACCHARIDES

The Agar-agar producing plants are called Agarophytes. Wall galactans of the agarophytes contain disaccharide repeating units of 3-linked, β-D galactose and 4-linked, α-galactose. In the carrageenans, the galactose units are substituted in a characteristic manner with sulfate groups.

Agar-Agar

The word agar-agar is of Malay origin and is used for the food-gel obtained from *Eucheuma* sp. The polysaccharide from this alga is also used to prepare media for the growth of bacteria. Later, all gelling polysaccharides are called Agar (Yaphe, 1984). Polysaccharides extracted from *Gelidium, Ahnfeltia, Pterocladia* and *Gracilaria* have identical structure.

The final products are called as Danish (*Furcellaria*) agar, British (*Chondrus*) agar, Russian (*Phyllophora*) agar and *Hypnea* agar. These preparations contain different types of carrageenans.

Composition of Agar

It consists of two components (i) agarose which consists of (1-->4)-linked 3,6-anhydro-α-L-galactose alternating and (1-->3)-linked, β-D-galactose, forming the repeating disaccharides neoagaribiose. It is a neutral polysaccharide (ii) Agaropectin consists of the same repeating units as agarose with some of the 3,6-anhydro-α-L-galactose being replaced by L-galactose-sulfate.

Duckworth and Yaphe (1971), using experiments on DEAE-Sephadex A-50 revealed that Arake's (1965) interpretation of agar consisting of two fractions is an oversimplification and proved that agar consists of agarose, agarose molecule substituted with methoxyl sulphate; pyruvate groups and sulphated galactans.

A brief description of different components of agar are as follows:

(i) Agarose (neutral)

(ii) Charged agarose

(iii) Sulfated galactan.

(i) Agarose (neutral) - Pure neutral agarose is at a low ebb in agars (Duckworth and Yaphe, 1971). Agarose was defined as a mixture of agar molecules having lowest charge content and, therefore, greatest gelling ability. The repeating D-galactose and 3,6-anhydro L-galactose in agarose may be replaced by 6-0 methyl D-galactose and 2-0 methyl-3,6-anhydro-L-galactose which increases the gelling temperature (Guiseley, 1970).

(ii) **Charged agarose** : The concentration of charged agarose molecules gradually increases due to substitution of D-galactose by D-galactose-sulfate or 4,6-O-(1-carboxy-ethylidene)-D-galactose and of 3,6-anhydro-D-L-galactose by L-galactose sulfate. The gel strength decreases with the increase in concentration of sulfate and decrease in concentration of 3,6-anhydrogalactose.

Fractionation of mucilage from *Gracilaria verrucosa* on Dowex X-2 shows extremes in structure in nonsulfated, highly methylated and a nonmethylated, highly sulfated (10%) charged agarose (Izumi, 1972).

(iii) **Sulphated galactan** : This fraction of the sugar has no 3,6-anhydro-L-galactose units. Food grade agars obtained from *Gracilaria* sp. (Duckworth *et al.*, 1971) have lower gel strength and a major fraction of agarose molecules are substituted with sulphate or pyruvate. The highly viscous mucilage from a *Gloiopeltis furcata* (Hirase and Watanabe, 1972) represents an agar with highly sulphated agarose molecule.

Agar from *Gracilaria*

Fractionation studies by column chromatography on DEAE-Sephadex A-50 (Duckworth *et al.*, 1971; Young *et al.*, 1971) show that *Gracilaria* agar is composed of different molecules. On the basis of chemical structure they are divided into different groups.

Group I has high gel strength and neutral agarose molecules are a major fraction.

IA : This group has a low methoxyl concentration in agarose e.g. *G. verrucosa* (Friedlander *et al.*, 1981).

IB : Includes agarose with high methoxyl content as reported in *G. deblis* (Duckworth *et al.*, 1971).

Group II Gel strength of agar, of this group, is lower than group I. Agarose molecules substituted with sulphate and pyruvate groups are major units in the agar.

II-A : Low concentration of pyruvate and methoxyl groups. Charged density is due to sulphate as in *G. verrucosa* (Friedlander *et al.*, 1981).

II-B : Sulphate is responsible for charged density. Methoxyl present as 6-0 methyl-D-galactose and/or 4-0 methyl-3,6 anhydro-L-galactose in *G. verrucosa*.(Usov *et al.*, 1979).

II-C: Methoxyl present as 2-0 methyl-3, 6-anhydro-1-galactose as in *G. eucheumoids* (Yane *et al.*, unpublished).

11-D: Charge density is due to pyruvate as in the agar of *G. compressa* (Duckworth *et al.*, 1971)

Group III has low gel strength and charged density is due to sulphate eg. *G ferox* Dockworth *et al.*, 1971).

Extraction and determination of heterogeneity of agar polymers

An extraction procedure was developed based on the differential solubility of agar polymers in ethanol and water at different temperatures.

The heterogeneity of agar polymers in the cell walls of *Gracilaria tenuistipitata, G. eucheumoides,* and *G. blodgetti* was demonstrated by sequential extraction of the algae in water at 22° C; in boiling with 80%, 60% 40% and 20% ethanol-water solutions; and in water in 100° C and 121° C (Lahaye *et al.*, 1985). Agar polymers in each extract analysed by ^{1}H and ^{13}C nuclear magnetic resonance spectroscopy (NMR). The agar polymers extracted from the three algae with water at 22° C had a high concentration of repeated units substituted by alkali-labile sulfate. Methylated agar polymers were extracted with boiling 80% and 60% ethanol from *G. tenuistipitata* and *G. eucheumoides.* The former having a high content of 6-0-methyl-D-galactose and a ion concentration of 2-0-methyl-3-6-anhydro-L-galactose and the latter consisting mainly of 2-0-mythyl-3, 6-anhydro-1-galactose. Sulfated agar polymers (alkali-stable sulfate) were extracted from *G. tenuistiptata* and *G. blodettii* with boiling 20% ethanol and water (100° C and 121° C). Floridean starch was detected by ^{13}C nuclear magnetic resonance in the extracts from *G. tenuistipitata* and *G. eucheumoides.*

Variation in Yield

Penniman (1977) reported little difference in agar yield between vegetative, cystocarpic and tetrasporic plants. Agar extracted from tetrasporic and male gametophyte plants of *G. bursa-pastoris* and *G. coronopifolia* show no difference in yield and gel strength (Hoyle, 1978). Kim and Henriquer (1979), however, noted in *G. verrucosa* distinct agar gel strength between cystocarpic and tetrasporic plants.

Carrageenan

Carrageenan is the term used to denote the polysaccharides extracted through hot water from member of the family Gigartinaceae and specially *Gigartina stellata, Chondrus ocellata, C. crispus.* O'Neill *et al.* (1955) showed that k-carrageenan is composed of D-galactose, 2 6-anhydro-D-galactose and sulfate groups in the ration of 6:5:7. They proposed a structure for k-carrageenan consisting of a long chain of alternating β-D-galactospyranse-4-sulfate and 3,6-anhydro-α-D-galactopyranose unity. The galactose residues are linked through C-1 and C-3, whereas the hydrogalactose is lined through C-1 and C-4.

(i) Chondrus ocellata - Araki and Hirase (1956) observed that *Chondrus ocellata* extracts are composed of the same 1:3-linked-dissaccharides units as k-carrageenan, but they obtained no evidence concerning the position of sulfate groups.

(ii) Furcellaria fastigiata - Kylin (1943) found that furcellarin (the mucilage from this plant) is potassium magnesium salt of galactan containing pentoses. It was suggested that mucilage is like k-carrageenan, albeit with a lower sulfate : hexose ratio of approximately 2:5.

(iii) Hypnea spicefera - Clingman and Nunn (1959) obtained a polysaccharide through hot water extraction. Analysis of the extract showed 47.4% galactose; 31.4%, 3,6-anhydro-D-galactose and 21.2% So_3Na. This polysaccharide was found to resemble k-carrageenan but it has a different infrared spectrum Yaphe (1959) also found 97% k-carrageenan like polysaccharide in extracts of *Hypnea musciformis.*

(iv) Eucheuma species : Two types of mucilages are found which resemble k-carrageenan, containing galactose, 3,6-anhydrogalactose and sulfate. Both mucilages have sugar and sulfate ratio of 1:1 (Zaneevelodg, 1959).

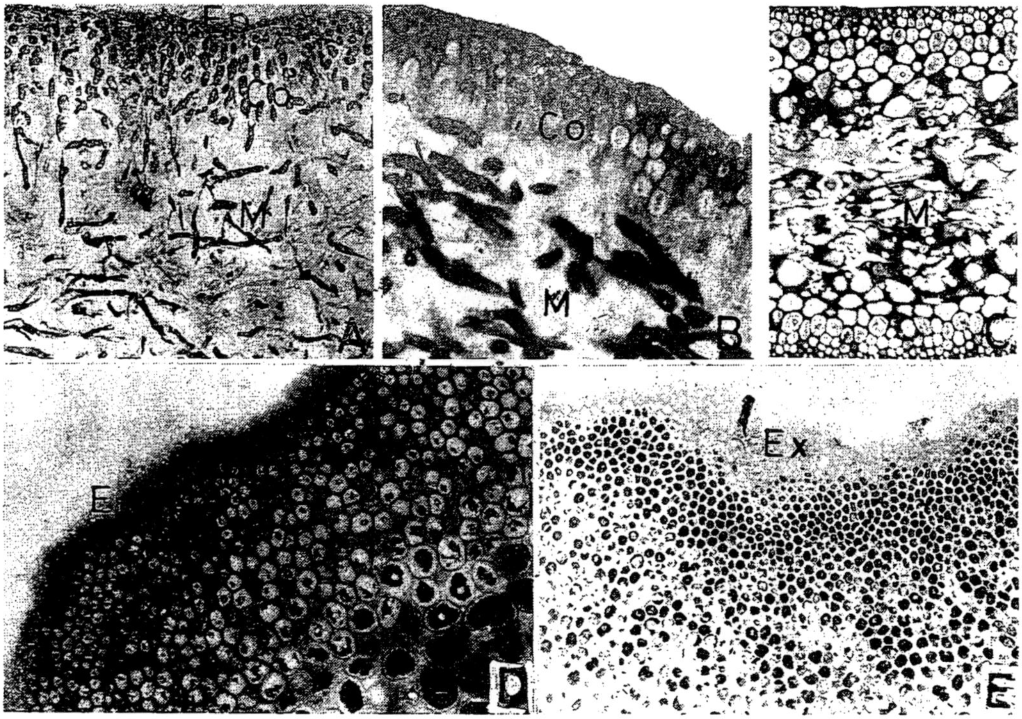

Fig. 80. A-E, Histochemical studies of *Nizymenia australis.*

(Co, cortex; Ep, epidermis; Ex, extracellular layer; M, medulla)

A. Transverse section of holdfast showing epidermis, cortex, medulla. Cortex is undifferentiated whereas medullary cells shwo loose arrangement. The holdfast cells are rich in proteins. (after Vijayaraghavan and Bhatia, 1991.)

B. T.S. of holdfast showing distribution of insoluble polysaccharides in cortex and medulla. (after Bhatia and Vijayaraghavan, 1990b).

C. T.S. of thallus showing distribution of polysaccharides and polyphenols in inner cortex and medulla. (After Bhatia and Vijayaraghavan, 1990b).

D. L.S. of thallus showing extracellular secretions at the growing apex. The extracellular covering stains deeply with PAS reagent. (after Vijayaraghavan, and Bhatia 1990a).

E. L.S. of thallus to show a veneer of extracellular proteins. (after Bhatia and Vijayaraghavan, 1990a).

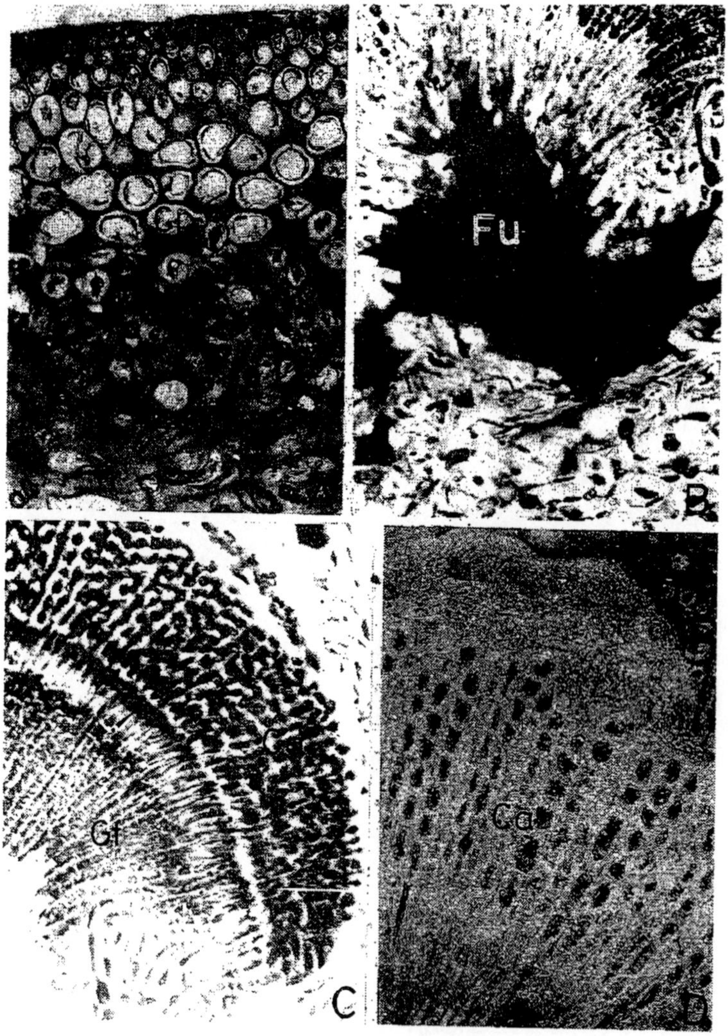

Fig. 81. A-D Histochemistry of carposporophyte of *Nizymenia australis.*

(after Vijayaraghavan and Bhatia 1990b)

(Ca, carposporangia; Cp, carpogonium; Cs, carpospores; Fu, fusion cell; Gf, gonimoblast filament)

A. Post fertilization carpogonium showing lysis of trichogyne (arrow).

B. Portion of cystocarp showing finger-like projections of large fusion cell which is rich in protein .

C. Portion of cystocarp showing carposporangium and carpospores which stain intensely for proteins and polyphenols.

D. Portion of cystocarp showing carposporangia which contain insoluble polysaccharides.

Histochemical Localization of Polysaccharides

In *Nizymenia australis*, the cell walls of epidermis, cortex and medulla of both the thallus and the holdfast are rich in wall polysaccharides. The epidermis is bereft of polysaccharide grains. Outer cortical cells contain a few centrally located grains; whereas the inner cortical and medullary cells are filled with this metabolite. The cortical cells of the axis contain sulphated and carboxylated polysaccharides (Fig. 80 B, C; Bhatia and Vijayaraghavan, 1990b). Intercellular secretions are rich in polysaccharides.

Female Reproductive System

Cytoplasm of carpogonium, hypogynous and supporting cells, contain a few peripheral polysaccharide grains. The cytoplasm of these cells shows honeycomb-like structures and takes purple stain with TBO (Fig. 65B). In cystocarp, the fusion cell cytoplasm has modicum of polysaccharides grains. The cell wall of gonimoblast filaments are rich in wall polysaccharides. Carposporangia are filled with polysaccharide grains but their walls are poorly stained (Fig. 81 D). pericarp cells are rich in wall polysaccharides.

Spermatangia

The thick outer covering that overlayers the spermatangial sorus stains intensely for insoluble, sulphated and carboxylated polysaccharides (Fig. 63 B,C; Vijayaraghavan and Bhatia, 1991). The cells of spermatangial filaments are rich in wall polysaccharides and their cytoplasm contains polysaccharide grains. The spermatium has a PAS positive membrane but cytoplasm is devoid of starch grains (see Bhatia and Vijayaraghavan, 1990b).

PROTEINS AND POLYPHENOLS

In *Nizymenia australis*, the epidermis and outer cortical cells are filed with cytoplasmic proteins. In the inner cortex only the peripheral layer of cell cytoplasm and refractive bodies stain intensely for proteins. The cytoplasm of holdfast cells are also protein-rich (see Vijayaraghavan and Bhatia 1991).

Carpogonium and Cystocarp

The cytoplasm of the carpogonium, hypogynous, supporting and auxiliary cells stain positively for proteins. In the cystocarp, the fusion cell, gonimoblast filaments, carposporangia and pericarp show differential staining for proteins. The cytoplasm of the large fusion cell is rich in proteins and polyphenols (Fig. 81 B). Both carpospores and carposporangia stain extensively for proteins (Fig. 81 C) as well as for insoluble polysaccharides (Fig. 81 D). In the pericarp cells cytoplasmic proteins show polarized distribution Bhatia and Vijayaraghavan, 1990b).

EXTRACELLULAR SECRETION IN *NIZYMENIA AUSTRALIS*

The extracellular secretions in this taxon contour the outer tangential walls of the epidermis. These are present in copious amount at the growing apex. The extracelluar covering stains intensely with PAS reagent (Fig. 80 D) and shows deep purple metachromasy with TBO. These histochemical localization suggests that the secretions contain a mixture of sulphated and carboxylated polysaccharides. The extracellular covering is also rich in proteins (Fig. 80 E; Bhatia and Vijayaraghavan, 1990a). These extracellular secretions form a veneer over the spermatangial sorus (Fig. 80 E). These extracellular mucilage covering has a dual function namely: (i) protection from algal epiphytes and pathogens; and (ii) help in spermatia release.

BIOMINERALIZATION

The plants of *Nizymenia australis* are firmly attached to the rocks by distinctive holdfast which shows an upper cylindrical, uncalcified portion and a lower calcified portion. The noncalcified portion shows the typical anatomy of a mature axis. The calcified portion shows a network of hallow tubes of calcium carbonate formed by holdfast cells in the intercellular spaces between them, resulting in the formation of a calcareous skeleton. After attachment to the rocks holdfast cells appear to be withdrawn into these calcified tubes which grow further additional deposition of calcium carbonate. The growth of calcareous tubes creates a partial vaccuum inside the skelton. Due to this vaccuum, the holdfast is firmly attached to the rocky substratum (Bhatia et al., 1990).

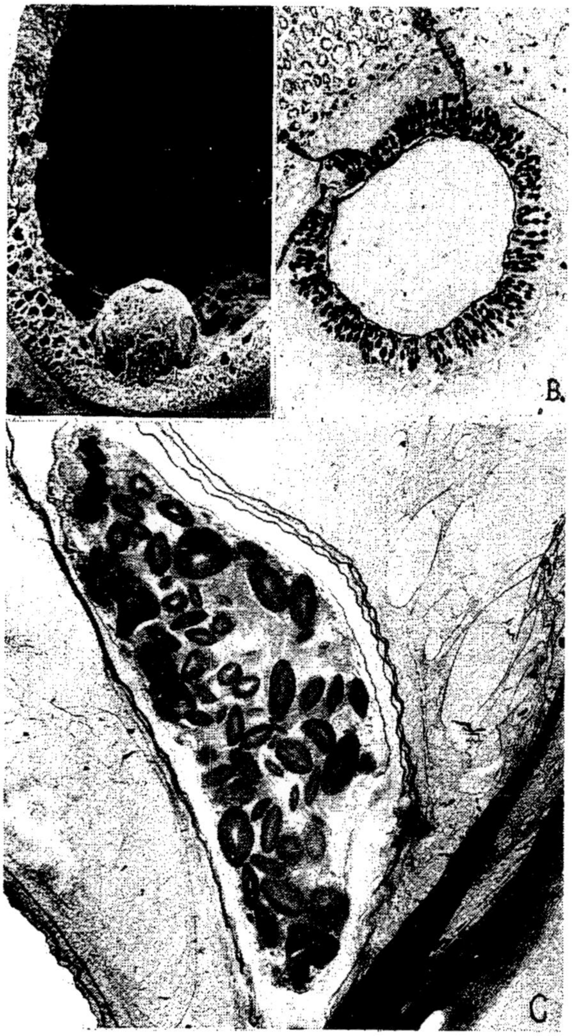

Fig. 82 A-C. Cystocarp in *Soleiria robusta*
(after Nashier, 1996)

(csg, carposporangium; cys, cystocarp; ep, epidermis; fs, floredian starch grain; fu, usion cell; ic, inner cortex; im, intercelluar mucilage; me, medulla; mf, medullary filament; oc, outer cortex).

A. Scanning electron micrograph of fractured thallus to show mature cytocarp embeded in the medulla. The medullary filaments are elongate and loosely arranged around the cystocarp.
B. Longitudinal section of thalli to show mature cystocarp. embedded in the medullary region. Carposporangia are gorged with floredian starch grains and are arranged in chains around the fusion cell.
C. Mature carposporangium showing seven layered thick cell wall. The cytoplasm is replete with floredian starch grains that are of various sizc and shape.

6 Rhodymeniales

CLASSIFICATION

Schmitz (1892) originally proposed the order Rhodymeniales comprising of six families Sphaerococcaceae, Rhodymeniaceae, Delesseriaceae, Bonnemaisoniaceae, Rhodomelaceae and Ceramiaceae.

Oltmanns (1904) created a new order Ceramiales to separate those taxa where the auxiliary cell is not formed until after fertilization, and accommodated the families Delesseriaceae, Rhodomelaceae (including Dasyaceae) and Ceramiaceae.

Kylin (1923) removed the family Bonnemaisoniaceae from Schmitz's Rhodymeniales to the order Nemaliales (as Nemalionales) because its members lacked a typical (generative) auxiliary cell and had an assumed haplontic type of life history. Sjostedt (1926) removed the family Sphaerococcaceae to a new order Sphaerococcales, which Kylin (1928) placed in the order Gigartinales.

Bliding (1928) distinguished two families for the remaining Rhodymeniales:

1) **Champiaceae** : Comprising taxa in which the thallus (or most of it) is hollow with longitudinal filaments bordering the cavity, and with tetrahedrally divided tetrasporangia.

2) **Rhodymeniaceae** : Comprising taxa where the thallus is solid, or if hollow, lacks longitudinal filaments, and with cruciately divided tetrasporangia.

Kylin (1931) placed all taxa in the two families of Bliding (1928), but subdivided them into subfamilies.

1. **Champiaceae:**
   a) Lomentarieae.
   b) Champieae.
2. **Rhodymeniaceae:**
   a) Faucheae
   b) Rhodymenieae
   c) Hymenocladieae.

Sparling (1957) regarded the order as consisting of only two families, the Rhodymeniaceae (with two subfamilies Rhodymenieae and Hymenocladieae) and the Champiaceae.

Lee (1978) accepted Sparling's (1957) Rhodymeniaceae with the two subfamilies; accepted Guiry's (1974) Palmariaceae as a separate family (in which carposporophyte generation is absent); and elevated Kylin's (1931) two subfamilies Lomentarieae and Champieae to the family status as Lomentariaceae and Champiaceae. In Lomentariaceae he included the taxa with 3-celled carpogonial branches, central cavities septate by multirowed cellular diaphragms, almost all gonimoblast cells functioning as carposporangia, and tetrasporangia produced terminally from cortical cells. In Champiaceae he placed those taxa in which the carpogonial branch is 4-celled; central cavity is septate by single row cellular diaphragms, carposporangia develop only from the surface gonimoblast layers, and development of intercalary tetrasporangia in dispersed sori from the cortical cells.

Guiry (1974) segregated the Palmariaceae from the Rhodymeniaceae on the basis of tetrasporangium formation (presence of a generative stalk cell), as well as presumed absence of carpogonia and carposporangia and formation of spermatangia in pairs on the spermatangial parent cell. Guiry and Irvine (1978, 1981) subsequently removed the Palmariaceae from Rhodymeniales and established a new order Palmariales. Further justification for placing this family in a separate order was provided by culture studies of the life history and sexual reproduction in members of Palmariaceae, including report of microscopic female gametophyte in some; and ultrastructural study of pit-plugs (Pueschel and Cole, 1982) distinguishing members of Rhodymeniaceae (pit-plugs bereft of caps) from the Palmariales (pit-plugs with cap-layers) (Hawkes and Scagel, 1986).

In this chapter the classification of Hawkes and Scagel (1986) is followed:

(a) **Champiaceae** — Tetrasporangia tetrahedrally divided, hollow portions always present in the thallus and provided with longitudinal filaments around the periphery of the lumen; carpogonial branch 4-celled; carposporangia formed only from superficial gonimoblast cells.

(b) **Lomentariaceae** — Tetrasporangia tetrahedrally divided; thallus as in Champiaceae: carpogonial branch 3-celled; carposporangia formed from almost all gonimoblast cells.

(c) **Rhodymeniaceae** — Tetrasporangia cruciately divided (except *Hymenocladia* and *Hymenocladiopsis*, which have tetrahedrally divided tetrasporangia); thallus solid or hollow, hollow portions lack longitudinal filaments around the periphery of the lumen.

MORPHOLOGY

The thalli are generally free living or epiphytic; vary in length from about 10 mm to over 1 m; soft to cartilaginous or occasionally mucilaginous (Hawkes and Scagel, 1986). Thalli may be unbranched, or regularly branched. They may be peltate (*Maripelta* sp.), stellate (*Sciadophycus* sp.), foliose or ligulate. The blades may be sessile or arise from branched or unbranched stipe. Some members consist of hollow sacs as in *Botryocladia*, or extensive hollow axes divided by septa as in *Lomentaria*, *Champia* and *Fryeella* (Fritsch, 1945). *Minium* is a crustose form (Hawkes and Scagel, 1986).

Parasitic forms like *Faucheocolax*, *Rhodymeniocolax*, *Gloiocolax* and *Champiocolax* grow on closely related genera and are extremely reduced and tuberculate, vary in hue from red to white, with filaments of the parasite penetrating the host cells.

In **Champiaceae** the thalli are multiaxial, cylindrical or compressed, and branched with all or some portions being hollow. Thalli are septate through single layered, cellular, transverse partitions called as transverse diaphragms. In *Gastroclonium subarticulatum* (Fig. 83A) the thallus arises from a disc-shaped holdfast and is irregularly to subdichotomously branched. Terete upright axes produce rhizomes, which form a mat of intertwined axes. Main axes are solid, terete and bear hollow, cylindrical branchlets toward their apices (Hawkes and Scagel, 1986). *Champia parvula* forms dense, bushy tufts arising from a number of creeping threads attached by multicellular haptera. The branches often arise singly. Steele *et al.* (1986) carried out genetical characterization of spontaneous mutants in *C. parvula* and suggested that both the morphology and pigmentation of the plant is affected by mutant genes. The mutant gene for pigmentation is designated as *Yell* and acts as a recessive lethal during carposporophyte development. In *Champia somalensis,* (Bhalla, 1996 unpublished) several hollow, erect axes arise from a multicellular haptera (Fig. 84 C) and cystocarps arise as tuberculate protrusions (Fig. 84 D). *Chylocladia kaliformis* has a jointed habit in which the fistular thallus is constricted at the points of origin of whorled branches. Thallus is attached to the substratum by a minute disc.

In **Lomentariaceae,** thallus is multiaxial, slightly constricted at septa, cylindrical or flat, branched, and at intervals hollow with multilayered cellular septa. In *Lomentaria hakodatensis,* the brownish-purple and cylindrical thallus is opposite to irregularly branched and forms a procumbent intertangled turf or clum (Fig. 83 B). The thallus is attached by a stoloniferous system of creeping branches, fastened by disc-shaped holdfast. *L. clavellosa* shows distichous branching on an unconstricted thallus which is throughout hollow. In *L. articulata,* the thallus is solid and constricted at the point of branching (Fritsch, 1945). Lee *et al.* (1988) described *Binghamiopsis caespitosa* gen. et sp. nov. from Cypress Point, California and Baja California, Mexico. The cylindrical to compressed thalli bear dichotomous, trichotomous or alternate to opposite branches with multiaxial meristems.

The family **Rhodymeniaceae** includes both free living and parasitic genera. The thalli exhibit diverse habits and may be crustose, foliose, cylindrical or peltate. The thallus may be solid or hollow, with the hollow portions lacking longitudinal filaments around the periphery of the lumen that are characteristic of Champiaceae and Lomentariaceae (Price and Kraft, 1991). The free living genera are *Botryocladia, Fauchea, Fryeella,* and *Rhodymenia. Botryocladia* consists of a solid, terete and branched axis which terminates in swollen, mucilage filled sacs of varying size (Fig. 83 F). The solid stipe arises from a dichotomously branched rhizoidal base in *B. ardreana* (Brodie and Guiry, 1988), or from a small, discoid holdfast as in *B. pseudodichotoma* (Hawkes and Scagel, 1986). *Botryocladia leptopoda* is one of the stately marine plants of Indian coast and ranges between 30 to 50 cms. in height. The thallus is purplish red and is attached to the substratum by a well defined holdfast (Fig. 84 A). The cystocarps are seen as small dots on the surface of mucilage-filled sacs (Fig. 84 B; Bhalla, 1996 unpublished).

In *Fauchea* (Fig. 83 C), the thallus is flat, flabellate, dichotomously or irregularly branched and lubricous. In *Fryeella,* the erect or prostrate, flat, foliaceous and dichotomously lobed thallus has thin, mucilage filled internal cavities that are segregated by curved internal septa (Hawkes and Scagel, 1986).

In *Rhodymenia,* the thallus habit shows great variations. In *R. pertusa,* it may be flat and blade-like or in *R. californica* dichotomously to irregularly branched, erect or spreading, with a disc-shaped holdfast. The thallus is usually stipitate, with or without cylindrical basal stolons. In *R. pacifica,* the thallus is often broad and blunt at apices. In *R. pertusa,* from sheltered coasts, the thallus is typically large and membranous, whereas in those from exposed areas it is small and coriaceous. The thallus shape varies from obovate to elliptical, and is typically perforate over most of the surface (Fig. 83 D).

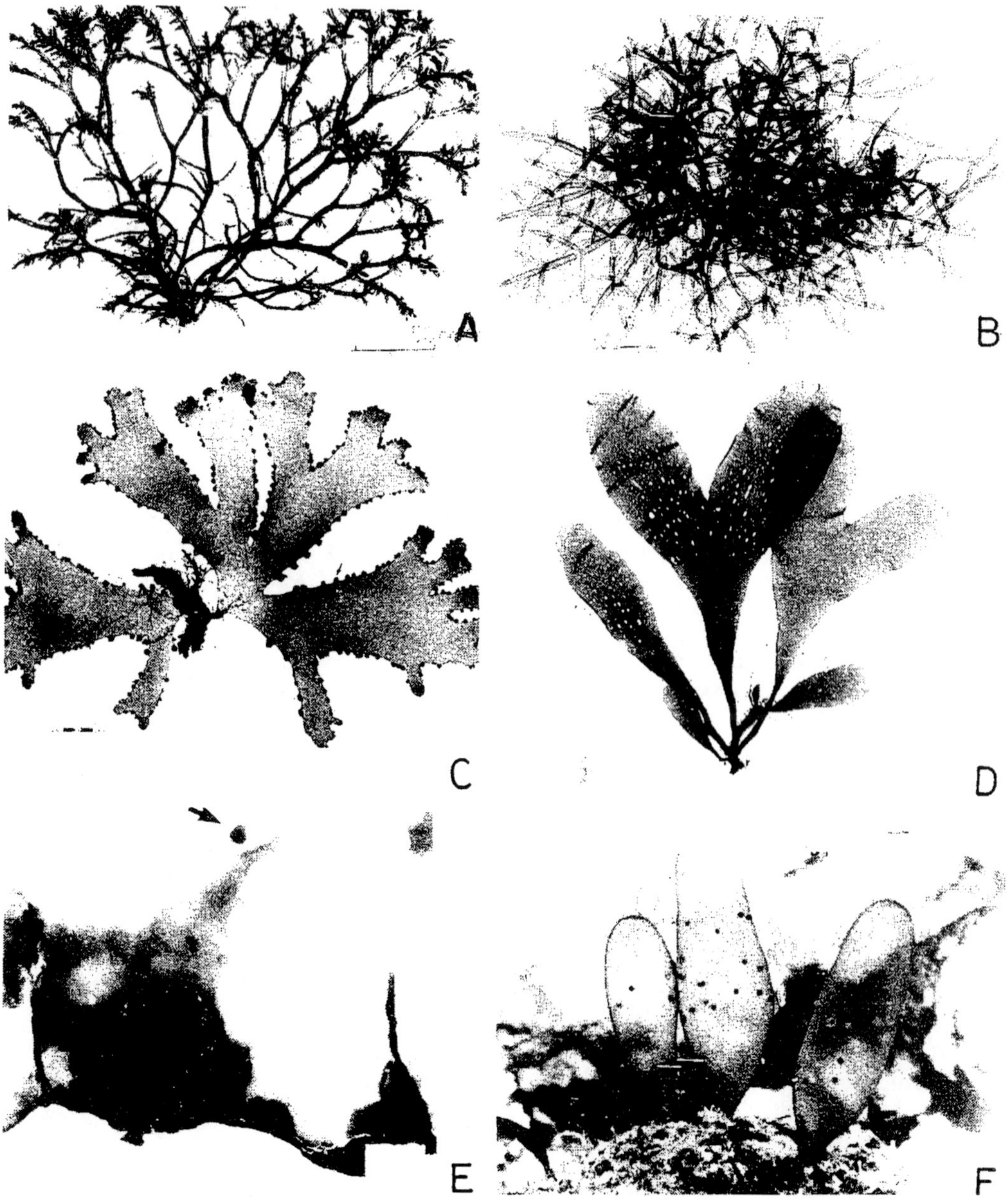

Fig. 83. A-F. Morphology.

(A-E : after Hawkes and Scagel, 1986b; F : after Ballantine, 1985).

A. *Gastroclonium subarticulatum* : Habit of vegetative plant.

B. *Lomentaria hakodatensis* : Habit of vegetative plant.

C. *Fauchea fryeana* : Gametophyte bearing marginal cystocarps (arrow).

D. *Rhodymenia pertusa* : Young vegetative thallus.

E. *Minium parvum* : Gametophytic crust on an intergeniculum of *Calliarthron tuberculosum*. Cystocarp (arrow) prominently raised above the crust.

F. *Botryocladia wynnei* : Habit of three cystocarpic plants each with solitary vesiculate branches growing on the substrate.

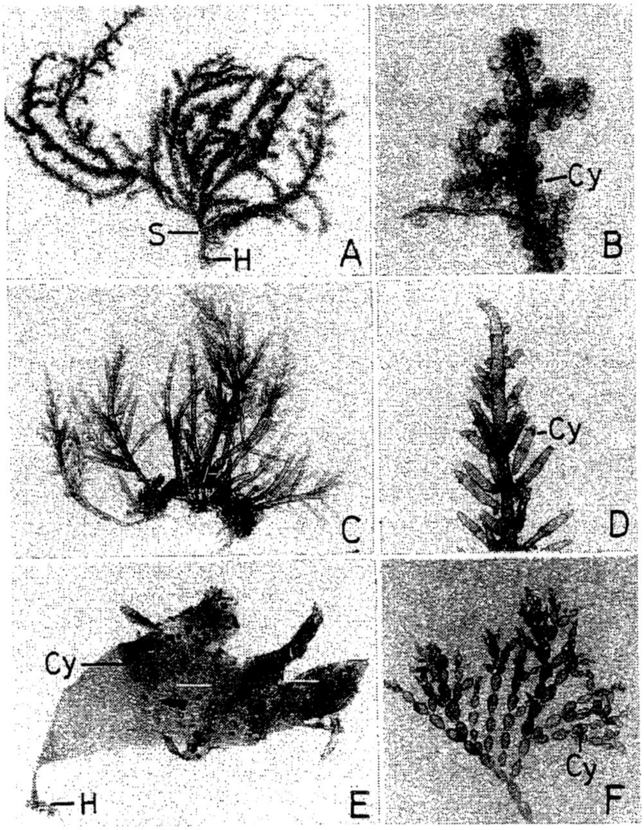

Fig. 84. A-F. Morphology. (after Bhalla, 1996, unpublished).
(Cy, cystocarp; H, holdfast; S, stipe)

A. *Botryocladia leptopoda* : Thallus showing a well developed holdfast and a solid main axis. The dichotomously branched stipe bears several branched axes, each of which is surrounded by several mucilage-filled, sac-like structures.

B. *B. leptopoda* : Portion of the thallus showing cystocarps.

C. *Champia somalensis* : Thallus is bushy in appearance with several hollow, erect axes arising from a multicellular haptera. Each axis is filled with mucilage and shows distinct diaphragms.

D. *C. somalensis* : Portion of the thallus showing cystocarps.

E. *Rhodymenia australis* : Thallus is blade-like, dichotomously branched and has a small holdfast. Cystocarps, of various sizes, occur as protrusions on the thallus surface.

F. *Coelarthrum opuntia* : Portion of the thallus showing regular constrictions. The presence of distinct nodes and internodes gives a beaded appearance to the thallus. Cystocarps are seen as dots on the thallus surface.

In *Rhodymenia palmata f. mollis,* the nature of holdfast produced in culture was unlike that of the plants observed in nature. In culture, numerous blades were produced from each holdfast whereas in naturally occurring tetrasporophytes, only a single blade on each holdfast was observed (Sparling, 1961). *Rhodymenia australis* occurs in inconspicuous habitats that are cut off from direct sunlight. They are usually seen in large and deep rock pools. The thallus is blade-like and cystocarps are scattered over the thallus as dense, deep-red irregular spots (Fig. 84 E; Bhalla 1996, unpublished).

Coelarthrum opuntia is a beautiful purplish red alga of the Indian coast and reveals minute basal disc that gives rise to a cluster of beaded and jointed fronds. Internodes are separated by small, slender filiform nodes. Cystocarps are minute and scattered irregularly on the internodes (Fig. 84 F; Bhalla 1996, unpublished).

In *Cordylecladia erecta,* the fronds are terete or slightly compressed or sparingly branched. Tetrasporangial plants are recognized by the formation of a swollen apical pod or spindle-shaped structure. The male plants have spermatangia occurring in elongated pod-like structures that gives a pointed or aristate appearance to the plants. Hairs are totally absent from spermatangial and tetrasporangial plants. Sterile hairs associated with depression or pits are sites of procarp formation. They are of variable length and develop from cortical cells (Brodie and Guiry, 1988).

In *Maripelta atlantica,* the plant shows a lobed holdfast, and an erect, unbranched stipe bearing a terminal peltate blade. In old plants, the stipe is 2-5 mm in diameter, 60 mm tall, irregularly cylindrical and with annular scars. The blade is nearly circular, upto 70 mm in diameter, 0.25-0.50 mm thick and often slightly depressed in the centre (Eiseman and Moe, 1981).

Parasitic forms exhibit extremely reduced development. In *Faucheocolax attenuata,* the thallus consists of a basal stalk that supports short, thick branches or lobes and grows on the surfaces and margins of *Fauchea spp. Rhodymeniocolax* parasitizes *Rhodymenia* spp. forming diminute (2-4 mm diameter) tuberculate outgrowths and short basal stalk and many short, thick branches (Hawkes and Scagel, 1986).

Ceratodictyon is a symbiotic taxon. *C. spongiosum,* in nature, grows symbiotically with the sponge *Sigmadocia symbiotica.* The thallus consists of a reticulate network of algal axes surrounded and covered by the sponge symbiont. The symbiotic association forms irregular cushions, 15-30 cm in diameter and 5-15 cm in height. The surface of most symbionts are punctuated by scattered, and prominent holes (called 'oscula') that are the exit pores for water currents set up by the sponge Choanocytes (Price *et al.*, 1984). The specimens are attached to the solid substrate by slightly splayed but otherwise unmodified bases or lateral extensions of major axes, there being no expanded disc or thickened holdfast pads (Price and Kraft, 1991). *Minium* is a crustose genus (Fig. 83 E) and the thallus is composed of basal layer of pseudodichotomously branched short and upright filaments.

ANATOMY

The thallus is multiaxial and pseudoparenchymatous, differentiated into an outer cortex and an inner medulla. The cortex is composed of either a single layer or a few anticlinal rows of small pigmented cells. Unicellular hair cells are sometimes scattered over the thallus surface. The medulla is usually filamentous, or occasionally pseudoparenchymatous, and it may be 1 to 8 cell layers thick, The medullary cells are large, isodiametric, hyaline and frequently vacuolate. The change in cell size from cortex to medulla can be abrupt or gradual. Internal diaphragms are composed of small cells or filamentous multicellular septa. Small internal secretory gland cells (vesicle cells) are frequently present in the medulla (Hawkes and Scagel, 1986).

CHAMPIACEAE

The multiaxial thallus may be wholly or partially hollow and septate through transverse diaphragms which is composed of a single layer of cells. Longitudinal filaments line the periphery of the lumen which is filled with mucilage. In *Gastroclonium subarticulatum,* abundant colourless hair cells are produced from the cortex of young branchlets. In *Chylocladia kaliformis,* the central cavity at the base of the branches is bridged by diaphragms and elsewhere is filled with mucilage. The entire thallus is besieged by a mucilaginous envelope, veenered by a well marked cuticle (Fig. 85 E). In the older thallus, the surface consists of a single layer of radially elongate cells, with 16-20 little branched threads that run longitudinally on the inner side at more or less even intervals. These longitudinal filaments converge towards the round apex and terminate in apical cells (Fritsch, 1945).

LOMENTARIACEAE

The hollow portions of the multiaxial thallus is septate through multilayered cellular diaphragms. In *Lomentaria hakodatensis,* the hollow axis is filled with mucilage and the wall is differentiated into a cortex and a medulla. The cortical layer consists of a single row of highly pigmented cells. The medulla is composed of 4-6 layers of round to elliptical and less pigmented cells. The inner wall bordering the hollow centre is lined with scattered hyphal filaments. The hollow cavity is divided at intervals by internal septa (Hawkes and Scagel, 1986). The multilayered septa apparent in *Lomentaria* and the lacunae enclosed by medullary cells evident in *Binghamia* are absent in *Binghamiopsis (*Lee *et al,* 1988*).*

RHODYMENIACEAE

The members exhibit medullar structures that are throughout pseudoparenchymatous or completely hollow. A few taxa have hollow laterals borne on solid stipes of lateral axes (Price and Kraft, 1991). When hollow, the longitudinal filaments are irregularly spaced around the central cavity. Only a few genera have vesicle cells.

In *Botryocladia leptopoda,* (Bhalla 1996, unpublished) the stipe shows pseudoparenchymatous construction and is differentiated into a two or three layered cortex and several layered medulla(Fig. 86B). In *B. ardreana* (Fig. 85 A) the stipe consists of small cortical cells and axially elongate medullary cells (Brodie and Guiry, 1988). In *B. wynnei* (Fig. 85 C) the outer cortical cells are pigmented (Ballantine, 1985). In *Fauchea fryeana,* there is a distinct 3 or 4 celled thick cortical zone and a medulla of large cells (Fig. 85 B, Fritsch, 1945).

In *Botryocladia pseudodichotoma,* the mucilage-sac wall is composed of two zones - an outer cortex of 2 or 3 layers of small, pigmented cells, and an inner cortex of 1 or 2 layers of large, colourless and vacuolate cells. Vesicle cells, in groups of 6-20, project into the mucilage-sac lumen from the surface of the inner cortical cells (Hawkes and Scagel, 1986). In *B. ardreana,* the vesicle wall is 80-130 μm in thickness and composed of three layers. Cortical layer forms an irregular, incomplete network; secretory cells are formed in groups of one to six on the secretory initial cells that are derived from the inner faces of medullary cells. The secretory cells frequently proliferate to form filaments upto three cells long (Brodie and Guiry, 1988). In *Botryocladia*

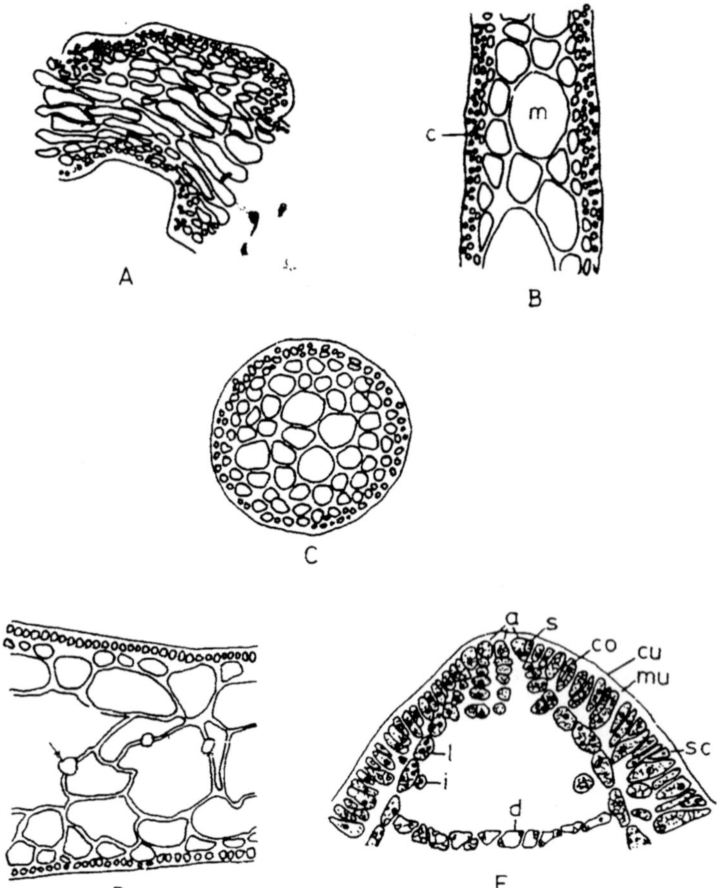

Fig. 85. A-E. Anatomy.

(a, apical cells; co, primary cortical cells; cu, cuticle; d, diaphragm; i, bulb-cells; m, medullary cells; mu, mucilage; s, segment; sc, secondary cortical cells).

A. *Botryocladia ardreana* : Longisection of stipe to show axial elongation of medullary cells (after Brodie and Guiry, 1988a).

B. *Botryocladia wynnei* : Transverse section of the stipe. Pigmented cortical cells are stippled (after Ballantine, 1985).

C. *Fauchea fryeana* : Section of the thallus (after Fritsch, 1945).

D. *Fryeella gardneri* : Transection of vegetative thallus consisting of small-celled cortex and large celled medulla. Medullary cells form diaphragms which traverse the hollow, mucilage-filled central cavity. Mucilage producing vesicle cells (arrows) are present (after Hawkes and Scagel, 1986b).

E. *Chylocladia kaliformis* : Longitudinal section through apex of the thallus (after Fritsch, 1945).

wynnei, vesicle cell wall consists of an inner layer of large, polygonal cells and an outer layer of small cells. Vesicle cells in clusters of two to six are formed from the medullary cells and project toward the vesicle cavity (Ballantine, 1985).

In *Fauchea lacinata,* the cortex is filamentous and occasionally proliferates around the medullary cells. In *Faucheocolax,* the cortex comprises of small cells which are arranged into anticlinal filaments. The medulla consists of large, isodiametric and colourless cells. *F. attenuata* is a putative parasite and shows varying degrees of penetration towards the host tissue. Medullary cells of the host and parasite are contiguous, show secondary pit-connection, with no apparent modification in the host cell structure. In some instances this host-parasite interface is filled with a mass of small cells that grow between and within the host cells (Hawkes and Scagel, 1986).

In *Minium parvum,* the crustose vegetative thallus consists of a basal layer of pseudodichotomously branched filaments which lack rhizoids and vesicle cells. From this basal layer, arise upright and occasionally branched filaments that are composed of one to several, nearly spherical cells.

In *Rhodymenia,* the thallus is pseudoparenchymatous, without any central cavity. The cortex consists of 2-4 rows of small, pigmented cells and the medulla consists of large, and thin walled cells. In *Coelarthrum opuntia* (Fig. 86 A), the holdfast shows well differentiated several layered cortex and medulla (Bhalla 1996, unpublished).

In *Rhodymeniocolax,* a parasite on *Rhodymenia* spp., the thallus is divided into a few celled, pseudoparenchymatous cortex and a medulla of both large and small cells. In *Ceratodictyon spongiosum,* a symbiont with sponge, each algal reticulum consists of a pseudoparenchymatous outer cortex and a central medulla of aggregated narrow filaments. In vegetative axes, the outer cortex has a single layer of cuboidal to rectilinear cells that are subtended by progressively large and elongate cells, which in turn surround the medullary core of compactly arranged filamentous cells (Price and Kraft, 1990). In *Cenacrum,* inner membranes of subcortical cells elongate to produce short-running filaments which basipetally fill the medullary filaments. The filaments are not continuous from the apex. Densely staining gland cells occur irregularly on isolated medullary filaments (Ricker and Kraft, 1979).

In *Fryeella gardneri* (Fig. 85 D) the medullary cells form diaphragms which traverse the hollow, mucilage-filled central cavity. Mucilage secreting vesicle cells are present on these diaphragms (Hawkes and Scagel, 1986).

REPRODUCTION

SPERMATANGIA

The gametophyte in Rhodymeniales is mostly dioecious, but monoecious gametophyte has been reported in *Botryocladia wynnei* (Ballantine, 1985), and *Champiocolax sarae* (Bula-Meyer, 1985), where the spermatangia are produced much early than the carpogonium. The spermatangia are scattered over the whole thallus surface in *Hymenocladiopsis crustigena* (Moe, 1986), *Cenacrum subsutum* (Fig. 87 A, Ricker and Kraft, 1991) and *Botryocladia ardreana* (Brodie and Guiry, 1986); or appear localized in nemathecia as in *Fauchea lacinata, Minium* spp., *Rhodymenia* spp. and *Lomentaria* spp. (Hawkes and Scagel, 1986). Nemathecia may be scattered over the whole surface; or localized either apically or subapically as in *Rhodymenia. In Champiocolax sarae* (Fig. 87 B), the spermatangial mother cell originates from the outer cortical cells (Bula-Meyer,

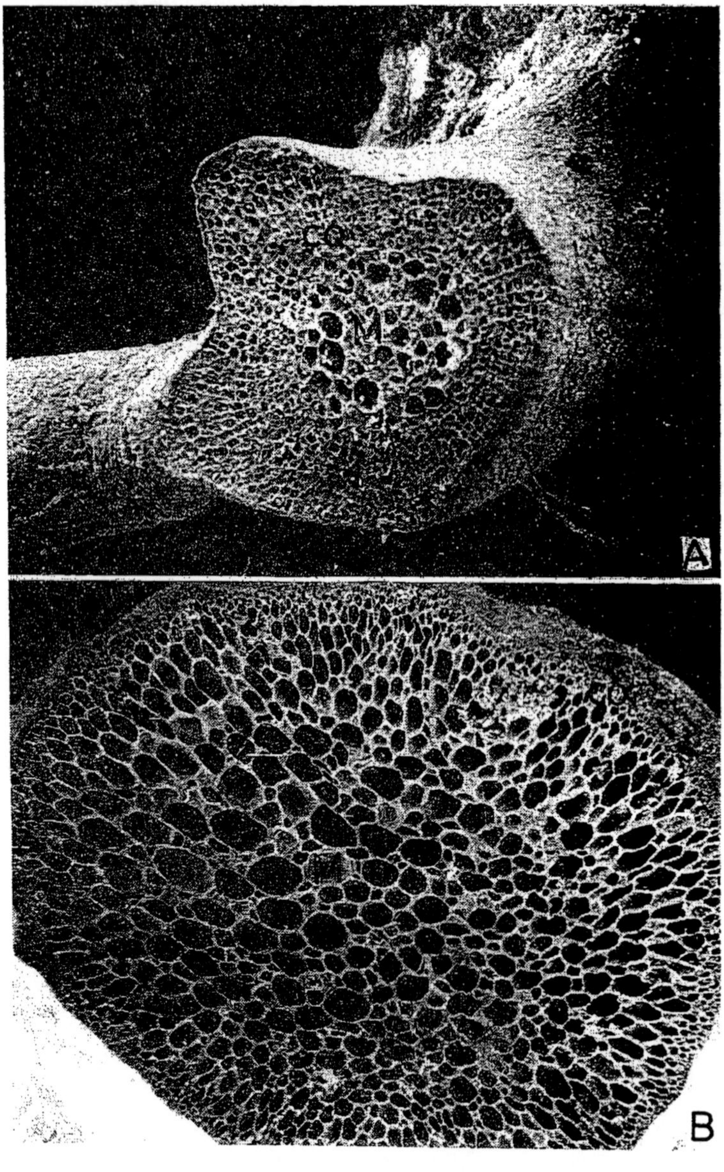

Fig. 86 A, B. Anatomy.

(after Bhalla, 1996, unpublished).

(Co, cortex; M, medulla)

A. *Coelarthrum opuntia* : SEM of section of holdfast showing well differentiated cortex and medulla.

B. *Botryocladia leptopoda* : SEM of section of stipe showing pseudoparenchymatous construction with 2 or 3 layered cortex and several layered medulla.

1985). In *Ceratodictyon spongiosum* (Fig. 87 C), the spermatangia occur either singly or in chains of two (Price and Kraft, 1991).

All taxa exhibit a distinct uniformity in spermatangial development. In all the three families namely Champiaceae, Lomentariaceae and Rhodymeniaceae, the superficial cortical epidermal cells cut off either terminally or laterally two or three primary spermatangial mother cells. Each mother cell successively produces apically or subterminally two or three spermatangia. In *Hymenocladiopsis crustigena* only one spermatangium is produced per mother cell (Moe, 1986).

Interestingly spermatangial plants have not been reported in *Fryeella, Rhodymeniocolax, Faucheocolax* and *Fauchea fryeana* from the British Columbia populations (Hawkes and Scagel, 1986).

CARPOGONIAL BRANCH

The carpogonial branch is usually 2-, 3-, or 4-celled, with the carpogonium at the apex of the filament and close to the auxiliary cell (Procarpic). The auxiliary cell is formed prior to fertilization (diagnostic character for the order) and is borne at the apex of 1-, 2-, or 3-celled branch called auxiliary cell branch, that arises from the same supporting cell as the carpogonial branch. The carpogonial branch is usually curved and has a short, somewhat laterally emerging trichogyne (Fig. 88 A).

Champiaceae: Carpogonial branch is 4-celled. The auxiliary cell is formed before fertilization from the supporting cell in *Gastroclonium subarticulatum* (Hawkes and Scagel, 1986), *Champiocolax sarae* (Fig. 88 A; Bula-Meyer, 1985) and *Champia* sp. (Fig. 88 C; Fritsch, 1945).

Lomentariaceae: In this family, the carpogonial branch is 3-celled. One or two auxiliary cells are borne on a 2-celled branch, which develops from the supporting cell as seen in *Lomentaria* (Fig. 88D, Fritsch, 1945).

Rhodymeniaceae: The carpogonial branch is 3-celled in *Botryocladia, Fauchea, Faucheocolax, Minium, Ceratodictyon* (Fig. 88 B) or 4-celled in *Rhodymeniocolax* and *Botryocladia ardreana* (Brodie and Guiry, 1988). In *Cenacrum subsutum* and *Rhodymenia* sp. it may be either 3- or 4-celled. The auxiliary cell is produced before fertilization on a 2-celled auxiliary cell branch. The supporting cell is conspicuously enlarged and has dense protoplasm. It differentiates either from inner cortical cells or from outer medullary cells.

Brodie and Guiry (1988) reported procarps from cultured female plants of *Cordylecladia erecta*. Procarps are formed in small depressions surrounded by sterile hairs. The carpogonial branch is 4-celled and the carpogonium forms a strongly emergent trichogyne that protrudes amongst the sterile hairs. Carpogonial branches are formed on a supporting cell that also bears a 2-celled auxiliary cell branch which is formed before fertilization. The development of carposporophyte occurs outwards from a fusion cell.

In *Minium parvum*, the carpogonial branches arise from surface cells of a female nemathecium. The carpogonial branches are found embedded at various depths beneath the surface as the vegetative filaments of the nemathecium continue to grow. The carpogonial branch initial cytoplasm is denser than the surrounding vegetative cells including the supporting cell. An oblique division of the initial produces two cells. The distal cell divides immediately into two, of which the upper daughter cell differentiates into the carpogonium (Moe, 1979).

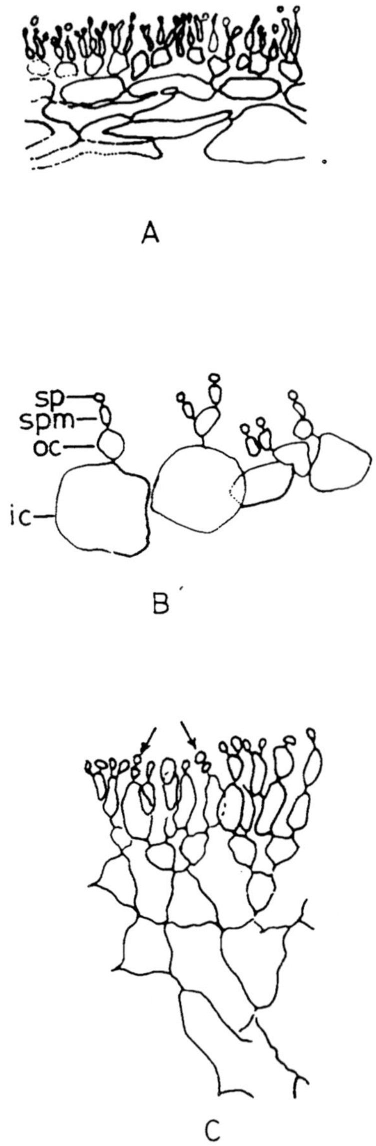

Fig. 87. A-C. Male reproductive structure.

(ic, inner cortical cell; oc, outer cortical cell; sp, spermatangium; spm, spermatangial mother cell).

A. *Cenacrum subsutum* : Spermatangial blade showing spermatangia in various stages of budding from spermatangial mother cell (after Ricker and Kraft, 1979).

B. *Champiocolax sarae* : Thallus showing development of spermatangia (after Bula-Meyer, 1985).

C. *Ceratodictyon spongiosum* : Outer cortex of a spermatangial axis to show that spermatangia occur either singly or in chains of two (arrows) on spermatangial mother cells (after Price and Kraft, 1991).

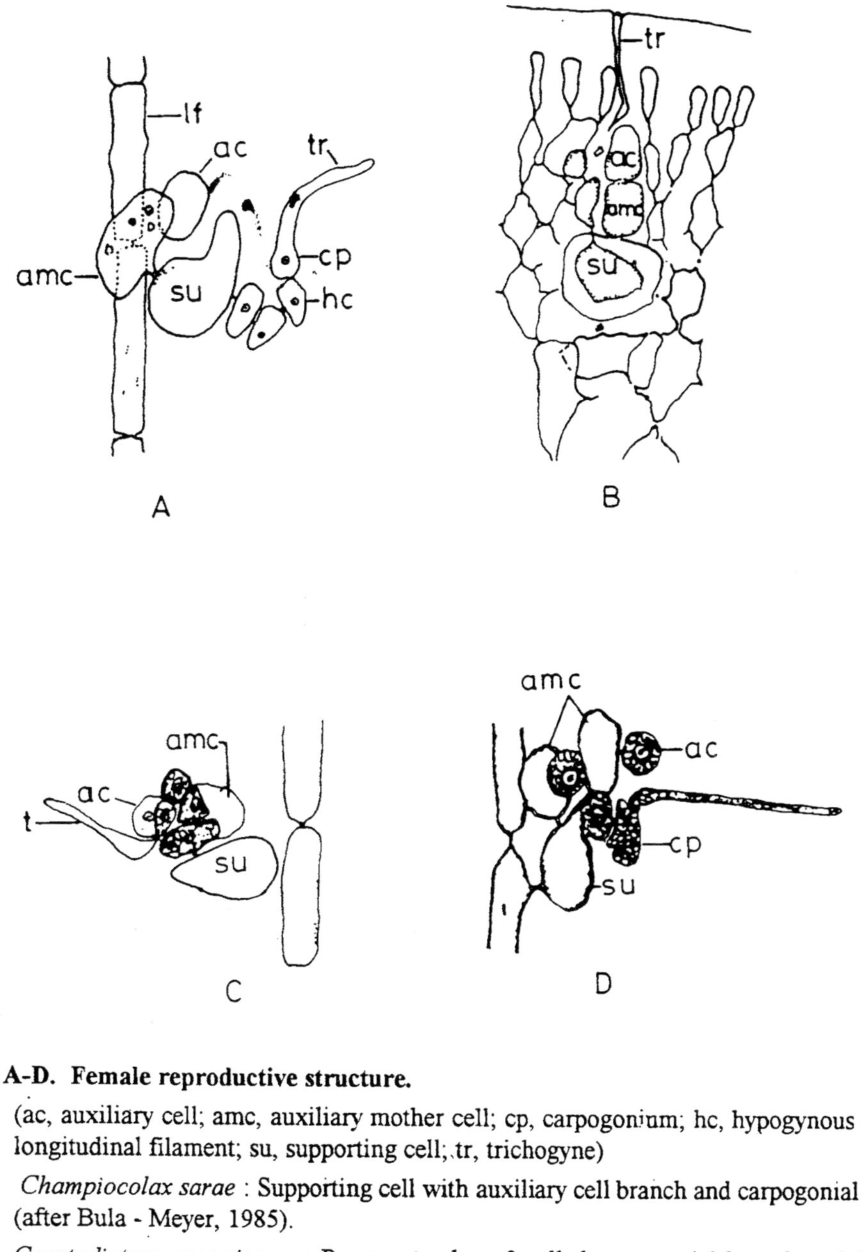

Fig. 88 A-D. Female reproductive structure.

(ac, auxiliary cell; amc, auxiliary mother cell; cp, carpogonium; hc, hypogynous cell; lf, longitudinal filament; su, supporting cell; tr, trichogyne)

A. *Champiocolax sarae* : Supporting cell with auxiliary cell branch and carpogonial branch (after Bula - Meyer, 1985).

B. *Ceratodictyon spongiosum* : Procarp to show 3-celled carpogonial branch and 2-celled auxiliary cell branch. The auxiliary cell is not yet linked to adjacent cells through secondary pit-connections (after Price and Kraft, 1991).

C. *Champia parvula* : Procarp (after Fritsch, 1945).

D. *Lomentaria clavellosa* : Procarp with two auxiliary cells (after Fritsch, 1945).

Moe (1986), in *Hymenocladiopsis crustigena,* observed that the supporting cell is a cell of the outer medulla or inner cortex that is at first not distinguishable from surrounding cells. The carpogonial branch initial divides anticlinally and the derivative cells are joined by pit-connections. The derivative without a pit-connection to the supporting cell divides periclinally and unequally into two, of which the upper cell differentiates into the carpogonium with laterally curved trichogyne. The auxiliary mother cell later forms secondary pit-connections with the surrounding cortical cells.

In *Cenacrum,* the auxiliary cell is large, conspicuous and formed towards the thallus surface from the auxiliary mother cell. It has one or more dark staining, proteinaceous inclusions (Ricker and Kraft, 1979) which are common in many genera of the Rhodymeniaceae (Sparling, 1957). In *Ceratodictyon spongiosum,* however, the proteinaceous inclusions are absent in the auxiliary cell (Price and Kraft, 1991). In *Maripelta atlantica,* the auxiliary cell branch is curved so that the auxiliary cell is juxtaposed to the carpogonium (Moe, 1981). In addition to the auxiliary cell branch, a sterile branch has been reported in *Minium parvum* (Moe, 1979) and *Maripelta atlantica* (Moe, 1981).

CARPOSPOROPHYTE

After fertilization, the trichogyne degenerates and the carpogonium cuts off a small connecting cell which makes contact with the auxiliary cell and transfers the diploid nucleus. Alternatively, a short connecting filament without any connecting cell establishes direct contact with the auxiliary cell. Subsequently, the carpogonial and the auxiliary cell branches fuse rapidly and the diploidized auxiliary cell from its upper surface gives rise to gonimoblast cells.

Concomitantly the cortical cells adjoining the carpogonium divide rapidly and form a thick, dome-shaped pericarp. Cystocarps are large, protruding and are either scattered over the surface or confined to the margins of the thallus. They may be with or without an ostiole.

The gonimoblast cells arising from the diploidized auxiliary cell may cut off secondary gonimoblast cells. The primary and secondary gonimoblast cells produce profusely branched gonimoblast filaments. Most cells, or sometimes only the terminal cells of these gonimoblast filaments produce carposporangia. The mature carposporophyte is large (upto 1-2 nm in diameter), comprises on the fusion cell, which results from swelling of the diploidized auxiliary cell, and fusion of auxiliary cell with cells of the carpogonial branch and adjacent cells, the gonimoblast filaments and in some genera even with vegetative filaments. The cystocarp is lined by the procarpic wall (Hawkes and Scagel, 1986).

Champiaceae: The cystocarps lack an ostiole. Only the terminal gonimoblast cells form the carposporangia. In *Champia somalensis,* (Bhalla 1996, unpublished) cystocarps occur as protrusions on the thallus and are bereft of ostioles (Fig. 90 C). Carpospores are distinctly wedge-shaped with narrow basal regions (Fig. 91 D). In *Chylocladia* and *Gastroclonium* extensive fusion results in a large fusion-cell, occupying the middle of the mature cystocarp. The fusions involve the inner cells of the gonimoblast, the auxiliary cells, the auxiliary mother cells and a few adjacent nutritive cells (Fritsch, 1945). In *Champia* (Fig. 89 B), there is comparatively less fusion. In *Champiocolax sarae,* the inner cells of the pericarp become stellate, forming a net-like tissue termed 'tela arachnoidea' (Bliding, 1928). Similar development is seen in *Champia* (Fig. 89 C, Fritsch, 1945).

Lomentariaceae: The cystocarps are ostiolate and elevated. After fertilization a column-like fusion cell is formed. Almost all gonimoblast cells form carposporangia. In *Lomentaria* (Fig. 89 D), the richly branched gonimoblast gives rise successively to many gonimolobes where most of the cells form the carposporangia (Fritsch, 1945).

Rhodymeniaceae : The cystocarp is large, ostiolate and protuberant. Most of the gonimoblast cells form carposporangia. In *Rhodymenia australis,* (Bhalla 1996, unpublished) cystocarps of various sizes are scattered over the thallus surface and each cystocarp has an ostiole (Fig. 90 B). In *Coelarthrum opuntia,* (Bhalla 1996, unpublished) although the cystocarps do not protrude on the thallus surface yet the surface view of the thallus shows well developed ostioles (Fig. 90 A). Carpospores are spherical to triangular in appearance (Fig. 91 C).

Brodie and Guiry (1988) in culture study of *Botryocladia ardreana* found the presence of a flat, plate like placental tissue composed of deeply staining cells at an early stage of cystocarp formation. This tissue persisted during the formation of pericarp and was present at the base of the mature cystocarp. The vegetative filaments (tela arachnoidea) derived from inner cells of procarpic wall are present in *Botryocladia ardreana* but absent in *Fryeella, Minium* and *Rhodymeniocolax* (Hawkes and Scagel, 1986). In *B. leptopoda,* (Bhalla 1996, unpublished), the cystocarp is ostiolate and shows several deeply staining cells lining the ostiole (Fig. 91 A). Carpospores are spherical to elongate in shape (Fig. 91 B).

Ceratodictyon spongiosum in culture shows the development of fusion cell by enlargement and fusion of auxiliary cell and auxiliary mother cell. The fusion cell ultimately incorporates the supporting cell and the bulk of the earliest formed gonimoblast cells. Individual clusters of carposporangia develop on the ramifications of the fusion cell and merge into a synchronously maturing whole. Gonimolobes are not formed on the fusion cell (Price and Kraft, 1991). In *Faucheocolax* (Fig. 89 A), the multinucleate gonimoblast initials cleave to form lobes of young carpospores. The fusion cell continues to incorporate other cells which also includes the supporting cell.

In *Maripelta atlantica,* simultaneous with the development of pericarp, a periclinal rupture takes place through the filamentous tissue just above the young carposporophyte. The tissue above the rupture arches outward creating a cavity in which the carposporophyte develops (Eiseman and Moe. 1981). Similar cavity formation has been observed in *Hymenocladiopsis crustigena* (Moe, 1986), *Cenacrum subsutum* (Ricker and Kraft, 1979), *Minium parvum* (Moe, 1979) and *Ceratodictyon spongiosum* (Price and Kraft, 1991).

In Rhodymeniaceae there is a distinct pattern of the division which separates the auxiliary cell from the gonimoblast initial. In *Gloioderma, Gloiocolax, Fauchea, Epymenia* and *Rhodymenia* (Sparling, 1957), the primary gonimoblast is formed by a transverse division which cuts the auxiliary cell into a large upper cell (the gonimoblast initial) and a considerably narrow lower one (the remainder of the auxiliary cell).

TETRASPOROPHYTE

The tetrasporangia usually originate from the outermost cortical cells of the tetrasporophyte thallus, and at maturity are embedded beneath the surface. In *Maripelta,* tetrasporangia are intercalary (Eiseman and Moe, 1981). Tetrasporangia are either scattered throughout the cortical layer or are produced in distinct nemathecia (sori). The nemathecia develop either apically or subapically on the thallus branch or are scattered over the thallus. Cruciately divided tetrasporangia are encountered in most members of Rhodymeniaceae. In *Hymenocladiopsis* (Moe, 1986) and *Hymenocladia* (Rhodymeniaceae) and all the members of Champiaceae and Lomentariaceae, the

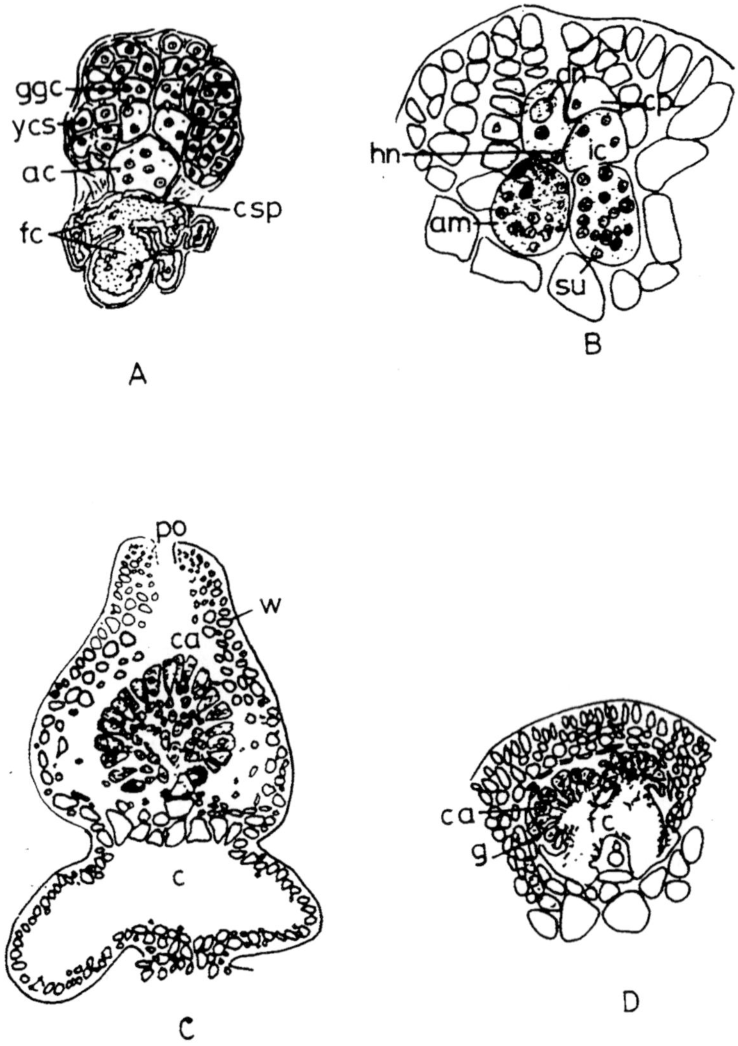

Fig. 89 A-D. Post-fertilization development.

(ac, auxiliary cell; ca, carposporangia; cp, carpogonium; dn, diploid nucleus; fc, fusion cell; ggc, generative gonimoblast cell; gi, gonimoblast initial; p, protein mass of auxiliary cell; po, aperture of cystocarp; su, supporting cell; w, wall of cystocarp; ycs, young carpospores).

A. *Faucheocolax* : After fertilization, multinucleate gonimoblast initials have cleaved to form lobes of young carpospores. The fusion cell continues to incorporate cells, including the supporting cell (after Kugrens and Delivopoulos, 1985).

B. *Champia parvula* : Procarp after fertilization (after Fritsch, 1945).

C. *C. parvula* : Vertical section of cystocarp (after Fritsch, 1945).

D. *Lomentaria clavellosa* : Procarp after fertilization (after Fritsch, 1945).

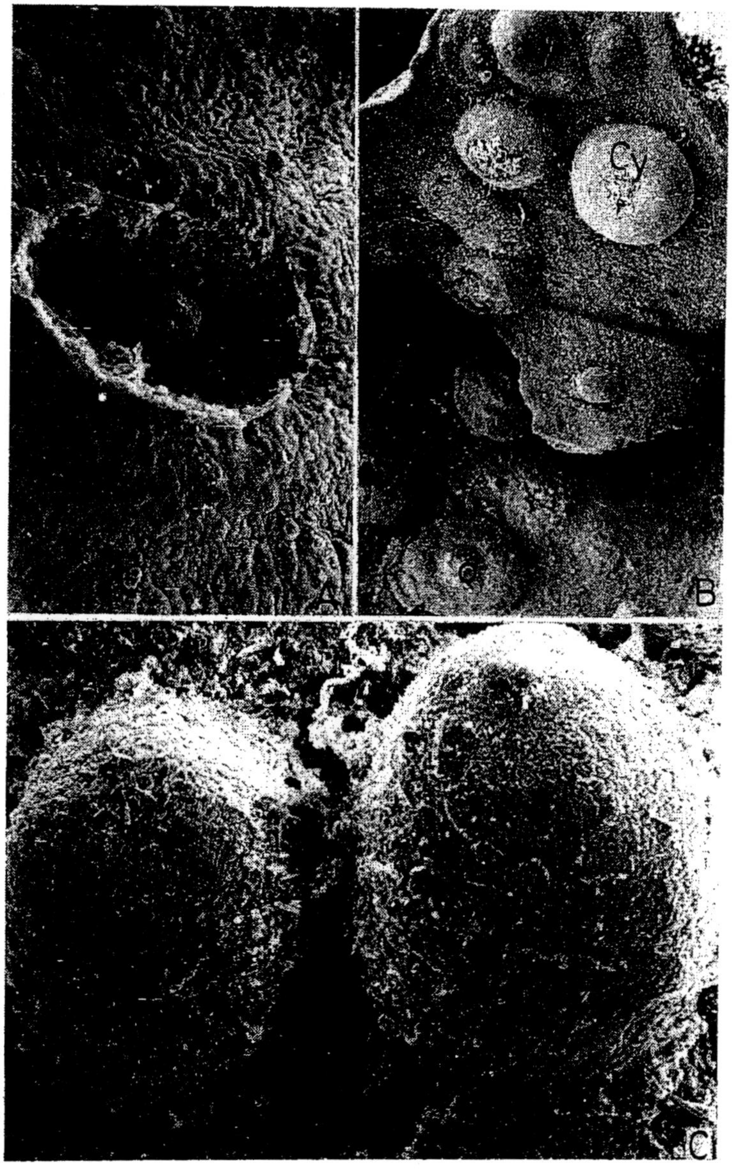

Fig. 90 A-C. Cystocarp.

(after Bhalla, 1996, unpublished).

(Cy, cystocarp; O, ostiole)

A. *Coelarthrum opuntia* : SEM showing well developed ostiole of cystocarp.

B. *Rhodymenia australis* : SEM of thallus with several ostiolate cystocarps.

C. *Champia somalensis* : SEM of cystocarps bereft of ostioles.

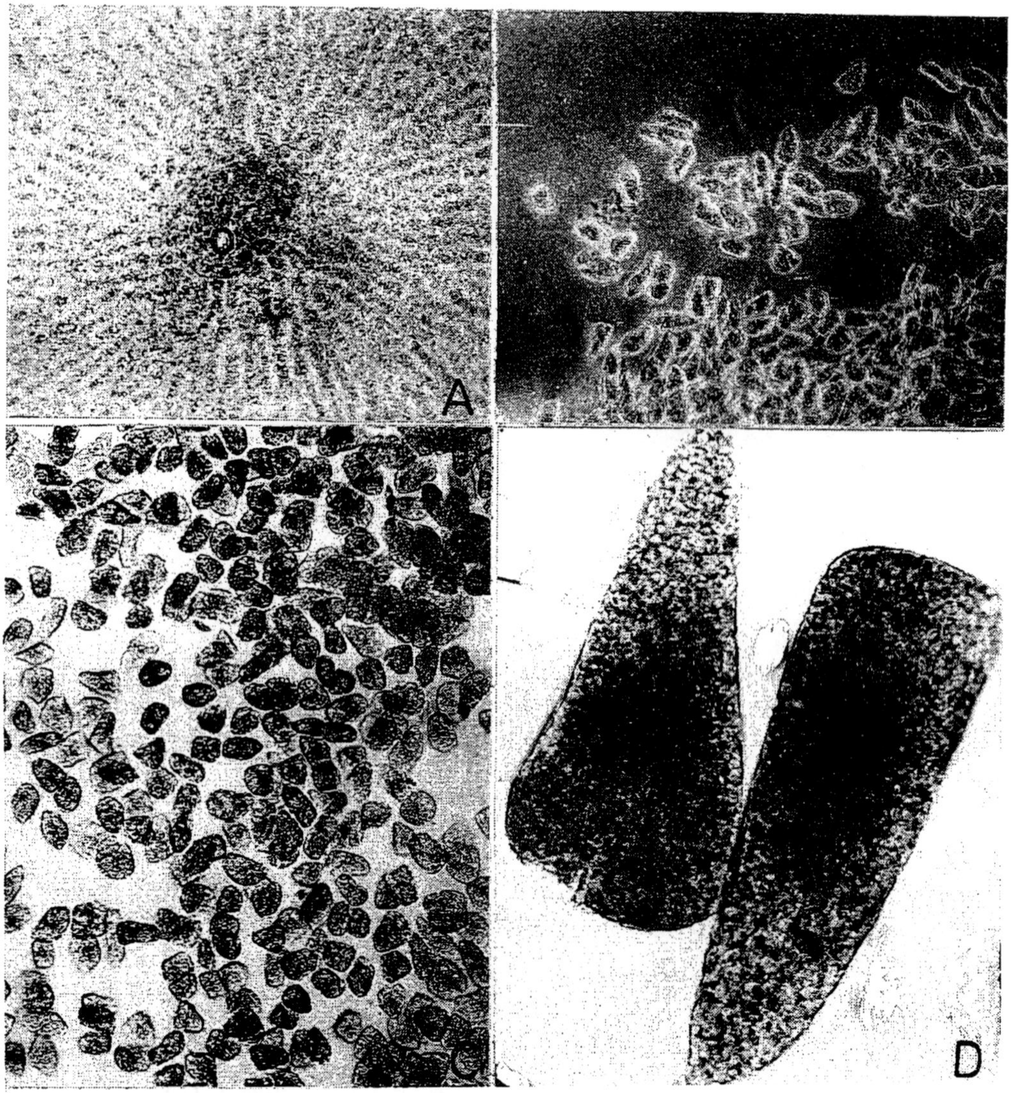

Fig. 91 A-D. Carpospore.

(after Bhalla, 1996, unpublished).

(O, ostiole)

A. *Botryocladia leptopoda* : Portion of the cystocarp showing deeply staining cells lining the ostiole (squash preparation).

B. *B. leptopoda* : Carpospores (phase contrast).

C. *Coelarthrum opuntia* : Spherical to triangular carpospores (squash preparation).

D. *Champia somalensis* : Wedge-shaped carpospores. (squash preparation).

tetrasporangia are tetrahedrally arranged and the division is simultaneous. This is in contrast to successive division and cruciate tetrasporangia found in Rhodymeniaceae.

Champiaceae : The tetrasporangia are tetrahedrally arranged and occur in distinct sori in the cortical layer. In *Chylocladia kaliformis,* they usually originate from the uninucleate intercalary cells of the secondary cortex. In *Gastroclonium,* they project in the central hollow of the thallus (Fritsch, 1945). In *Champiocolax,* they develop from the intercalary cortical cells (Bula-Meyer, 1985). *Gastroclonium* sometimes forms polysporangia with 8-16 polyspores (Hawkes and Scagel, 1986).

Lomentariaceae : The tetrahedrally divided tetrasporangia are terminal in the cortical layer and formed in aggregate sori. In *Lomentaria hakodatensis,* the sporangia develop on small swollen branchlets, form sori on branch intersepta and are embedded (Hawkes and Scagel, 1986).

Rhodymeniaceae : In most of the members tetrasporangia are cruciately divided except *Hymenocladiopsis* (Moe, 1986) and *Hymenocladia* where they are tetrahedrally divided. Kylin (1931), on the basis of type of division in tetrasporangia (cruciate or tetrahedral), divided the family Rhodymeniaceae into subfamilies Faucheae and Rhodymenieae (cruciately divided tetrasporangia) and Hymenocladieae (tetrahedrally divided tetrasporangia). Sparling (1957) and Lee (1978) accepted the above classification.

In *Hymenocladiopsis crustigena* (Moe 1986), the tetrasporangia are scattered over one or both surfaces of primary segments and pinnae, and are not grouped into sori. Tetrasporangial initials are cut off from outer medullary cells, or from inner or outer cortical cells. They enlarge, and the nucleus divides meiotically into four, and the wall formation takes place simultaneously and centripetally. The sporangia are arranged in a tetrahedral pattern.

In *Maripelta,* the tetrasporangia are produced in continuous nemathecia, which are formed by filaments cut off from the superficial cells. Tetrasporangium mother cell differentiates from intercalary cortical cells. The mother cell divides sequentially. Initially a periclinal division occurs and the derivatives divide anticlinaly. The spores are arranged in cruciate manner (Eiseman and Moe, 1981).

According to Price and Kraft (1991), in culture, the tetrasporangia of *Ceratodictyon* are borne in terminal nemathecia. The spores are usually decussately/ cruciately arranged, albeit occasionally many sporangia with irregularly arranged spores are encountered (Fig. 92 A). The cells borne in positions homologous to tetrasporangia remain undivided and elongate considerably and like hairs, protrude beyond the surface of the cuticle (Fig. 92 A). Their function is, however, unknown.

In *Botryocladia ardreana,* the tetrasporangia are produced by intercalary cells in slightly modified cortical sori, and are successively divided. Tetraspores are cruciately or slightly irregularly arranged (Fig. 92 C, Brodie and Guiry, 1988).

Sterile filaments densely interspersed among the tetrasporangia have been observed in the nemathecia in *Minium parvum* (Moe, 1979), *Maripelta atlantica* (Eiseman and Moe, 1981), and *Fryeella gardneri* (Hawkes and Scagel, 1986).

Tetrasporangia in *Fauchea lacinata* and *Fryeella gardneri* (Fig. 92 B) are produced in anastomosing nemathecia and only on the ventral surface of the thallus (Hawkes and Scagel, 1986). In *Rhodymenia,* the cruciately divided tetrasporangia are terminal, and are aggregated in subapical sori in *R.californica* and *R. pacifica*, or dispersed over the thallus surface in *R. pertusa* (Hawkes and Scagel, 1986).

LIFE CYCLE STUDIES

Based on field collected specimens, most members appear to have a sexual, **Polysiphonia** type life-history (West and Hommersand, 1981) which is karyologically trigenic (i.e., one haploid and two diploid phases) and morphologically triphasic having a gametophyte-carposporophyte-tetrasporophyte alternation with the gametangial (monoecious or dioecious) and meiosporangial (tetrasporophyte) generations being free living and isomorphic, and the carposporophyte being parasitic on the female gametophyte (Hawkes and Scagel. 1986). Asexual life-history phenomena such as apomeiosis and vegetative propagation have been reported in some species.

LOMENTARIACEAE

In *Lomentaria* spp., Svedelius (1937) reported apomeiosis and Lee and West (1980) observed vegetative propagation, suggesting an asexual phase in life-history in addition to sexual, **Polysiphonia** type of life-history. In *L. hakodatensis,* both sexual and asexual populations have been observed in the field. This was confirmed by *in vitro* studies on the life-history of Korean populations where one isolate exhibited a **Polysiphonia** type of life-history and another isolate only vegetative reproduction (Lee and West, 1980).

CHAMPIACEAE

In *Gastroclonium subarticulatum,* the life-history has not been studied in laboratory conditions, In the field, isomorphic tetrasporophytes and gametophytes have been observed, indicating a sexual, **Polysiphonia** type of life-history. In British Columbia populations only tetrasporophytes are known which suggests an asexual life-history via apomeiotic cycling of the tetrasporophyte (Hawkes and Scagel, 1986).

RHODYMENIACEAE

A **Polysiphonia** type sexual life-history pattern has been suggested in most of the members. This is based on the presence of isomorphic, female gametophytes (bearing cystocarps) and tetrasporophytes. But apparent absence of male gametophytes in British Columbia populations of *Fauchea lacinata,* and *F. fryeana* (Sparling, 1961); *Botryocladia pseudodichotoma, Faucheocolax attenuata, Fryeella gardneri* (Hawkes and Scagel, 1986), suggests the existence in addition to sexual life cycle, an asexual life cycle through apogamous development of the carposporophyte.

The British Columbia populations of *Rhodymenia* spp., in addition to the **Polysiphonia** type of life-history pattern, exhibit asexual life-history pattern. In *R. californica,* only tetrasporophyte has been observed suggesting apomeiotic cycling of tetrasporophytes. *R. pacifica* comprises of only vegetative populations suggesting life cycle via vegetative apomixis.

ULTRASTRUCTURAL STUDIES

PLASTID

In *Botryocladia botryoides,* the actively growing cells have plastids which comprise of a girdle lamella enclosing parallel, single thylakoids on which phycobilisomes are scattered. Plastoglobuli (involved in lipid storage) and occasionally starch grains are visible between the thylakoids. In the plastidial matrix, several electron-transparent regions are present in which

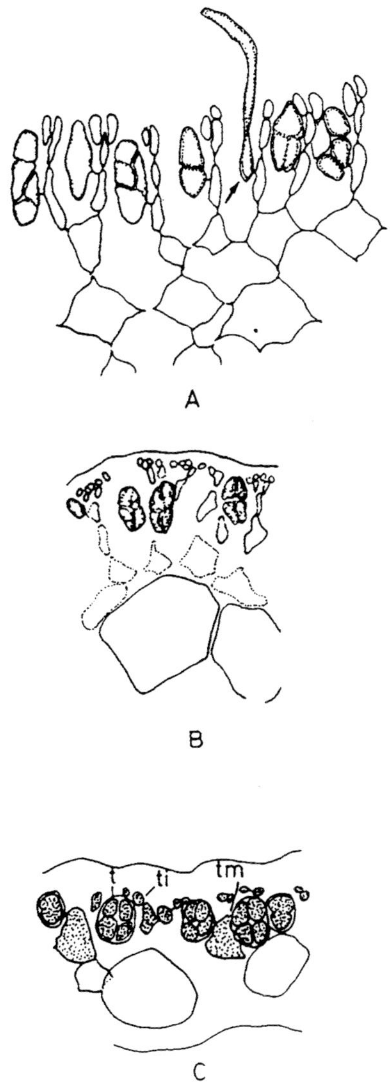

Fig. 92 A-C. Tetrasporangial development.

(t, tetrasporangia; ti, tetrasporangial initial; tm, tetrasporangial mother cell).

A. *Ceratodictyon spongiosum* : Various stages of initiation and differentiation of tetrasporangia, including regularly and irregularly divided types. A hair-like extension (arrow) of unknown function occurs in the position of a tetrasporangium (after Price and Kraft, 1991).

B. *Fryeella gardneri* : Tetrasporangial nemathecium (after Hawkes and Scagel, (1986b).

C. *Botryocladia ardreana* : Formation of tetrasporangia and tetrasporangial initial in pairs from tetrasporangial mother cell (after Brodie and Guiry, 1988a).

fibrils, representing the genophore are located. The fibrils are frequently in contact with a membranous body (Fig. 93 B), some of which seem to connect these bodies to the thylakoids (arrow in Fig. 93 A). The inner lumen of the membranous bodies (MB) is typically more electron-dense than the surrounding matrix. Occasionally, a pair of electron-dense core may be present. But the MBs in plastids of the inner layers of the cortex have an electron-transparent inner space and such MBs have poorly evident structures of the membranes (Fig. 93 B), whereas the connection with the fibrils remain. In oblique sections the MBs reveal a stalk suggesting a membranous connection with the nearest thylakoid (Tripodi *et al.*, 1986).

In the gonimoblast cells of *Botryocladia botryoides,* during the early stages of differentiation, the plastids completely lack thylakoid membranes whereas they are fully differentiated in carposporangia (Tripodi *et al.*, 1986).

Delivopoulos and Kugrens (1984) studied the successive development of plastids in the carpospores of *Faucheocolax attenuata.* In the young carpospores, the proplastid has a peripheral thylakoid (Fig. 93 C). Concentric lamellae are initially observed at the periphery which later move to the centre (Fig. 93 C, 93 D) and give rise to internal thylakoids of the plastid. Phycobilisomes are present on fully developed thylakoids. The concentric lamellae, which eventually become small are seen attached to some thylakoids in older plastids (Fig. 93 D). Delivopoulos and Kugrens (1985) observed the presence of membranous bodies in the plastids of *Faucheocolax* and opine that the function of the MBs is storage of molecules to be later used in the expansion of thylakoid membrane.

The chloroplasts in the gland cells of *Botryocladia pseudodichotoma* are irregular in shape and have a peripheral thylakoid encircling the internal thylakoids which have a concentric orientation parallel to the peripheral thylakoid (Fig. 94 D). Osmiophilic stromal droplets and electron-transparent zones are present in the plastids (Young, 1978).

GLAND CELL

Clusters of small, pyriform, gland (vesicular) cells protrude from the inner cortex into the inflated bladders in *Botryocladia pseudodichotoma* (Fig. 94 A; Young, 1978). In *B. wynnei* (Fig. 94 B), the gland cells occur in clusters of 2 to 6 and are produced from normal medullary cells (Ballantine, 1985). *B. ardreana* has gland (secretory) cells which often proliferate to form filaments upto three cells long (Brodie and Guiry, 1988). Young (1978), in *B. pseudodichotoma,* showed that the gland cell is multinucleate, and contains a large, central vacuole surrounded by dense cytoplasm (Fig. 94 C). The chloroplasts are irregular in shape, and have a peripheral thylakoid encircling the internal thylakoids which have a concentric orientation parallel to the peripheral thylakoid (Fig. 94 D). The two most prominent organelles in the gland cell are osmiophilic bodies and Golgi bodies. The osmiophilic bodies (OB) may be spherical, subspherical or irregular (Fig. 94 C). They are abundant in the cytoplasm and are also released in the vacuole. Their composition is unknown.

The abundant Golgi apparatus appear to form by coalescence of vesicles and tubules derived from endoplasmic reticulum. The Golgi apparatus produces large, electron-transparent vesicles that contain an osmiophilic, fibrous material. This material is secreted into a diffuse wall layer which appears to be continuously sloughed.

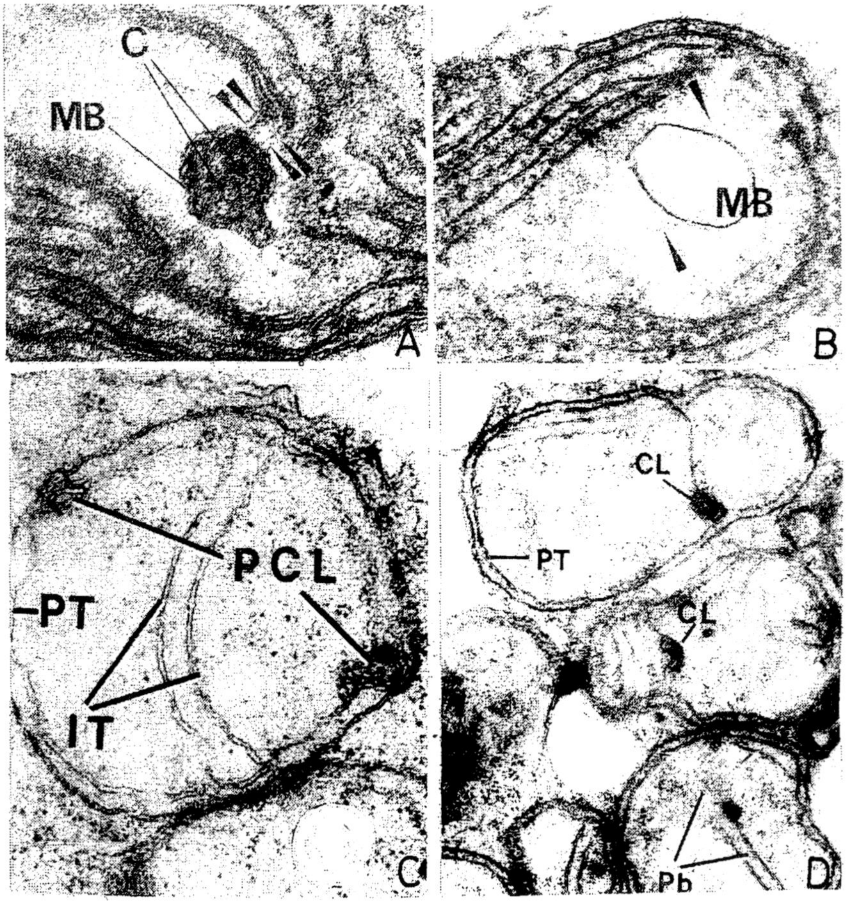

Fig. 93. A-D. Ultrastructure of plastids.

(A-B : after Tripodi *et al.*, 1986; C-D : after Delivopoulos and Kugrens, 1984).

(C, core; CL, concentric lamellae; IT, internal thylakoids; MB, membranous body; PCL, peripheral concentric lamellae; PG, plastoglobuli; PT, peripheral thylakoid)

A. *Botryocladia botryoides* : Membranous bodies in the electron transparent area of the plastidial matrix. Fibrils are connected with the surface of MB and that of nearby thylakoids (arrowheads). Two electron dense cores are visible in the MB.

B. *B. botryoides* : The plastids in the inner layers of the cortex show MBs with a poorly evident structure of the membranes, whereas the connection with the fibrils are still visible (arrowheads). The inner space of MB in these plastids is electron transparent.

C. *Faucheocolax attenuata* : Proplastid with peripheral thylakoid, two internal thylakoids and two peripheral concentric lamellae.

D. *F. attenuata* : Chloroplasts with fully formed internal thylakoids bearing phycobilisomes and small concentric lamellae with attachments to some thylakoids.

The gland cells of *Botryocladia* show many similarities with mucilage secreting cells of other algae like *Laminaria* (Phaeophyta) (Evans *et al.*, 1973), *Lomentaria* (Bouck, 1962), and *Leveringiella* (Kugrens and West, 1973). In general, these cells are characterized by the presence of many well developed mitochondria, abundant Golgi bodies, and the granulocrine discharge of Golgi derived vesicle-contents via exocytosis (Schuef, 1974).

The precise nature of the secretions in *Botryocladia* is unknown. Histochemical localization of the mucilage adhering to the vesicle cell indicates it to be an acidic polysaccharide (Young, 1978).

CARPOSPOROPHYTE DEVELOPMENT

Auxiliary Cell

Faucheocolax attenuata reveals an auxiliary cell that contains several nuclei during its early development but during progressive stages of differentiation more nuclei are produced. It however, produces uninucleate gonimoblast initials that are connected to it by small septal plugs. The auxiliary cell (Fig. 95 A) contains numerous small vacuoles. Plastids are in the proplastid stage, either lack thylakoids or have only portions of the peripheral thylakoid. The plasmalemma profile is irregular. As the auxiliary cell ages, this profile of plasmalemma and that of the nuclei become more irregular indicating perhaps cytoplasmic degeneration. Septal plugs between the auxiliary and gonimoblast cells increase in size during later stages of development (Kugrens and Delivopoulos, 1985).

Gonimoblast Cell

In *Faucheocolax attenuata,* the gonimoblast initials adjacent to the auxiliary cell are large while those that are further away are initially small but later enlarge during development. The terminal gonimoblast cells become meristematic and produce carpospores whereas the axiale internal ones assumes a sercretory function. The septal plugs between gonimoblast cells are small, but enlarge as development proceeds (Kugrens and Delivopoulos, 1985).

The multinucleate generative gonimoblast cells (GGC) (Fig. 95 B) remain meristematic and divide successively to form uninucleate cells which eventually differentiate into carpospores. GGC possesses numerous small vacuoles. Proplastids occur in the peripheral cytoplasm while most mitochondria are found in the central cytoplasmic area. Golgi vesicles, containing a single dark staining granule, appear to be exocytosing material, thereby giving the plasmalemma an irregular profile. Smooth endoplasmic reticulum and concentric membrane bodies are common in these cells. Concentric membrane bodies are tightly coiled. The generative gonimoblast cell is devoid of floridean starch.

The secretory gonimoblast cells (SGC) (Fig. 95 C) are also multinucleate, lack vacuolation, but have considerable Golgi activity. The proplastids have unusual internal membrane arrangements, some proplastids have concentric membrane bodies while others have reticulate membranes or extended concentric lamellae. Cytoplasmic concentric lamellae are continuous with the smooth endoplasmic reticulum and often become inflated by the secretion of granular materials within the lamellae. Numerous dictyosomes secrete mucilaginous material which is released from the cell and augments additional mucilage. Dark staining spherical masses are also common within these secretory cells. In addition, patches of dark staining granular material occur in indented portions of the nuclear envelope.

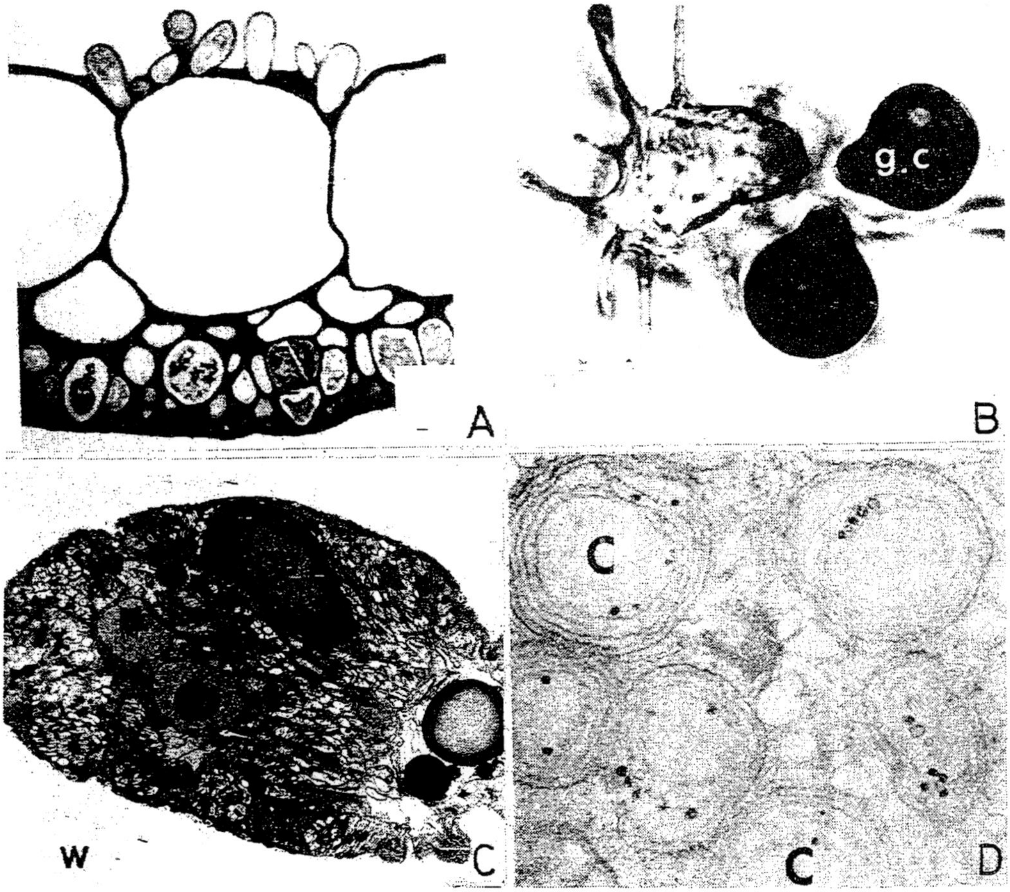

Fig. 94. A-D. Ultrastructure of gland-cells in *Botryocladia*.

(A, C and D : after Young, 1978; B : after Ballantine, 1985).

(C, chloroplast; gc, gland cell; N, nucleus; OB, osmiophilic body; W, wall).

A. *B. pseudodichotoma* : Cross-section of bladder illustrating small, outer cortical cells (bottom); large, vacuolate inner cortical cells, and a cluster of cells (top).

B. *B. wynnei* : Inner vesicle wall showing an irregular stellate gland supporting cell with two stained gland cells.

C. *B. pseudodichotoma* : Electron micrograph of longitudinal section of developing gland cell. Note the abundant electron-transparent Golgi-derived vesicles. Developing OBs occur within areas resembling inflated endoplasmic reticulum cisternae (right of centre).

D. *B. pseudodichotoma* : Concentric orientation of thylakoids in the chloroplasts of a gland cell.

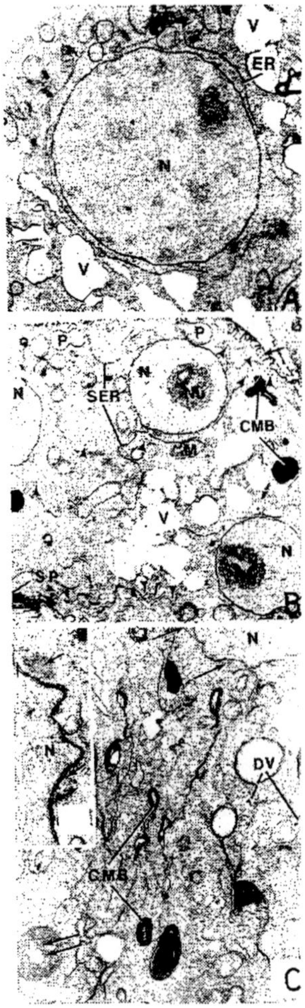

Fig. 95. A-C. Ultrastructure of auxiliary and gonimoblast cells in *Faucheocolax attenuata*.

(after Kugrens and Delivopoulos, 1985).

(CMB, concentric membranous body; DV, dictyosome vesicle; ER, endoplasmic reticulum; M, mitochondria; N, nucleus, Nu, nucleolus; P, plastid; SER, smooth endoplasmic reticulum; V, vacuole)

A. Portion of a typical young auxiliary cell which exhibits an irregular plasmalemma and contains proplastid, as well as numerous small vacuoles.

B. Portion of a generative gonimoblast cell having several nuclei, small vacuoles, concentric membrane bodies and proplastids.

C. Central cytoplasm of a secretory gonimoblast cell. Numerous concentric membrane bodies (apparently formed from endoplasmic reticulum) and a dark staining, non-membranous body (double arrows) are found in this region. Indentations of the nucleus contain dark staining granular material (arrows).

Carpospore

In *Gastroclonium clavatum,* the carpospores lie in radial arrangement in the cystocarp (Fig. 96 A). The carpospore is wedge-shaped and 80-100 μm at its longest diameter. The nucleus lies in the centre in uncondensed form, and has a regular outline and a single large nucleolus. The plastids are spread throughout the cytoplasm and lie preferentially in clusters and show either round, or elliptical or irregular profiles (Fig. 96 B). The thylakoids are present in separate rows and run across several peripheral matrix sites which probably represent centres of membrane proliferation. The matrix of the plastid is evenly opaque except in small, randomly scattered areas which are interpreted as DNA-containing areas. Osmiophilic globules, upto 150 μm in diameter are present inside the thylakoid matrix. They are found either scattered in the matrix or associated in small clusters. Starch-grains are present interspersed through the whole cytoplasm. Dictyosomes are closely associated with mitochondria, which are small in size but have well developed cristae. Periodic acidthiocarbohydrazide-silver proteinate test shows the presence of several aggregation of polysaccharides without any limiting membranes (Fig. 96 C). They comprise of compactly layered material arranged in complex patterns. Ellipsoidal, round, rod-like or less often irregular shaped multilayered bodies are present in the cytoplasm. They comprise of electron-transparent and opaque layers lying inside large electron-translucent areas (Gori, 1980).

The mature carpospore in *Faucheocolax attenuata* shows some compartmentalization in cytoplasmic constituents. The nucleus is situated in the centre with perinuclear, large starch grains. Mature chloroplasts besiege this ring of starch. In the peripheral region of the carpospore are present a few small starch grains, dictyosomes, adhesive vesicles produced by the dictyosomes; and fibrous vacuoles. Dictyosomes are inactive in mature carpospores. Fibrous vacuoles are large and press against the outer cytoplasm and plasmalemma. They often fuse with each other. The carpospore is surrounded by a two -layered carpospore wall (Delivopoulos and Kugrens, 1984).

Carposporogenesis

In *Faucheocolax attenuata* immediately after cleaving from the multinucleate gonimoblast cells, the young carpospores are embedded within confluent mucilage produced by gonimoblast cells. These young carpospores contain a large nucleus, a few starch grains, concentric lamellae, as well as proplastids with a peripheral and occasionally with some internal thylakoids. Concentric lamellar bodies are also present in the proplastids. Cytoplasmic concentric membranes produced mucilage in mucilage sacs (Fig 97 A) Mucilage sac has a reticular fibrous structure embedded within electron transparent matrix. Subsequently the mucilage sacs release their contents, forming an initial reticulate deposition of carpospore wall material (Fig. 97 A). Dictyosome vesicles with large, single, dark staining granules (Fig. 97 A) also contribute to wall formation and may create a separating layer between the mucilage and carpospore wall. Starch is polymerized in the perionuclear cytoplasmic area (Fig. 97 B) during the latter stages of young carpospores. They are in close contact with endoplasmic reticulum (Delivopoulos and Kugrens, 1984).

Intermediate-aged carpospores continue their starch polymerization. Dictyosomes deposit more wall material and continue to form fibrous vacuoles (Fig. 97 C). Thylakoids are formed in thc proplastids from concentric lamellar bodies (Fig. 97 B).

Mature carpospore is surrounded by a two-layered carpospore wall (Fig. 97 D). Large floridean starch granules, peripheral fibrous vacuoles, mature chloroplasts and curved dictyosomes are present in the cytoplasm. The dictyosomes produce cored vesicles that in turn transform into

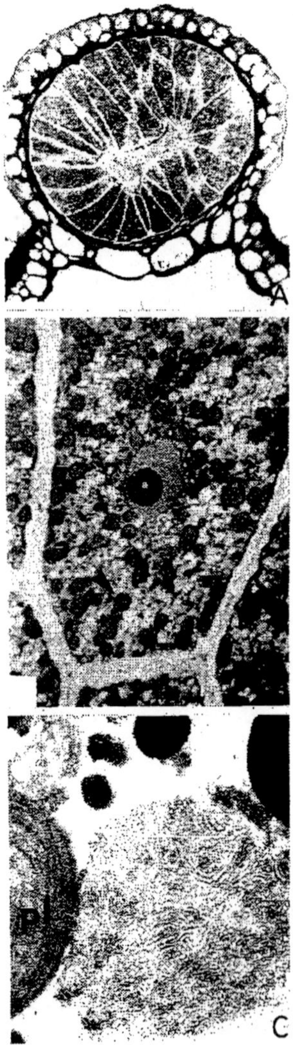

Fig. 96. A-C. Ultrastructure of carpospore in *Gastroclonium clavatum*

(after Gori, 1980).

A. Section of cystocarp crowded with wedge-shaped radially arranged carpospores. Multilayered pericarp and basal placenta can be seen. In the centre, residual portion of gonimoblast; bottom, portion of the large inner cavity typical of the mature branches.

B. Portion of a carpospore. The nucleus shows a large nucleolus; many plastids lie in the cytoplasm (arrow).

C. Labyrinthine polysaccharide aggregation as seen after periodic acid-thiocarbohydrazile-silver proteinate tests.

adhesive vesicles. Pit-connections between carpospores begin to lyse. This degeneration appears to be mediated by microtubules.

METABOLIC BYPRODUCTS

In *Chylocladia verticillata* (Champiaceae), numerous small cells are observed scattered on the cortex surface facing the cavity of the hollow thallus (Fig. 98 A). These small inner cells are characterized by many membrane-bound inclusions in the cytoplasm (Fig. 98 B). Staining with Coomassie Brilliant Blue R and digestion with Pepsin solution demonstrate the proteinaceous nature of these inclusions (Fig. 98 B) and clustering of ribosomes on their cytoplasmic face support their proteinaceous constitution, and indicate that they may originate from the endoplasmic reticulum. The proteinaceous inclusions are made up of about 20 nm thick, fibrillar, elements and appear either as successive arrays of arcs or as parallel rows of alligned dots when observed in different planes (Tsekos and Bosabalidis, 1988).

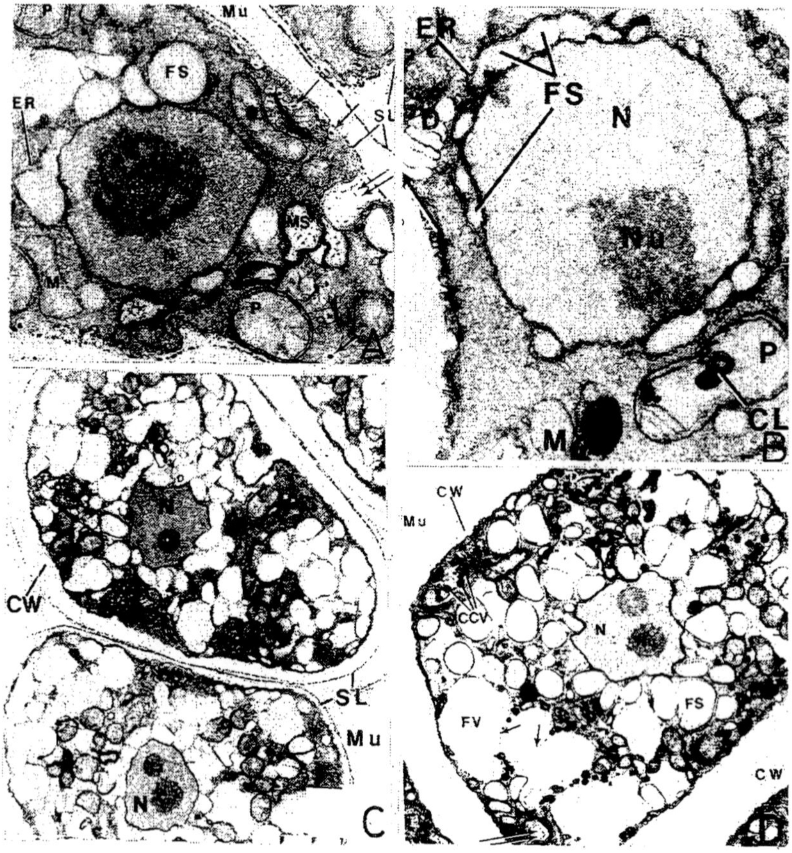

Fig. 97. A-D. Ultrastructure : Carposporogenesis

(after Delivopoulos and Kugrens, 1984).

(CCV, condense cored vesicles; CL, concentric lamellae; CW, carpospore wall; D, dictyosome; ER, endoplasmic reticulum; FS, floridean starch; fv, fibrous vacuole; M, mitochondria; Ms, mucilage sac; Mu, mucilage; N, nucleus; Nu, nucleolus; P, plastid; SL, separating layer)

A. Young carpospore showing discharging mucilage sacs (double arrows). Golgi vesicles (arrows) are also releasing their contents, thus forming the initial separating layer.

B. Young carpospore with perinuclear starch grain formation, ER is in intimate association with starch grains.

C. Low magnification of a medium-aged carpospore containing large starch grains, numerous fibrous vacuoles and well-developed plastids, Carpospore wall deposition has been completed and the wall consists of the outer separating layer and an electron-transparent fibrous layer.

D. Low magnification of a typical mature carpospore showing some compartmentalization of organelles.

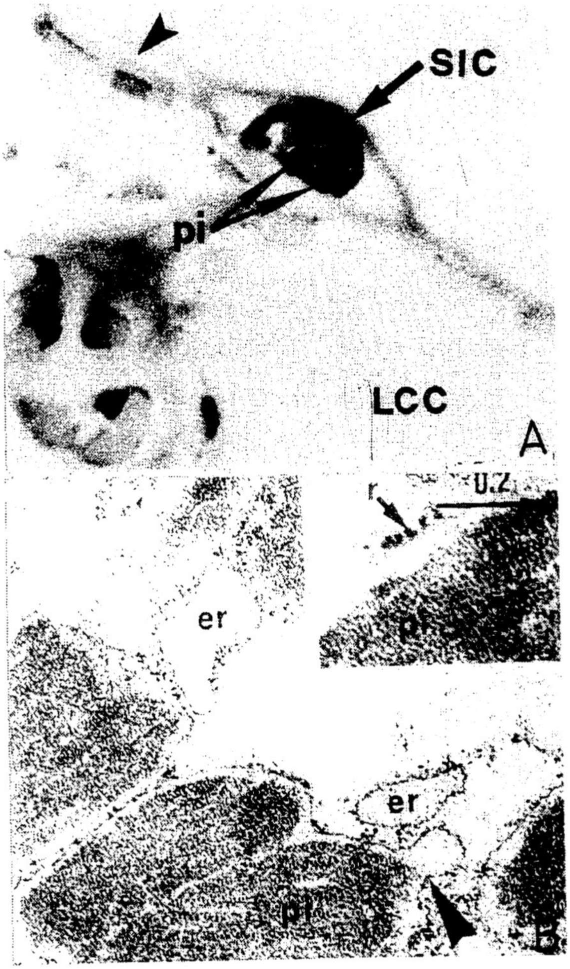

Fig. 98 A, B. Proteinaceous inclusions in *Chylocladia.*

(after Tsekos and Bosabalidis, 1988).

(cw, cell wall; er, endoplasmic reticulun; LCC, large cortical cell; pi, proteinaceous inclusion; r, ribosome; SIC, small inner cell).

A. Cross section of the thallus with large cortical cells and a small inner cell. Coomassie Brilliant Blue R staining. The fibrillar inclusions are strongly stained, while the outer wall of the small inner cell is weakly stained. No reaction is found in the walls of the large cortical cells (arrowheads).

B. Part of a small inner cell of the thallus, short dilated elements of the rough endoplasmic reticulum occasionally appear fused to the inclusion membrane (arrow head). Inset: ribosome clusters on the cytoplasmic face of the inclusion membrane.

7 Palmariales

CLASSIFICATION

The taxa currently placed in Palmariales have been of uncertain systematic position. *Palmaria, Halosaccion* and *Leptosarca* were previously placed in the family Rhodymeniaceae, order Rhodymeniales mainly because of apparent similarities in vegetative construction to Rhodymenalian taxa, despite a lack of information on the female reproductive system (Kylin 1931, 1956). The unique method of tetrasporangium formation in these genera, where the tetrasporangium is borne on a generative stalk cell; the presumed absence of carpogonia and carposporophyte; consequent biphasic life-history; and formation of spermatangia in pairs on the spermatangial mother cell, led to the removal of these taxa from the family Rhodymeniaceae to a new family Palmariaceae (Guiry, 1974) in the order Rhodymeniales. Subsequently, Guiry and Irvine (1978) created the order Palmariales to accomodate the single family Palmariaceae.

Supporting evidence to justify the erection of the order Palmariales has been provided by the ultrastructural study of pit-plugs. In Palmariales, two layers of materials (two cap layers) cover each end of the plug core. A cap membrane separates the two cap layers. The outer cap layer forms a thin plate (Pueschel, 1980; Pueschel and Cole, 1981, 1982).

The placement of additional two taxa *Rhodophysema* and *Halosacciocolax* in the Palmariales is still debatable (Silva, 1983). *Rhodophysema* seems to belong to Palmariales on the basis of it having pit-plugs with two cap layers (Pueschel and Cole, 1982) and a lack of carpogonial branch (DeCew 1982; DeCew and West, 1982). All the *Rhodophysema* species lack generative stalk cells, nor is it known whether these cells are homologous to those of the other Palmariales. Further *Rhodophysema* is monoecious with highly reduced tetrasporophyte and is parasitic on the gametophyte (Hawkes and Scagel, 1986).

Guiry and Irvine (1983) placed *Halosacciocolax* in the Palmariales on the basis of its two layered pit-plug caps, but Kraft (1981) concludes that its affinities are uncertain. Hawkes and Scagel (1986) provisionally included the genera *Rhodophysema* and *Halosacciocolax* in Palmariales. The classification of Hawkes and Scagel (1986) is as follows:

Order : **Palmariales**

Family : **Palmariaceae.**

Genera :

1. ***Halosacciocolax*** **:** Thallus a diminutive parasite on *Palmaria*.
2. ***Rhodophysema*** **:** Thallus a small epilithic or epiphytic crust.
3. ***Palmaria*** **:** Thallus a large foliose blade, often palmately branched.
4. ***Halosaccion*** **:** Thallus a single, hollow sac; without mucilage but filled with water.

MORPHOLOGY

Palmariales, in nature, are characterized by the absence of female gametophytes and presence of isomorphic, macroscopic male gametophytes and tetrasporophytes. Recent *In vitro* studies have shown the presence of small. microscopic, short-lived female gametophytes in many genera thereby exhibiting sexual dimorphism. *Rhodophysema* is an exception which is bisexual.

MALE GAMETOPHYTE/ TETRASPOROPHYTE

The thalli are multiaxial in construction and may be parasitic (*Halosacciocolax*); free-living or epiphytic; solitary or gregarious; erect (*Palmaria*) or crustose (*Rhodophysema*) and solid or hollow (*Halosaccion*) (Hawkes and Scagel, 1986).

Erect forms : They are membranous to coriaceous, and reach upto 1 metre in length. Blades are broadly flattened and branching is simple, palmate or irregularly proliferous. Thallus is attached by a small discoid or hapteroid holdfast. Stipe is absent or inconspicuously present in *Halosaccion* (Fig. 99D).

A variation in blade morphology and branching pattern is observed within the genus *Palmaria,* In *P. callophylloides,* the blades are dichotomously or trichotomously or irregularly branched (Fig. 99 C); in *P. hecatensis* (Fig. 99 A) single, unbranched blades are strap-shaped or occasionally broad and lobed, arising from extensive crustose base; and in *P. mollis* (Fig. 99 B), the typically membranous blades are either single or several arising from a discoid holdfast, often with short marginal proliferations near the base, giving it a frilly appearance.

In *Halosaccion* (Fig. 99 D), the thallus is erect, simple or branched hollow sac, stipitate and attached to the substratum with discoid holdfast.

Crustose forms : They are microscopic or upto 40 mm in diameter, epiphyte on other algae or sea grasses, epizoic on various invertebrates or epilithic being attached to rocks. *Rhodophysema* is a typical crustose genus with *R. elegans* showing a flat, noncalcified crust, 1-3 mm in diameter. The monostromatic margins display unlimited horizontal growth, spreading over the substratum. Central regions of large crusts are polystromatic perithallus comprising of erect filaments. In *R. georgii,* the thallus may be crustose or globose to pear-shaped pustule. The crusts often grow together to form patches that are several millimeters in diameter.

Parasitic forms : The thallus of *Halosacciocolax* is colourless, tuberculate, globose or cushion-like, within a mucilaginous envelope. *H. kjellmanii* is a putative parasite forming wart-like, colourless to brownish-yellow outgrowths (0.5-1.5 mm wide and 160-180 μm thick) on *Palmaria.* They parasitize only in those parts of the host where cortical thickening has occurred. The vegetative surface of the parasite is covered with a thick mucilaginous envelope that ruptures when reproductive structures (spermatangia and tetrasporangia in male gametophyte and tetrasporophyte respectively) are formed.

FEMALE GAMETOPHYTE

The female gametophytes in *Halosaccion glandiforme, Palmaria mollis* (Van der Meer and Bird, 1985), *P. palmata* (Fig. 99 E. Van der Meer and Todd, 1980) are microscopic and consist of a crustose disc of cells, reaching reproductive maturity in only a few days after tetraspore germination.

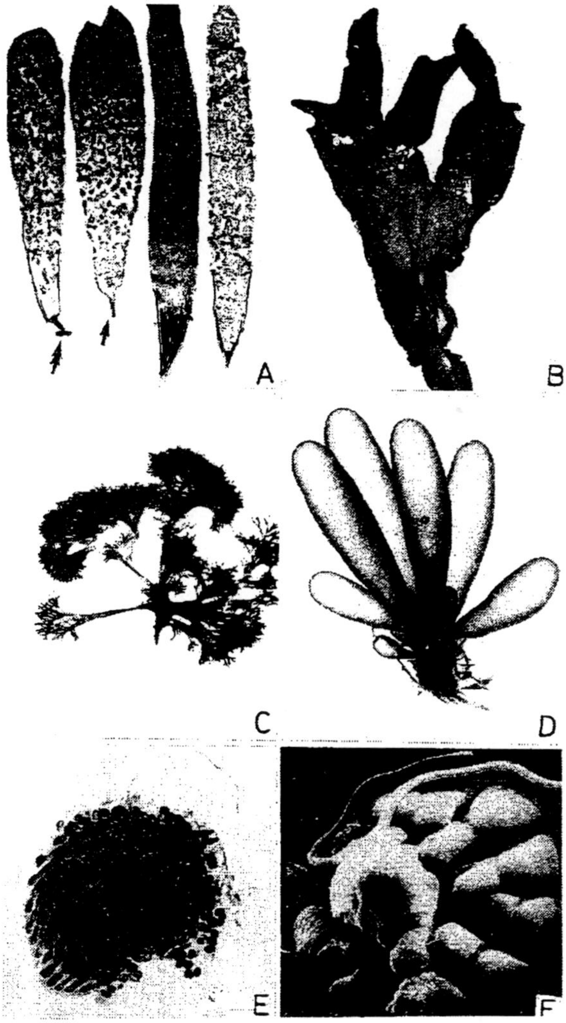

Fig. 99. A-F. Morphology

A. *Palmaria hecatensis* : Tetrasporangial thalli have mottled pattern and basal crust (double arrow) (after Hawkes and Scagel, 1986a).

B. *Palmaria mollis* : A sheltered water plant showing characteristic habit (after Hawkes and Scagel, 1986a).

C. *Palmaria callophylloides* : Vegetative thallus (after Hawkes and Scagel, 1986a).

D. *Halosaccion glandiforme* : Several young vegetative thalli growing epiphytically on *Prionitis sp.* (after Hawkes and Scagel, 1986a).

E. *Palmaria palmata* : A sexually mature, four week old female gametophyte with long slender trichogyne (after Van der Meer and Todd, 1980).

F. *Halosaccion americanum* : SEM of female gametophyte as a prostrate disc. Carpogonium has two trichogynes (after Mitman and Phinney, 1985).

In *vitro* studies in *Halosaccion americanum* show that the female gametophyte is initially only two-celled consisting of vegetative cell and carpogonium. Fertilization may take place at this stage. If the unfertilized female gametophytes are maintained for longer periods, they continue to grow. The initial single vegetative cell divides many times forming an eccentric disc of small cells around one side of the earlier formed carpogonium (Fig. 99 F), the enlarging disc produce numerous carpogonia. These gametophytes remain as small prostrate discs and never produce an upright thalli (Mitman and Phinney, 1985).

In *Rhodophysema,* the solitary carpogonium develops terminally on a vegetative filament. In Palmariales, carpogonium develops either terminally or intercalary on the vegetative filament and the carpogonial branch is absent.

ANATOMY

The thalli in Palmariales are multiaxial and distinguishable into an outer cortex and an inner medulla. The cortex consists of 1-3 (-18) layers of small, elongate, pigmented cells often arranged in anticlinal rows. The cells in surface layers contain more number of chloroplasts. The medulla comprises of 4-6 (-8) layers of cells which are comparatively large and are devoid of chloroplasts. They are isodiametric to oblong in shape and are either compact or loosely adherent. In hollow forms, a central cavity is present that is surrounded by medullary cells (Hawkes and Scagel, 1986).

In *Palmaria, Devaleraea* and *Leptosarca,* there is an abrupt transition from the small-celled cortex to the large-celled medulla. In *Halosaccion,* there is a more gradual transition in cell size.

In *Halosacciocolax kjellmanii* (Fig. 100 A) the thallus is composed of upright, parallel filaments issuing from a basal layer which is partially endophytic in the host cortex. These parasitic filaments branch between the host cortical cells. Cells near the margin of basal layer are elongate and are arranged in radiating rows, forming secondary pit-connections with the host.

In *Halosaccion americanum,* the cortex comprises of 1-4 simple or branched row of small anticlinally elongate cells bereft of pit-conections. The cells of the thallus become gradually large towards the central cavity, the innermost 3-10 layers made up of isodiametric medullary cells (Mitman and Phinney, 1985).

In *Halosaccion glandiforme* (Fig. 100 B), the thallus wall varies in thickness from 170 to 460 µm. The central cavity is surrounded by large, isodiametric medullary cells with stellate protoplast. These undergo fusion with the surrounding cells to form an anastomosing meshwork. Medullary cells become small towards the thallus surface and gradually merge into a cortex composed of one to several elongate cells arranged in vertical files.

In *Palmaria callophylloides,* the pseudoparenchymatous thalli are 120-580 µm thick, without any hollow cavity. The medulla consists of 1 or 2 or more large cells with smaller cells packed around them. The cortex is typical 2 or 3 or more cells thick in the basal and marginal regions. Fusion between cells occur in the transition region from medulla to cortex. In transverse section, the medulla of *P. callophylloides* shows more cells than of either *P. mollis* or *P. hecatensis* (Hawkes and Scagel, 1986).

In *P. mollis* (Fig. 100 C), the blades are 140-480 µm thick. Large medullary cells form bulk of the thallus with small medullary cells filling the space between them. The cortex is one cell thick, but in basal portion of older thalli it becomes four or five or more cells thick (Van der Meer and Bird, 1985).

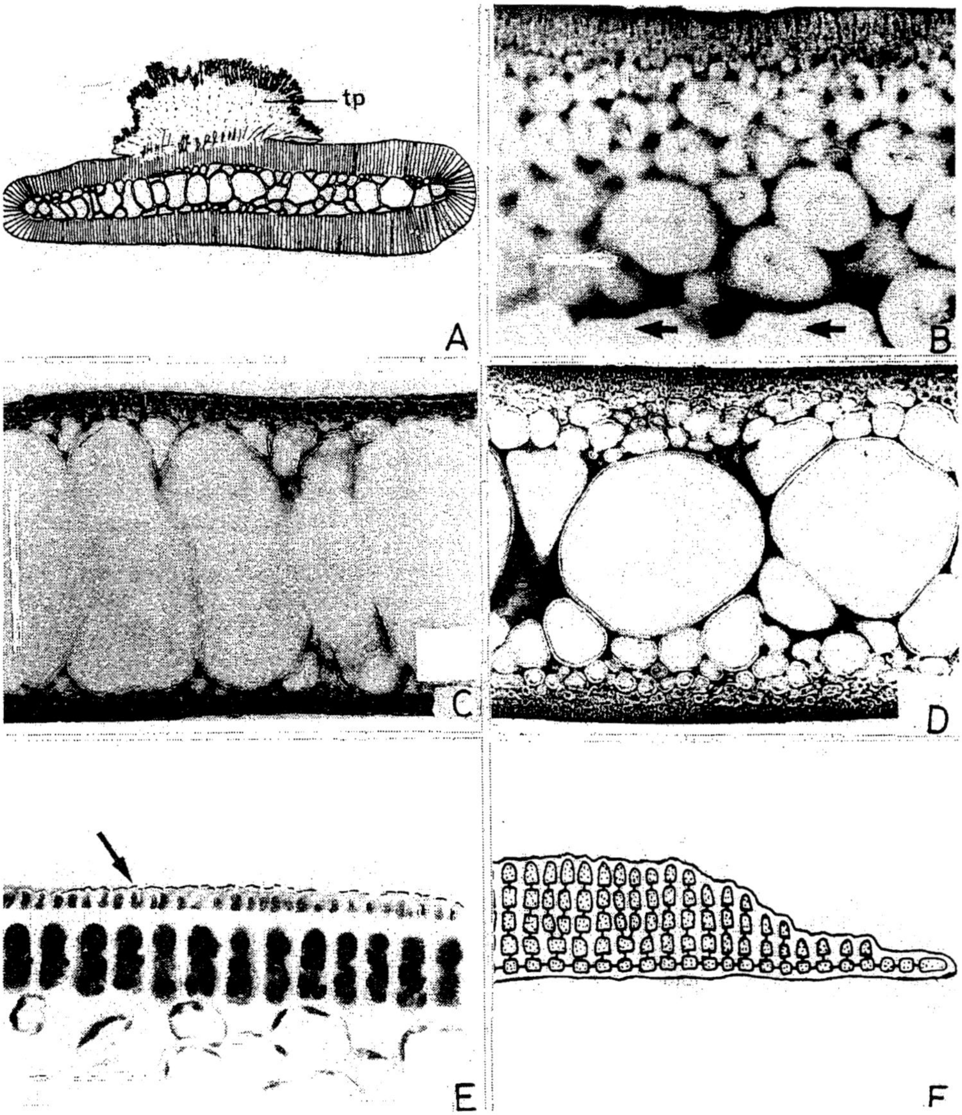

Figure 100 A-F. Anatomy.

(after Hawkes and Scagel, 1986a).

(hc, host cortex; tp, tetrasporangial pustule)

A. *Halosacciocolax kjellmanii* : Tetrasporangial pustule attached to host cortex.

B. *Halosaccion glandiforme* : Young vegetative thallus to show hollow cavity (two arrows).

C. *Palmaria mollis* : Vegetative thallus showing large celled medulla and small celled cortex.

D. *Palmaria hecatensis* : Mature vegetative thallus with well developed cortical layer.

E. *Rhodophysema georgii* : Monostromatic portion of crust (arrow) growing on *Phyllospadix* sp.

F. *Rhodophysema elegans* : Vegetative crust with monostromatic, distromatic and polystromatic portions.

In *P. hecatensis* (Fig. 100 D), the thallus is typical of **Palmaria** type where the cortical cells are arranged in closely appressed anticlinal filaments in 4 or more rows. Fusion between cells in the cortical layers is of common occurrence (Hawkes, 1985).

In *Palmariales,* the crustose form namely *Rhodophysema* (Fig. 100 E, F), is frequently confluent, monostromatic at the margin, polystromatic towards the centre where they are upto 13-20 cells thick and made up of erect, tightly coalesced filaments of elongate cells. In *Rhodophysema elegans* (Fig 100 F), the young crusts consist of a monostromatic hypothallus composed of closely adhering radiating filaments which lack rhizoids. With age the hypothallus produces a polystromatic perithallus (Fig. 100 F), consisting of tightly packed, 2-20 cells long, erect filaments which are usually simple or sometimes dichotomously branched. Only the central region of the crust becomes multistratified to any extent; there is usually an outer region 3 or 4 cells thick and occassionally extensive monostromatic regions are present at the growing margins. In surface view, cells of the hypothallus are rectangular to quadrate. Frequent lateral fusion occurs between adjacent hypothallus cells as well as between cells of the perithallus (Hawkes and Scagel, 1986).

In *R. georgii* (Fig. 100 E), the thallus development is similar to *R. elegans.* Here the cells of the perithallus contain several parietal and plate-like chloroplasts. In addition to crustose morphology, this species can form globose pustules, which form when the central cells swell and elongate to form a colourless tissue (Hawkes and Scagel, 1986).

REPRODUCTION

SPERMATANGIUM

Spermatangia are formed in slightly raised sori from the surface layer of cortical cells of the thallus. They may cover almost the whole surface of the gametophyte, except for the extreme apex and base. In parasitic forms, the spermatangia develop in a continuous soral area extending to the entire surface of the gametophyte.

In *Halosacciocolax kjellmanii,* spermatangia are known only in plants from Greenland, where they are reported to be "hyperparasitic" in small spermatangial pustules composed of erect parallel filaments. In *Halosaccion glandiforme* and *H. americanum,* spermatangia are abundantly formed in a continuous sorus from the cortical layer.

In *Palmaria hecatensis* and *P. mollis.* spermatangia initially occur in discrete soral patches; these soon coalesce to form a continuous sorus over the blade surface. The spermatangial sori are less pigmented than surrounding vegetative tissue, giving the spermatangial thallus a mottled appearance. During progressive development the individual sori coalesce to form a continuous sorus that cover the entire blade except at the margins and basal region.. In *P. callophylloides* (Fig. 101 B), spermatangial sori develop over most of the blade surface including the margins. In *Rhodophysema elegans* (Fig. 101 A) and *R. georgi,* one or two spermatangia are formed in sori on spermatangial mother cells, which are borne terminally on the perithallial filaments.

In Palmariales, the superficial cortical cells of the thallus give rise to spermatangial mother cells, with each cortical cell cutting 3-4 (-8) spermatangial mother cells at the upper corner. Each spermatangial mother cell cuts off subterminally a pair of spermatangia. Multicellular, elongate paraphyses are absent among spermatangia. Both spermatangia and spermatangial mother cell are beseiged by a common cuticular wall. Prior to spermatangial release, the entire cuticle is

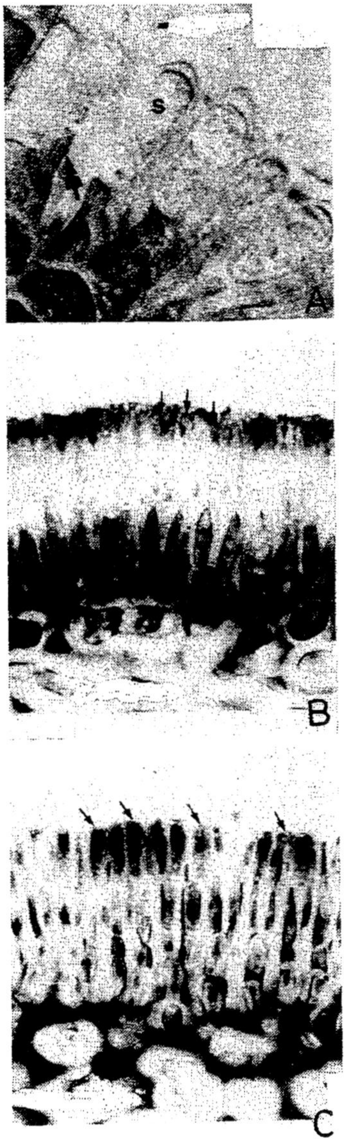

Fig. 101 A-C. Reproductive structure: spermatangium

(A: after Saunders *et. al.*, 1989; B-C; after Hawkes and Scagel, 1986a).

(sp, spermatangium).

A. *Rhodophysema elegans* : Mature spermatangial mother cells bearing spermatangia on long extensions (arrow), spermatangia surrounded by nests of walls.

B. *Palmaria callophylloides* : Long and narrow spermatangia (arrows).

C. *Halosaccion glandiforme* : Spermatangia (arrows).

shed. Regeneration of secondary and tertiary spermatangia occurs within the empty spermatangia. In *Palmaria, Halosacciocolax, Halosaccion glandiforme* and *Rhodophysema,* the spermatangial mother cell produces two spermatangia, whereas in *H. americanum,* it gives 3 or 4 spermatangia. The shape of the spermatangia varies from narrow (*H. glandiforme*) (Fig. 101 C), to elongate (*Palmaria sp.*) and ovoid (*Rhodophysema elegans*). In *Halosaccion glandiforme,* the spermatangium is 10-11 μm long; in *Palmaria callophylloides* it is 6-9 μm long, in *P. hecatensis,* 7-10 μm long and 1.5-2.0 μm wide, and in *P. mollis,* 5-10 μm long.

In vitro studies reveal that in tetrasporangiate type of *Rhodophysema elegans,* the spermatangial mother cells bearing ovoid spermatangia (6.5-14 μm x 4-9 μm) are associated with developing, immature sporangial sori in all isolates (Saunders *et al.,* 1989). Repeated regeneration of spermatangia within old walls resulted in nest of walls around spermatangia (Fig.101 A) and the formation of long, horn-like extensions from the spermatangial mother cells. Released spermatia are observed singly or in mucilaginous clouds and are spherical in shape.

In bisporangial cutures of *R. elegans* (Saunders *et. al.,* 1989), spermatia are reported to be elongate (8-13x4-6 μm), borne in pairs from spermatangial mother cells that terminate the erect filaments scattered among sterile filaments and bisporocytes in immature sori. Shed spermatia, 5-8 μm in diameter, are observed to be spherical. Spermatangia regenerated into empty sheaths of previously discharged spermatia.

CARPOGONIUM

The carpogonium is sessile and terminal on any vegetative filament (not a specialized branch) usually towards the margin of the crustose thallus. There is no auxiliary cell or carpogonial branch and no carposporophyte develops after fertilization.

In *Halosaccion ramentaceum,* the carpogonia are observed only in culture and they closely resemble those of *Palmaria palmata* (Van der Meer and Todd, 1980), consist of a single pigmented cell bereft of any obvious carpogonial branch or auxiliary cell. They resemble the surrounding vegetative cells, except for the long and occasionally branched trichogynes. The carpogonia are concentrated near the growing margins of the disc, and none have been observed on the upright fronds (Van der Meer, 1981).

In *Palmaria palmata, in vitro* studies reveal that the carpogonia appear to be solitary cells without any obvious carpogonial branch. No evidence for auxiliary cell has been obtained. Carpogonia develop usually near the surface, but sometimes are buried within the disc. Buried trichogynes are irregular in diameter, their shape perhaps resulting from the influence of surrounding cells (Fig. 102 C, Van der Meer and Todd, 1980).

In *Palmaria mollis* the carpogonium is solitary and each female thallus produces as many as 15 to 60 carpogonia which are typically borne away from the centremost portion of the disc. In *Halosaccion americanum* (Fig. 102 A; Mitman and Phinney, 1985), and *Devaleraea ramentacea* (Van der Meer, 1981), on the small and one week old female thallus, a large carpogonium develops on the margin of the disc. In both the taxa, as the disc expands additional carpogonia developed with only their trichogynes protruding and produced from internal cells of the vegetative disc (Fig. 102 B).

In *Rhodophysema elegans* the solitary carpogonium is produced only in young crusts (2 or 3 weeks old) or in areas or marginal meristems on older crusts (DeCew and West. 1982).

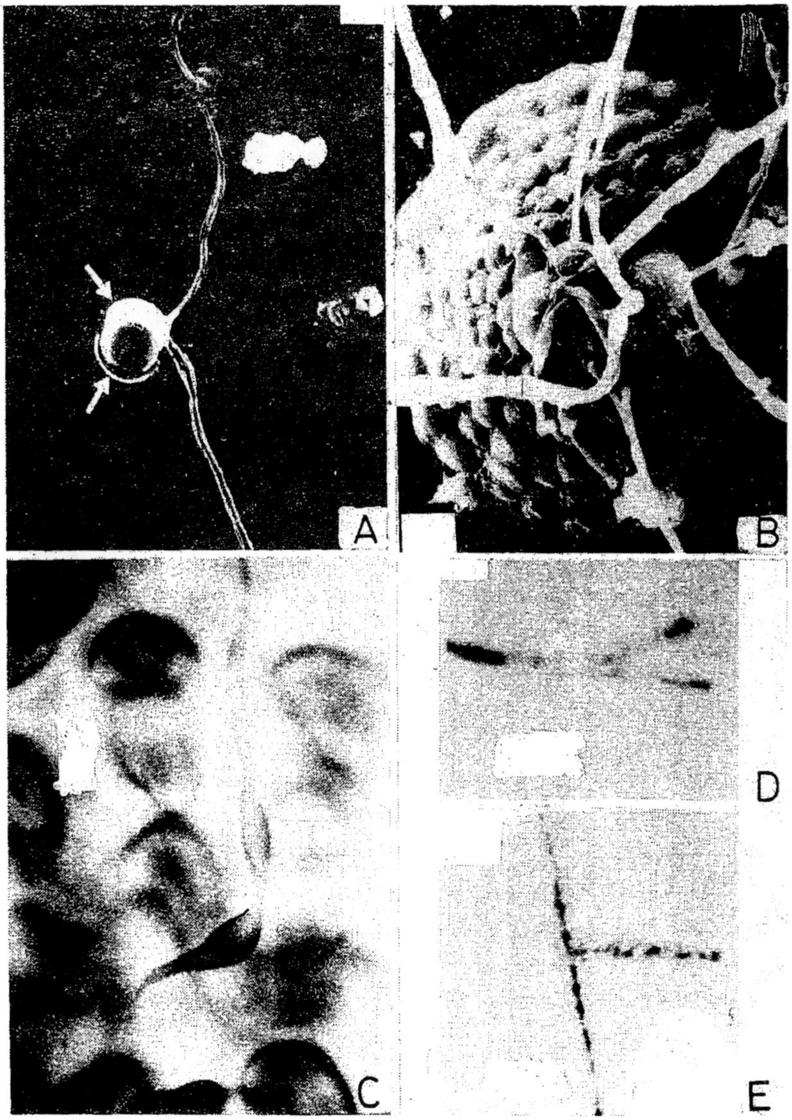

Fig. 102 A-D. Carpogonium and trichogyne.

(A-B : after Mitman and Phinney, 1985; C-D; after Van der Meer and Todd, 1980)

A. *Halosaccion americanum* : SEM of 24 hrs old, 2-celled female gametophyte. The fertilized carpogonium bears two trichogynes with spermatia still attached.

B. *Halosaccion americanum* : SEM of one month old female gametophyte to show many large carpogonia.

C. *Palmaria palmata* : Englarged view of buried trichogyne.

D. *Palmaria palmata* : Trichogyne branching, one is near the tip, whereas the other is distant from the tip.

Branching trichogynes are observed in *Plamaria plamata* (Van der Meer and Todd, 1980; Fig. 102 D)., *Devaleraea ramentacea* (Van der Meer, 1981) and *Halosaccion americanum* (Mitman and Phinney, 1985). In *H. americanum,* multiple trichogynes are sometimes reported (Fig. 102 A), a unique feature not reported for any other red algal taxon (Mitman and Phinney, 1985).

After fertilization, the trichogyne degenerates in *Palmaria, Halosaccion* and *Devaleraea,* and the zygote directly gives rise to the terasporophyte, which overgrows the microscopic female thallus and forms an independent holdfast on the substratum. In *Rhodophysema,* the trichogyne degenerates and the fertilized carpogonium elongates and divides into two unequal cells, the lower one is the stalk cell (presumably diploid) and the upper is the tetrasporangial initial. No carpogonial branch, carposporophyte or free living tetrasporophyte are observed.

TETRASPOROPHYTE

The carposporophyte generation is absent in Palmariales and the fertilized carpogonium (or the first division product of the zygote) directly gives rise to the tetrasporophyte (or only to tetrasporangial initial). The tetrasporophyte is macroscopic and morphologically and anatomically indentical to the male thallus.

The tetrasporangia are scattered or aggregated into sori, which can be confluent and cover almost the entire surface of the tetrasporophyte (except for the extreme apex and base). In parasitic forms (*Halosacciocolax;* Fig. 103 A), tetrasporangia may develop in a continuous soral area over the entire surface of the tetrasporophyte. Elongate multicellular (3-8 cells) paraphyses or vegetative filaments may or may not occur among the tetrasporangia.

The tetrasporangial initial usually originates from a superficial cortical cell of the tetrasporophyte. This cortical cell enlarges and divides periclinally to form two cells of unequal size, the inner of which is the stalk cell and the outer is the tetrasporocyte.

The tetrasporangium is arranged cruciately (or tetrahedrally in some populations of *Rhodophysema elegans*). The stalk cell is generative and usually regenerates secondary and frequently tertiary tetrasporangia. The presence of a generative stalk cell is the unique, diagnostic feature of the Palmariales (Guiry, 1978). A distinctive wall, called tetrasporangial wall is laid down around the tetrasporangial initial, so that the stalk cell is completely included, together with the tetrasporocyte, within the tetrasporangial wall. Considering this, Pueschel (1979) suggests that the product of the divided tetrasporangial initial should be collectively referred to as the tetrasporangium. The tetrasporangium will thus consist of the stalk cell and the tetrasporocyte. New walls are formed around successive tetrasporocytes (See Hawkes and Scagel, 1986).

The tetrasporocyte enlarges and divides cruciately (or tetrahedrally in some populations of *R. elegans)* to form four haploid tetraspores. After release of the tetraspores, the stalk cell expands and protrudes into the empty tetrasporangium, enlarges, divides periclinally and produces a new tetrasporocyte initial which develops identically to that leading to formation of the primary tetrasporangium.

In the genera *Halosaccion* (Fig. 103 B) and *Palmaria* (Fig. 103 C) the cruciately divided tetrasporangia with the generative stalk cell are terminal. Tetrasporangia occur generally in sori tetrahedrally divided, on both the surfaces of the tetrasporophyte thallus. The tetrasporangia with the generative stalk cell are terminal. Tetrasporangia are surrounded by vegetative filaments (*Halosaccion;* Fig. 103 B). or 3-5 celled paraphyses (*Palmaria,* Fig. 103 C).

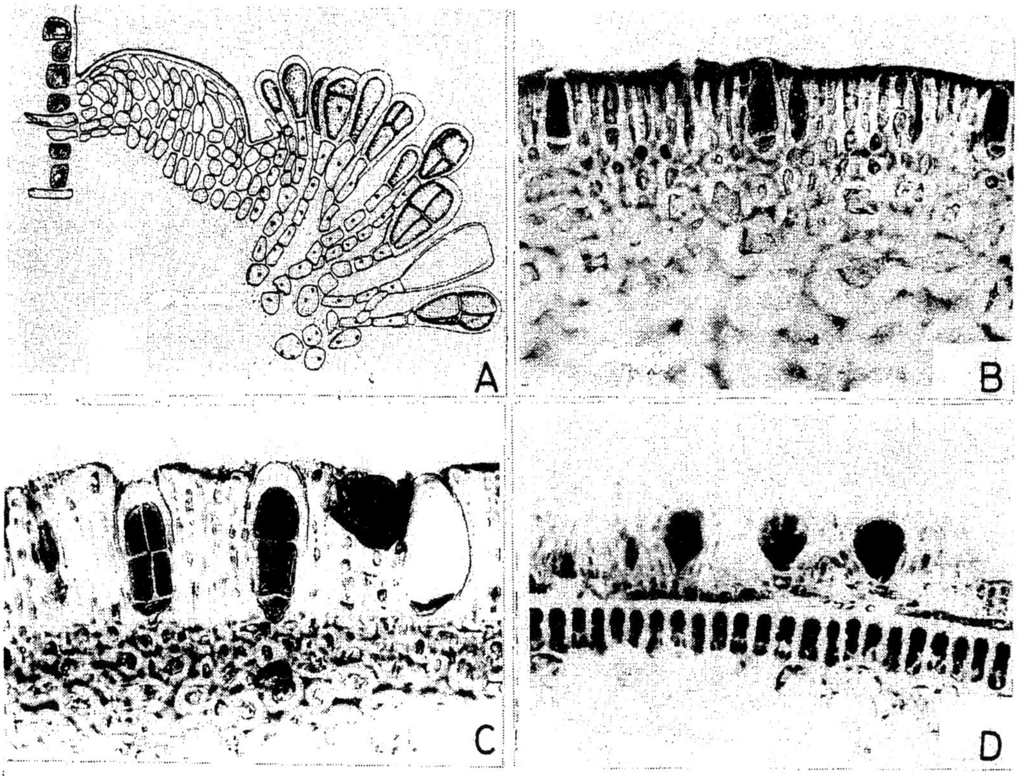

Fig. 103. A-D. Tetrasporangium.

(after Hawkes and Scagel, 1986a)

A. *Halosacciocolax kjellmanii* : Tetrasporangia at edge of pustule. Secondary filaments (arrow) penetrate the host cortex.

B. *Halosaccion glandiforme* : Tetrasporangia surrounded by vegetative filaments.

C. *Palmaria hecatensis* : Tetrasporangia with stalk cell separated by paraphyses.

D. *Rhodophysema georgii* : Irregularly tetrahedral tetrasporangia surrounded by paraphyses.

In *Halosaccion ramentaceum* (Van der Meer, 1981) the diploid tetrasporophyte initiates development on the female thallus but almost immediately forms an independent holdfast on the substratum beside the female thallus. After the holdfast is well established the first diploid fronds arise.

In *Palmaria palmata* (Van der Meer and Todd, 1980), the diploid tetrasporangial plants develop directly on female thallus. As they enlarge, they completely overgrow the smaller female gametophytes and form their own attachments with the substratum.

In *Halosacciocolax kjellmanii*, tetrasporangia develop terminally on the pustules which are composed of a base of irregularly round, colourless and elongate, thick walled cells which penetrate and form secondary pit-connections with the host cortex (Fig. 103 A). Tetrasporangia are cruciately divided (Lund, 1959).

In *Rhodophysema elegans*, (Fig. 104 A-G), a crustose form, the zygote elongates, later divides transversely into two unequal cells. The lower, small cell becomes the diploid stalk-cell and the terminal, long cell forms the tetrasporangium. The tetrasporangium divides cruciately. The tetrasporangium and stalk cell are borne directly on the gametophytic thallus in discrete and wart-like sori. The stalk cell is generative. Paraphyses are present and are composed of 5-7 pale cells. Kornmann and Sahling (1983) reported in this taxon irregular and tetrahedral terasporangia.

The tetrasporangia in *R. georgii* (from the type locality Isles of Scilly, England) are reported to be cruciately divided. The specimens from British Columbia, however, have tetrasporangia that are irregularly tetrahedrally divided (Fig. 103 D).

BISPORANGIUM

In *Rhodophysema elegans*, in addition to sexual life-history in tetrasporangial populations a direct, asexual, mitotic bispore to bisporophyte life-history pattern is observed in bisporoangial populations from the North Atlantic Ocean. (Saunders *et al.*, 1989). South and Whittick (1976) noted that bisporangial production predominated in the Newfoundland isolate of *R. elegans*. Fletcher (1977) presented morphological evidence of a direct life-history involving mainly bisporangia, without sexual structures in a British isolate of *R. elegans* (see Saunders *et. al.*, 1989).

Bisporangia in *R. elegans* have usually been reported as occurring on the same plants as tetrasporangia (South and Whittick 1976; Fletcher, 1977). However, separate bisporangial and tetrasporangial thalli have been found along Irish Coasts (Hawkes and Scagel, 1986) and in Japan (Masuda and Ohta, 1981).

Observations by Saunders *et al*, (1989) on field collected material of *R. elegans* from Irish coast show that like tetrasporangia, bisporangia are also formed in sori among the curved and projecting 3 to 5 celled sterile filaments (Fig. 105 D, E). Two bisporangia are sometimes borne on a single generative stalk cell. Bisporangia also regenerate sporocytes within the empty sheath layers of previous sporangia. Bisporangia and tetrasporangia are produced by separate plants. Rarely, however, bisporangia and tetrasporangia seem to occur on the same plant but in separate sori. The indentations in crust margins represent a confluence of two of more individuals.

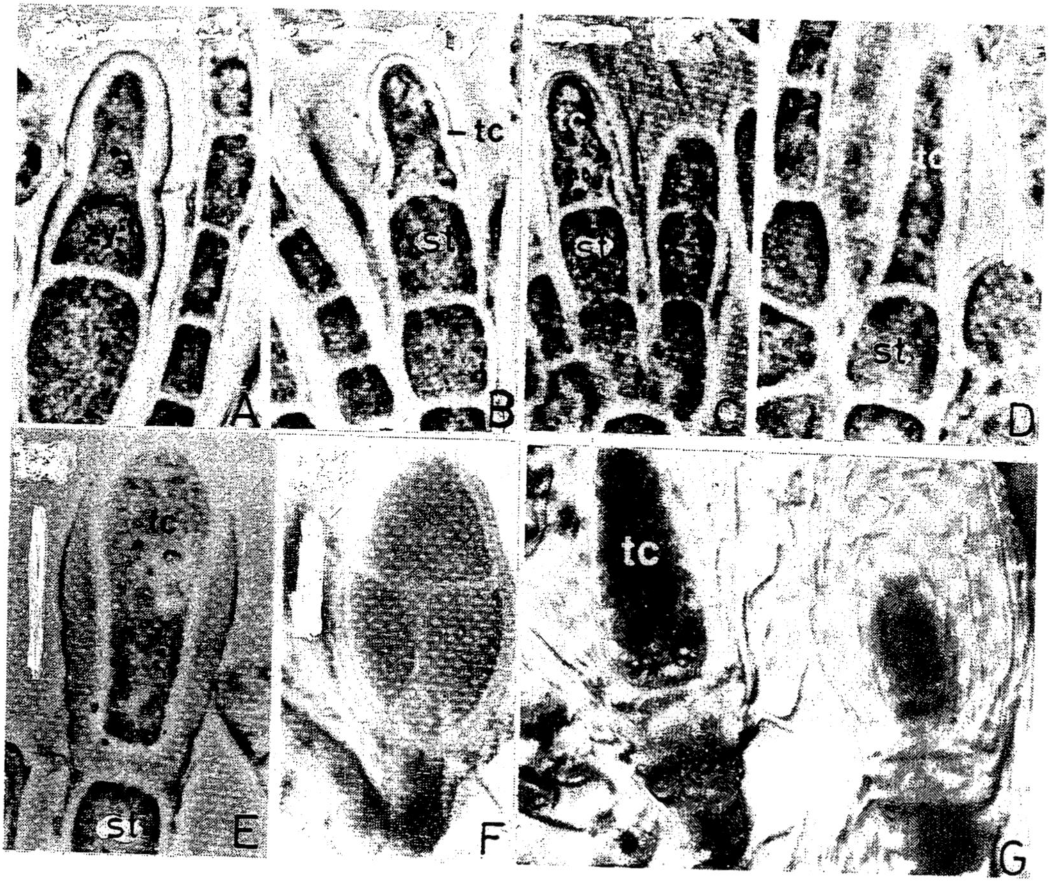

Fig. 104 A-G. Tetrasporogenesis in *Rhodophysema elegans.*

(after Saunders *et al.*, 1989)

A. Zygote after first division.
B. The initial division of zygote results in a primary tetrasporocyte and a generative stalk cell.
C. Tetrasporocyte has enlarged.
D. Elongate tetrasporocyte on stalk cell.
E. Tetrasporocyte has enlarged greatly, whereas the stalk cell remains the same.
F. Mature tetrasporangia, with spores in a regular, cruciate arrangement, inside the old walls of discharged and empty sporangium.
G. Tetrasporocytes inside nest of old walls attached at various levels to the stalk cell.

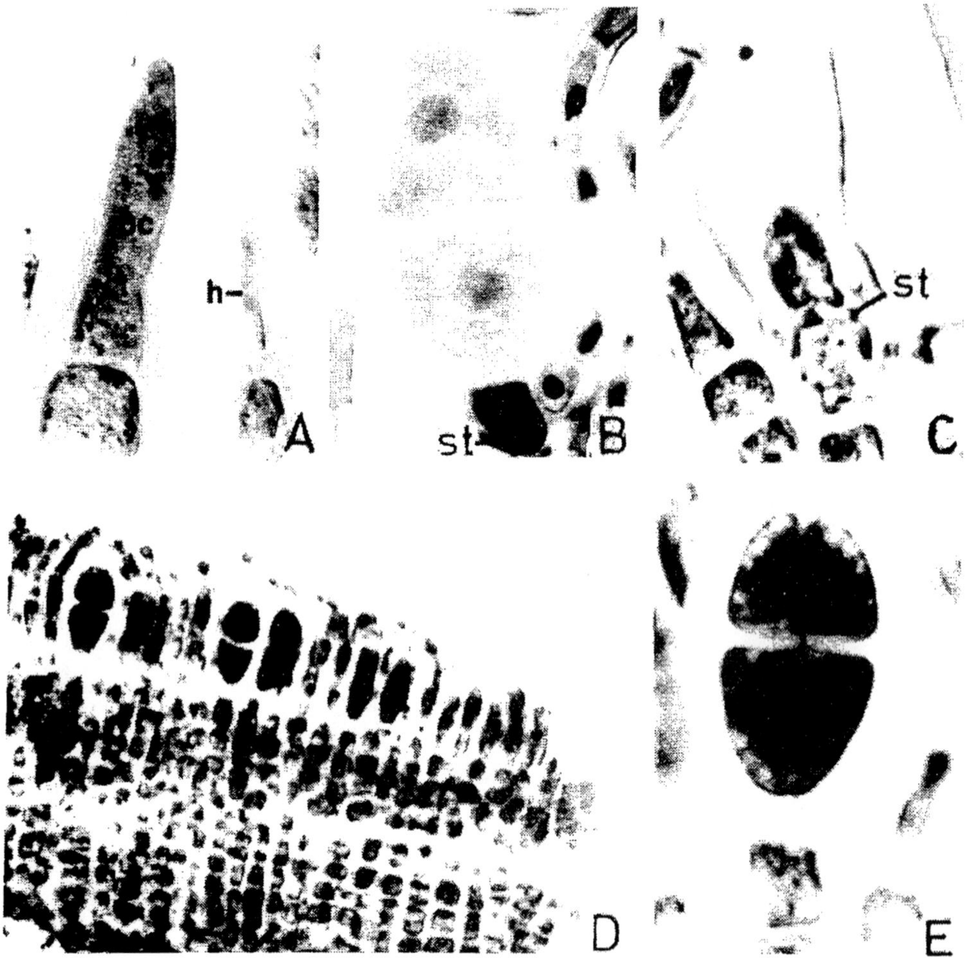

Figure 105. A-E. Bisporogenesis in bisporangial populations of *Rhodophysema elegans*.

(after Saunders *et al.*, 1989)

(A-C : Bisporangial culture; D and E : Field collected material)

(bc, bisporocyte; h, hair cell; st, stalk cell)

A. Elongate developing bisporocyte adjacent to narrow hair cell.
B. Mature binucleate bisporangium on intercalary stalk cell.
C. Stalk cell protruding into old sporangial wall prior to formation of a new sporocyte.
D. Thick crust with sorus of bisporangia surrounded by curved sterile filaments.
E. Bisporangium on stalk cell.

Bispores (from field collected bisporangial plants) reveal identical germination pattern as tetraspores in culture (Saunders *et al.*, 1989). The initial cleavage of bispore is followed by further divisions that result in formation of discs with marginal meristems. Vegetative thalli are identical to those developing from tetraspores and likewise form hairs (Fig. 105 A), Bisporangial initials are formed in sori among 3 to 5 celled curved sterile filaments. Sporangial initials are elongate (Fig. 105 A) and always mature into bisporangia (Fig. 105 B). Bisporocytes and mature binucleate bisporangia are frequently formed by intercalary sterile filaments. Bisporangial stalk cells are regenerative, with new sporangia developing within the sporangial sheaths of previous bisporangia (Fig. 105 A-E; Saunders *et al.*, 1989).

Morphological evidence of a direct life-history in bisporangial isolates is confirmed by the presence of 35-37 chromosomes in vegetative marginal cells and 34-36 in bisporocytes. Consequently, bisporangial plants are interpreted as polyploid gametophyte (2 x = 36) which undergo a direct mitotic biospore to bisporophyte life-history and also produce diploid spermatia. Examination of bisporangia with DAPI and mithramycin by Saunders *et al.*, (1989) demonstrated that the majority of bispores are uninucleate. Observed variations from the binucleate bisporangial pattern suggest *in situ* germination of mitotic bisporangia, rather than quadrinucleate (meiotic) bisporangia.

LIFE CYCLE STUDIES

The genera are characterized by the lack of the carposporophyte, with the dimorphic gametophyte (except in *Rhodophysema* where the gametophyte is bisexual) that alternate with the tetrasporophyte. Male thalli are macroscopic whereas the female gametophytes microscopic (when unisexual). Tetrasporophytes are independent and isomorphic to male thalli in *Palmaria, Halosaccion* and *Halosacciocolax;* whereas in *Rhodophysema* they are two-celled and parasitic on the gametophyte.

In *Palmaria* and *Halosaccion,* isomorphic and macroscopic male thallus and tetrasporophytes are known in the field for many years, but the female thallus is apparently absent. This led some investigators to speculate that in France populations the spermatia produced by the male thallus are functionless and the the tetraspores are apomeiotic because of chromosome number 14 for all phases (Magne, 1964). However, contrary cytological evidences indicate both haploid and diploid chromosome numbers (Yabu, 1976; Pueschel, 1979; Van der Meer and Todd, 1980), *In vitro* studies of *P. palmata* (Van der Meer and Todd, 1980), *P. mollis* (Van der Meer and Bird, 1985) and *Halosaccion ramentaceum* (Van der Meer and Chen, 1779; Van der Meer, 1981) indicate absence of sexual life histories in these taxa. These studies have demonstrated the presence of diminutive free living female gametophytes (which bear sessile carpogonia) but bereft of carposporophytes. The transitory nature and rapid maturation of the female gametophyte reveal that the spermatia available for fertilization originate from male gametophytes of the previous generation (Hawkes and Scagel, 1986).

In *Rhodophysema elegans* (from California), the life- history is haplo-diplontic, consisting of a haploid, monoecious (bisexual) gametophytic crust and diploid tetrasporangial initial and stalk cell borne directly on the gametophyte (DeCew, 1981; DeCew and West, 1982), Spermatangia and hair cells (presumably trichogynes) for *R. elegans* from Japan are known (Masuda, 1978; Masuda and Ohta, 1981), suggest that sexual reproduction in populations of this species may be widespread. According to South and Whittick (1986) and Fletcher (1977), some populations may have apomictic life histories as a result of apogamy and/or apomeiosis. *R. elegans* (from Germany) possess spermatangia, tetrasporangia and bisporangia, but lack carpogonium and fertilization. The spermatia are functionless and the tetraspores were formed apomeiotically (Karnmann and Sahling, 1983). Bisporangia in *R. elegans* have usually been reported as occurring on the same plants as tetrasporangia (South and Whittick, 1976; Fletcher, 1977). However, separate bisporangial and tetrasporangial thalli have been found along Irish Coasts (Hawkes and Scagel, 1986) and in Japan (Masuda and Ohta, 1981).

In *Rhodophysema georgii*, spermatangia and hair cells (trichogynes) are reported in Japanese population (Masuda and Ohta, 1975) but British populations were apomictic (Fletcher, 1975). It seems highly probable that both sexual and asexual populations coexist (Hawkes and Scagel, 1985). It was noted by Fletcher (1975) for *R. elegans* (from Germany) that the apomictic populations cycle asexually by means of bisporangia.

Magne (1982) suggests that the two major diphasic life history types of Palmariales : **Palmaria** type (Fig. 106 A) and **Rhodophysema** type (Fig. 106 B) may be derived from the triphasic **Polysiphonia** type life-history through elimination of the carposporophyte phase, followed by reduction of the female gametophyte (in the **Palmaria** type) or tetrasporophyte (in the **Rhodophysema** type). Gabrielson and Garbary (1986) also consider the life-history pattern in Palmariales an evolutionary reduction from **Polysiphonia** type.

Hawkes and Scagel (1986a), however, argue that "the absence of carposporophytes in Palmariales may not be due to a reduction and elimination of this phase because in the first place it may never have existed in these groups and thus derive it from biphasic ancestors." They stress a free-living tetrasporophyte may never have existed in the **Rhodophysema** type life-history.

Magne (1987) on the other hand, has reinterpreted the life-history of *Palmaria* as being triphasic. According to him "the generation formed by the development of the zygote is not a tetrasporophyte as was previously thought, but a carposporophyte; and tetrasporocyst mother cell is in fact a carposporocyst producing no carpospore but a two-cell parasitic tetrasporophyte composed of a tetrasporocyst and its stalk cell". Thus *Palmaria* has an independent, free-living carposporophyte. Compared to this, life-history of *Rhodophysema elegans* is biphasic; there is no carposporophyte, the zygote being transformed directly into the carposporocyst producing a two-celled tetrasporophyte.

ULTRASTRUCTURAL STUDIES

CHLOROPLAST

In Palmariales; the plastids are characterized by the lack of affiliation of the photosynthetic membranes. They typically have unstacked, evenly spaced thylakoids, and are known as chromatophores.

In *Palmaria palmata* (Fig. 107 A) two peripheral thylakoids, called girdle lamella or girdle band surround the central thylakoids which may be flat or coiled. The tetrasporocytes show rapidly proliferating chloroplasts with only one encircling thylakoid (Pueschel, 1979). The peripheral thylakoid is either complete and closed (Fig. 107 A), or it may be continuous with other thylakoids (Fig. 107 B; Cole and Sheath, 1990).

In a green pigmented mutant of *P. palmata* the peripheral encircling thylakoid is lacking and the tubular membranes show branching and resemble prolamellar bodies (Fig. 107 C). Amorphous inclusions are found in some chloroplasts of the mutant (Pueschel and Van der Meer,

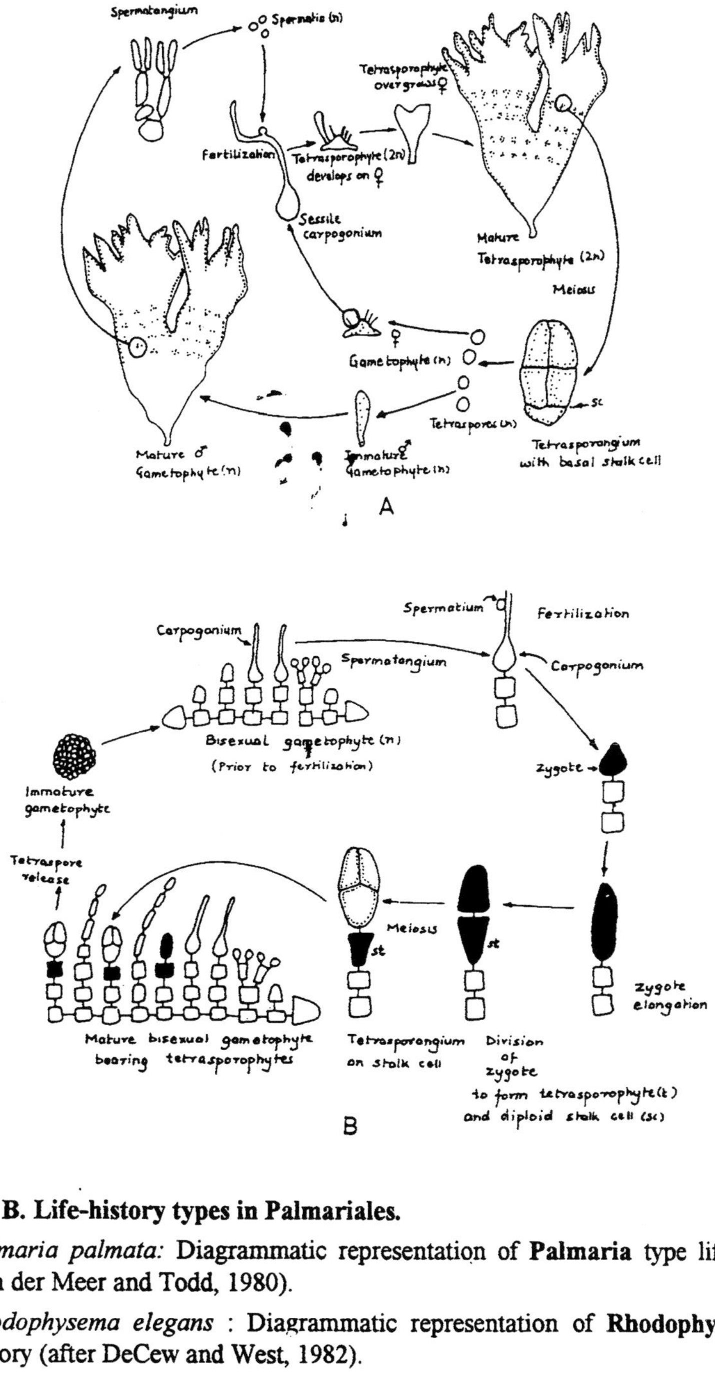

Fig. 106. A, B. Life-history types in Palmariales.

A. *Palmaria palmata:* Diagrammatic representation of **Palmaria** type life-history. (after Van der Meer and Todd, 1980).

B. *Rhodophysema elegans* : Diagrammatic representation of **Rhodophysema** type life-history (after DeCew and West, 1982).

Fig. 107. A-C. Ultrastructure : Chloroplast in *Palmaria palmata*.

(after Cole and Sheath, 1990):

A. Chloroplast showing two peripheral thylakoids surrounding a coiled thylakoid (ends marked by arrows) and flat central thylakoids. Genophore visible between central thylakoids as electron-transparent region containing fibrils.

B. Chloroplast showing continuity of outermost encircling thylakoids with other thylakoids (arrow).

C. Plastid of a green pigmented mutant. Branching tubular membranes resemble prolamellar body; peripheral encircling thylakoids are lacking.

1984). The DNA (genophore) is visible between central thylakoids of the chromatophore as electron-transparent region containing fibrils (Fig. 107 A).

PIT-CONNECTION

These are structures which are mostly found in florideophycean red algae, but also occur in the chantransia-stage of the Bangiophycean taxa. During cell division, incomplete septation results in cellular connection (called pit-connections) between the daughter cells. But cytoplasmic continuity is not maintained between cells for long, because shortly after the furrowing ceases, the septal apertures are occluded by plug-like structures called pit-plugs (Pueschel and Cole, 1982). Firstly tubular membranes appear in the septal apertures (Pueschel and Cole, 1982) and later a homogenously granular mass of protein-the plug-core is deposited around them (Pueschel, 1980). The tubular membranes eventually disappear.

The genera belonging to the order Palmariales are characterized by the presence of 2-layered plug-caps; the outer (cytoplasm-adjacent) and the inner (core-adjacent) cap layers, on either side of the plug-core (Fig. 108 B) The outer cap layer is formed by lateral coalescence of vesicles or plaques of electron-dense material about 50 nm from the surface of the plug-core (Fig. 108 A). In this order the outer cap layer forms a thin plate and completely covers the adjacent inner cap layer. This outer layer contains a carbohydrate component. The inner cap layer is electron-transparent and its origin is uncertain.

A cap membrane separates the inner cap layer from the outer cap layer. The cap membrane appears to be continuous with the plasmalemma by means of a junction that could be viewed as three phospholipid monolayers associated in paired permutations (Pueschel, 1977). The plasmalemma is continouus from one cell to the next along the sides of the pit-plug (Fig. 108 B; Pueschel, 1977). The origin of cap membrane is unknown.

In *Palmaria palmata,* treatment of section with periodic acid results in lack of contrast, especially in plug cap (Fig. 108 C). PA/PTA-chromic acid selectively stains outer layer (2) of plug cap (ca) (Fig. 108 C). Pit-plugs of *Rhodophysema* resemble those of *Palmaria* in the structure of the mature plugs, mode of formation of the plug caps and positive PTA-chromic acid reactivity of the outer cap layer. Some plugs of *R. georgii* have enlarged outer cap layers (Fig. 108 D) in comparison to thin, plate-like outer cap layer seen in other taxa of Palmariales.

TETRASPOROGENESIS

Ultrastructural study of tetrasporogenesis in *Palmaria palmata* reveals that the differentiation of tetrasporangium begins with the enlargement of the sporangial initial which are usually the apical cells of the cortical filaments, although intercalary initials are known. The sporangial initial deposits a layer of distinctive wall material followed by migration of nucleus from a central position to a more basal one and thus, foreshadowing an unequal cytokinesis of the premeiotic mitosis. Cytokinesis occurs by centripetal furrowing, forming a large, upper tetrasporocyte and a small, lower, stalk cell. Concomitant to division the sporangial wall becomes two-layered. The septum between the stalk cell and the tetrasporocyte is continuous with the less dense inner layer. Small vesicles with electron-dense contents lie close to the plasmalemma and adjacent to the indentations in the wall caused by a large nodule of amorphous electron-dense wall material. The position and contents of the vesicles suggest that they are the likely candidates of wall materials.

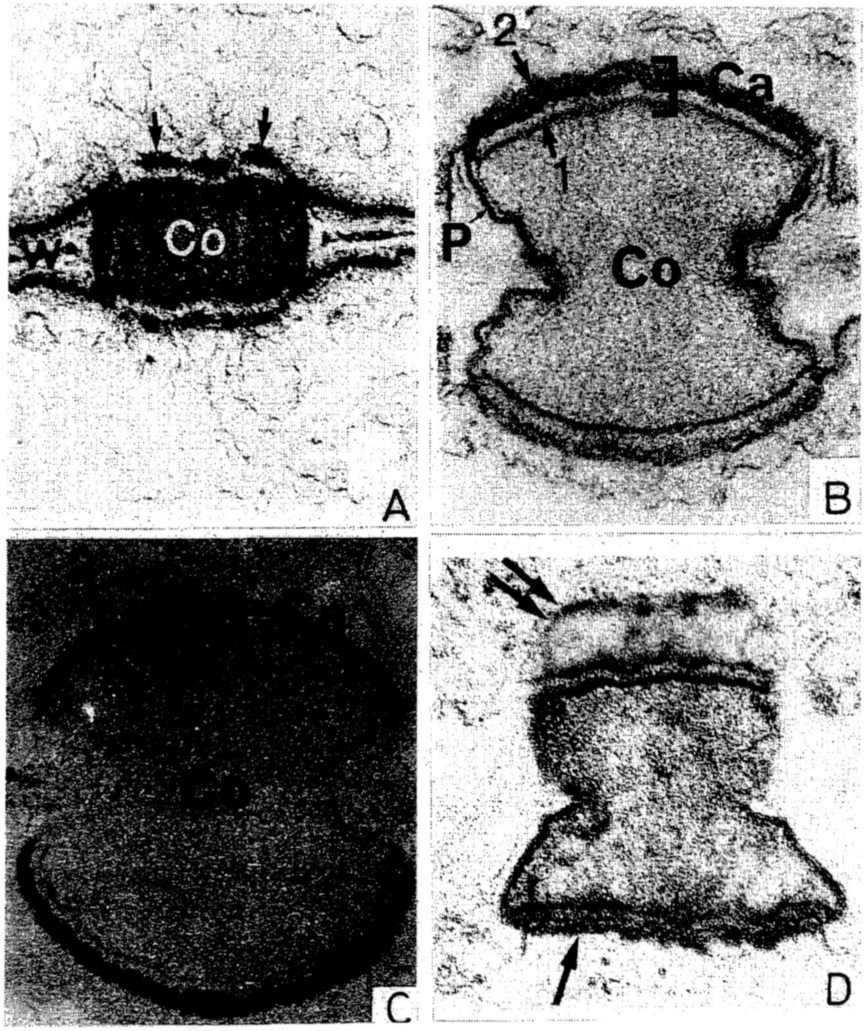

Fig. 108. A-D. Ultrastructure of pit-plugs in Palmariales

(after Pueschel and Cole, 1982).

(ca, plug cap; co, plug core; p, plasmalemma; w, septal wall).

A. *Palmaria palmata*: Plug-core is deposited around cisternae which invade the septal pore. Plaques of election-dense material (arrows) form the outer plug cap layer.

B. *Palmaria palmata*: Structurally complete pit-plug with plug-cap composed of inner (1) and outer (2) layers. The plasmalemma is continuous along the sides of the plug-core.

C. *Palmaria palmata*: PA/PTA-Chromic Acid selectively stains outer layer (2) of plug-cap.

D. *Rhodophysema georgii*: Exceptional pit-plug which has one normal (arrow) and one enlarged outer cap layers (double arrow).

The tetrasporocyte enlarges and its nucleus becomes considerably larger than that of the stalk cell nucleus. The stalk cell becomes cup-shaped and clasps the basal portion of the sporocyte. Onset of meiotic prophase 1 is marked by chromatin condensation and formation of synaptonemal complexes.Simultaneously, there is a deposition of floridean starch from the inner cortical cells in the tetrasporocyte cytoplasm. MLS (microbody like structures) and cluster of nonmembranous spherical bodies get associated with the nucleus.

During mid-prophase 1, chromatin continues to condense and the perinuclear ER sheath become more extensive. Mitochondria and dictyosomes are more abundant. A zone of exclusion from which organelles including ribosomes are absent, lay between the surface of the mitrochondria and the forming face of the associated dictyosome.

A "diffuse" or "resting" stage (Westbrook 1928; Magne 1964; Yabu 1971; Van der Meer; 1976) follows pachytene in which the chromosomes decondense. Presumably transcription occurs at this time. Nuclear surface area is increased by the finger-like projections and invaginations of the nuclear envelope and with increased density of the nuclear pores, may facilitate flow of materials in and out of the nucleus. Chloroplasts proliferate rapidly at this stage with only one encircling thylakoid in each chloroplast. Parallel cisternae of ER are commonly found among large clusters of starch grains. Scattered starch grains often occur between cisternal pairs. Abundant large vesicles are produced in which microfibrillar material is embedded in an electron-transparent matrix. Intercalation of mucilaginous material between the sporangial wall and the cell membrane results in massive convolutions of the plasmalemma. The mucilage layer thickens evenly around the tetrasporocyte, except at the apical end of the sporangium, where a lenticular thickening produces a broad indentation of the cell.

After the diffuse stage, the chromatin condensation occurs again and the first division of the tetrasporocyte nucleus (meiosis 1) takes place. One nucleus occupies the apical end of the cell and second nucleus the basal end. Second meiotic division takes place and the four post meiotic nuclei are positioned peripherally in the uncleaved cell.

The first cleavage of the tetrasporocyte occurs in the periclinal plane. Only mucilage is found in the furrow and the sporangial wall does not take part in cleaving the spores from the tetrasporocyte. The periclinal septation is arrested short of completion, and is resumed and completed during the formation of the anticlinal septa. Mucilage vesicles and mucilage sacs contribute to the formation of the septa and disppear soon after the completion of cytokinesis. The dictyosomes bud off adhesive vesicles to the cytoplasm of the nascent tetraspores. The adhesive vesicles contain a proteinaceous material. Another type of inclusion, the "fibrillar-bodies", without a bounding membrane and rich in acidic and sulfated polysaccharides is present in the cytoplasm. Their mode of origin and function is not clear. In the quatret, a clear zone which lacks stainable constitutents is distinguished around each spore.

At the time of spore discharge, the cuticle, vegetative cell wall and the sporangial wall are ruptured. As the tetraspores emerge, they lose their angular outlines and become spherical. The released tetraspores contain numerous adhesive vesicles, abundant starch grains, mitochondria, plastids, a few fibrillar bodies and a nucleus with an extremely irregular outline.

The chamber from which the tetraspores emerge maintains its integrity. The inner, less dense portion of the sporangial wall which lines the chamber apparently dissolves. The stalk cell begins to enlarge. A central vacuole develops, pushing the cytoplasm to a parietal position. The nucleus migrates from the basal position to the apex of the cell. The old sporangial wall of the stalk cell becomes thick. Functionally the stalk cell is now a tetrasporangial initial and will form a new stalk cell and a new tetrasporocyte by unequal mitotic division and the entire process of tetrasporogenesis is thus repeated. The regenerate sporangia have relatively less starch grains than the primary one.

In *Halosacciocolax kjellmanii,* fine structural investigations show that tetrasporangia either arise directly from an apical uninucleate cell terminating the vegetative upright filaments of the thallus or from cells regenerated from subterminal cells of filaments called 'stalk cells' Synaptonemal complexes are formed in the nuclei of young tetraspore mother cells. Following pachytene and during a prolonged diffuse stage mitochondria associated dictyosomes produce electron-transparent vesicles which presumably release mucilage-like substances which are believed to facilitate the liberation of spores. Large fibrous vesicles and small cored vesicles of uncertain function are also observed after the first cleavage. Floridean starch grains accumulate abundantly in the cytoplasm only towards the final phase of tetrasporogenesis. Cleavage of the TMC (tetraspore mother cell) is successive, cruciate and centripetal. At maturity, wall formation may not be complete in the spores. The stalk cell is generative. TMC and generative cell are never enclosed within a common wall (Chesnoy *et al.*, 1989).

METABOLIC BYPRODUCTS

Protein crystals are abundant in medullary and inner cortical cells of some specimens of *Palmaria. P. palmata* has slender, sheet-like crystals (Fig. 109 B), lack bounding membranes but develop in association with ER and are appressed to envelope of lobed (arrow) chloroplast. *P. mollis* has crystals that appear as polyhedron (Fig 109 A). Presumably the contrasting crystal morphologies reflect genetically based differences in the protein monomers (Cole and Sheath, 1990).

Osmiophilic inclusions, present in osmiophilic vesicles, have been demonstrated in *P. palmata* (Pueschel, 1977). The major low molecular weight organic metabolite in this species is heteroside floridoside (0 $-\alpha$ -D-galactopyranosyl - (1->2) glycerol) (Colin and Gueguen, 1930). In *P. palmata,* during late stages of tetrasporogensis fibrillar bodies are present in the cytoplam. They lack bounding membrane and their size and outline are highly variable . The matrix is composed of electron-dense fibrils which are swirled without any discernible pattern. The mode of origin and function of these bodies are obscure, but reactivity to Aniline Blue suggests they are rich in acidic and sulfated polysaccharides (Pueschel, 1979).

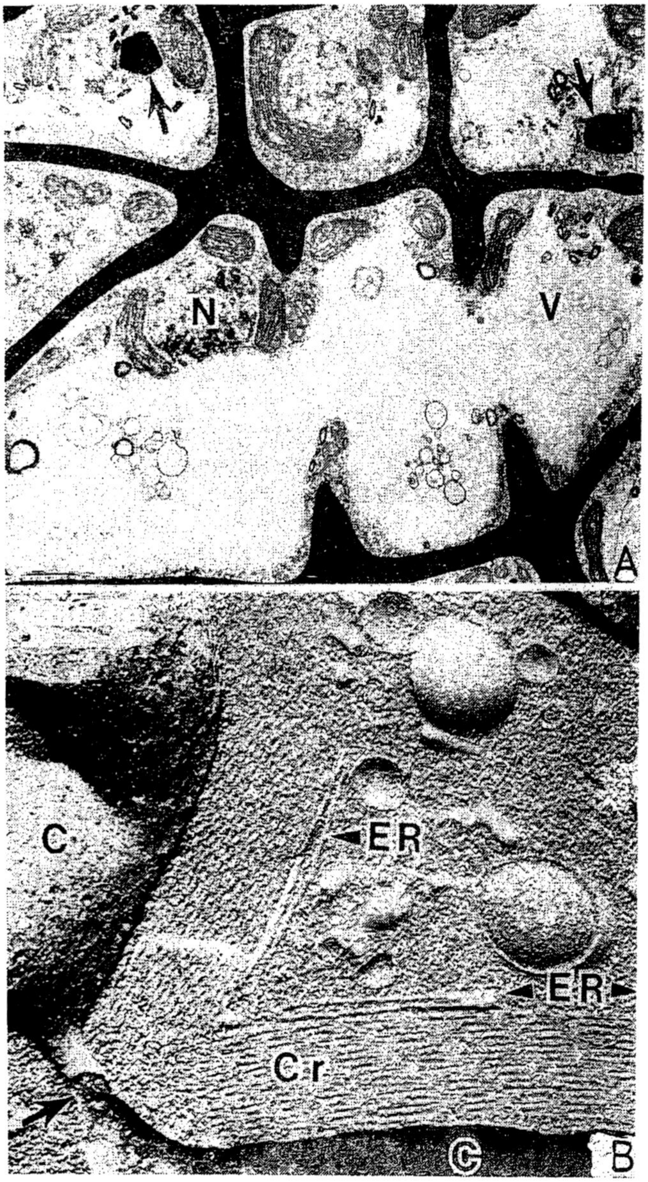

Fig. 109 A, B. *Palmaria* : Protein crystals.

(after Cole and Sheath, 1990).

(c. chloroplast; cr, protein crystals; er, endoplasmic reticulum)

A. *P. mollis* : Protein crystals (arrows) are located in the cytoplasm of inner cortical cells.

B. *P. palmata* : Protein crystals form sheets.

8 CERAMIALES

CLASSIFICATION

According to Bold and Wynne (1965) the order Ceramiales is monophyletic in origin. Papenfuss (1944) and Hommersand (1963) proposed relationship between Ceramiaceae, Delesseriaceae, Rhodomelaceae and Dasyaceae. Both the authors are unanimous on Ceramiales being monophyletic, Ceramiaceae being the most primitive of the four families and the three families being derived from Ceramiaceae. Papenfuss maintains that the families Delesseriaceae, Rhodomelaceae and Dasyaceae form a monophyletic line from Ceramiaceae whereas according to Hommersand the three families have arisen from different groups in Ceramiaceae. Moreover according to Hommersand (1963) the Ceramiales are derived from "one of the major independent lines of evolution that has originated from ancestors in Nemalionales".

For past forty years the basic classfication as proposed by Kylin (1956) for Ceramiales has remained much the same except for changes in minor details. Kylin divided Ceramiales into four families namely Ceramiaceae, Delesseriaceae, Rhodomelaceae and Dasyaceae.

A synopsis of classification by Fritsch (1945) with many new added genera is presented below.

CERAMIACEAE : *Antithamnion, Aristothamnion,*
Ballia, Bornetia,
Callithamnion,
Carpoblepharis,
Ceramium,
Ceramothamnion,
Compsothamnion, Corgynospora,
Crouania,
Cupilota, Dhorniella,
Gattya,
Griffithsia,
Haloplegma,
Halurus,
Lejolisia,
Microcladia,
Monospora,
Pleonosporium,

Plumaria,
Ptilocladia,
Ptilota,
Ptilothamnion,
Reinboldiella,
Seirospora,
Spermothamnion,
Sphondylothaminon,
Spyridia, Tiffaniella,
Vickersia,
Wrangelia.

DELESSERIACEAE

(*a*) Delesserieae : *Apoglossum,*
Caloglossa,
Claudea,
Delesseria,
Grinnellia,
Hemineura,
Hypoglossum,
Membranoptera,
Pantoneura,
Vanvoorstia.

(b) Nitophylleae : *Acrosorium,*
Cryptopleura,
Gonimophyllum,
Haraldia,
Martensia,
Myriogramme,
Nitophyllum,
Phycodrys,
Ploycoryne.
Polyneura

(c) Sarcomenieae : *Cottoniella,*
Platysiphonia,
Sarcomenia,
Taenioma

RHODOMELACEAE

(*a*) Polysiphonieae :	*Bryothamnion,*
	Digenea,
	Polysiphonia,
	Stromatocarpus
(*b*) Lophothalieae :	*Brongniartella,*
	Doxodasya,
	Lophocladia,
	Lophothalia,
	Murrayella,
	Pteronia,
	Spirocladia,
	Wrightiella.
(*c*) Bostrychieae :	*Bostrychia*
	Colacopsis
(*d*) Rhodomeleae :	*Odonthalia,*
	Rhodomela
(*e*) Chondrieae :	*Acanthophora,*
	Chondria,
	Coeloclonium,
	Herpochondria
(*f*) Laurencieae :	*Janczewskia,*
	Laurencia,
	Ricardia
(*g*) Pterosiphonieae :	*Dictymenia,*
	Pollexfenia,
	Pterosiphonia,
	Symphyocladia
(*h*) Herposiphonieae :	*Amplisiphonia,*
	Dipterosiphonia,
	Herposiphonia,

	Periphykon,
	Placophora
(*i*) Lophosiphonieae :	*Ctenosiphonia,*
	Lophosiphonia,
	Ophidocladus,
	Stictothamnion
(j) Polyzonieae :	*Cliftonaea,*
	Euzoniella,
	Leveillea,
	Polyzonia
(k) Amansieae :	*Amansia,*
	Halopithys,
	Kützingia,
	Lenormandia,
	Osmundaria,
	Rytiphloea,
	Vidalia.
DASYACEAE :	*Calacodasya,*
	Dasya,
	Dasyopsis,
	Dictyurus,
	Heterosiphonia,
	Thuretia.

MORPHOLOGY

All members in Ceramiales are uniaxial. The numerous taxa exhibit great diversity and attain a high degree of morphological complexity. All the four families Ceramiaceae, Delesseriaceae, Rhodomelaceae and Dasyaceae exhibit distinct morphological characteristics.

Ceramiaceae is the most primitive in the order and members of this family always show uniseriate axis with nodes. The cylindrical filaments are partially or wholly corticated, where the corticating cells are rarely as long as the cells they cover.

In *Compsothamnion* and *Sphondylothamnion* the laterals are arranged in a whorled manner and the axis is uncorticated. In *Griffithsia* the cells of the central axis are huge, uncorticated and multinucleate. The thallus appears to be copiously branched due to equal development of main

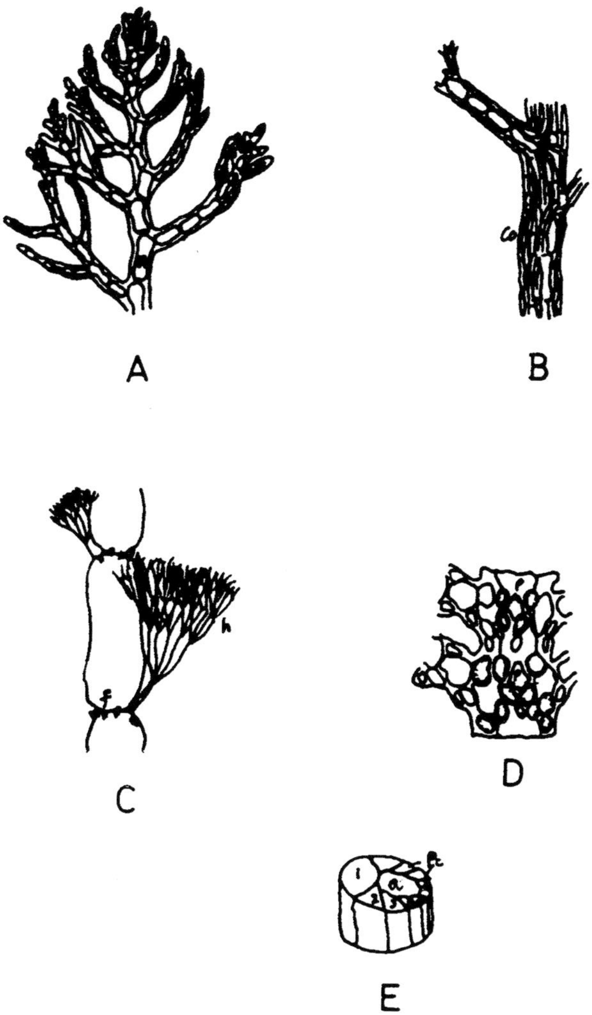

Fig. 110. A-E. Morphology.

(after Fritsch, 1945).

(a, axial cell; co, cortex; f, fertile shoots; h, hairs; p, cortex forming cells; pc, pericentrals).

A. *Callithamnion* : Apical part of a plant.

B. *Callithamnion* : Plant showing cortication.

C. *Griffithsia* : A few segments showing fertile shoots and hairs.

D. *Plumaria* : Portion of thallus enlarged to show the manner of cortication.

E. *Ceramium* : Diagrammatic representation of mode of formation of pericentrals.

axis and laterals. Reproductive branchlets are short and develop from the top of the young cells and are deciduous (Fig. 110 C). The entire thallus is encased in a thick, mucilaginous envelope.

In *Antithamnion,* the plant has a creeping thallus that consists of threads with upturned tips and bears many uncorticated filaments. Two kinds of branches are found (i) the branches of limited growth and (ii) the branches of unlimited growth. The branches of limited growth are borne either in whorls of three or four or opposite to each other; borne on nodes and formed in successive manner. The branches of unlimited growth alternate and take the place of short laterals. The cells of the thallus are uninucleate with many chromatophores. In *Plumaria elegans,* the thallus is pinnately branched and the younger parts of thallus resemble that of *Antithamnion.* A thick cortex envelopes the older parts of the thallus. The cortex comprises of two kinds of cells (i) cortical filaments formed from the basal cell of the lateral branch (ii) the axial cell gives rise to protrusions perpendicular to the main branch. These protrusions cut off four cells at four corners. These cells form short filaments covering the axial cell (Fig. 110 D). These two kinds of cells finally divide and form a thick envelope which consists of three parts (i) the outer small celled photosynthetic tissue (ii) an intermediate storage layer and (iii) a conducting system of long and narrow cells that surround the axial cells (Fritsch, 1945).

In *Ceramium,* the thallus is branched, shows transverse banding and the ultimate branches appear as a tuning fork. The portions of thallus uncovered by corticating filaments alternate with those portions that are covered. The thallus further shows regular dichotomy and each segment has a large cylindrical axial cell which successively cuts of a series of pericentral cells at the top of each segment (Fig. 110 E). These pericentrals make up the primary cortical band. Each of the pericentrals, at four corners cuts off secondary pericentrals identical to that seen in *Plumaria.* Each of these four cells produce branching threads which cover the axial cell. The cells of these branches are small and branches from successive segments are contiguous, thus, giving rise to alternating covered and uncovered areas of thallus (Fig. 112 A; Cormaci and Funari, 1991). In *Spermothamnion,* the laterals are arranged in a tetrastichous manner and are found in whorls (Gordon, 1972). In *Diplothamnion,* the plants are filamentous and uncorticated. Growth occurs by means of transverse divisions of the apical cell. The laterals are arranged in a distichous manner. The branches of indeterminate growth arise alternately from axial cells in place of determinate laterals. The cells of axis and indeterminate branch become short towards apex. The erect axis becomes secondarily prostrate and is anchored to the substratum through unicellular rhizoid (Huisman, 1991). The genus *Involucrana* is very much like *Diplothamnion,* except that the position of female axis is different. It is short, three celled and produces polysporangia instead of tetrasporangia.

Callithamnion has a filamentous habit and resembles in morphology the green alga *Cladophora.* The cells of the thallus may be uni, or multinucleate due to nuclear fragmentation and not division (Vijayaraghavan *et al;* 1995) ; have ribbon–shaped chromatophores; the branches arise below the septa and are either spirally or in some species distichously arranged. The principle axis is composed of broad, thick walled cells and show zig-zag disposition (Fig. 110 A). The older part of the thallus is covered by an investment of delicate and multicellular corticating threads that are formed either from the base of the lateral branches or from the base of the axial cell (Fig. 110 B). The cortex gives rise to filaments that help in anchorage (Fritsch. 1945). In *Callithamnion cordatum* the laterals are also branched and the female plants are bushier in appearance as compared to the male plants (O'kelly and Baca, 1984).

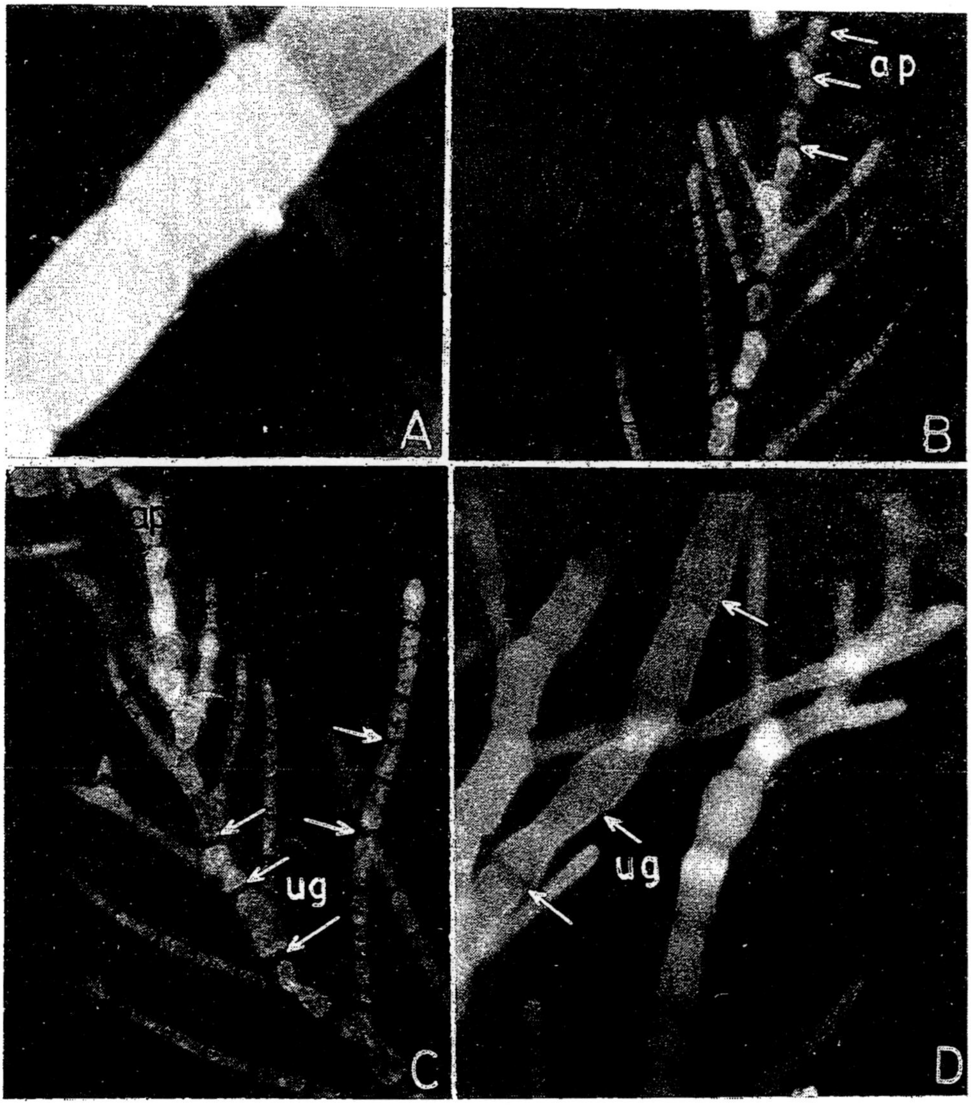

Fig. 111 A. D *Callithamnion* sp. in culture. Cell elongation. (Fluorescent microscopy) (after Vijayaraghavan et al., 1995)

(ap, apical cell; ug. unipolar band growth)

A. Main filament photographed immediately after staining with 0.01% Calcofluor.

B. Apical portion showing new cell wall material added at the apex . (arrows).

C. A portion of the main filament showing unipolar band growth in the lateral banches as well as the main filament (arrows).

D. Same, shwoing unipolar band growth (arrows).

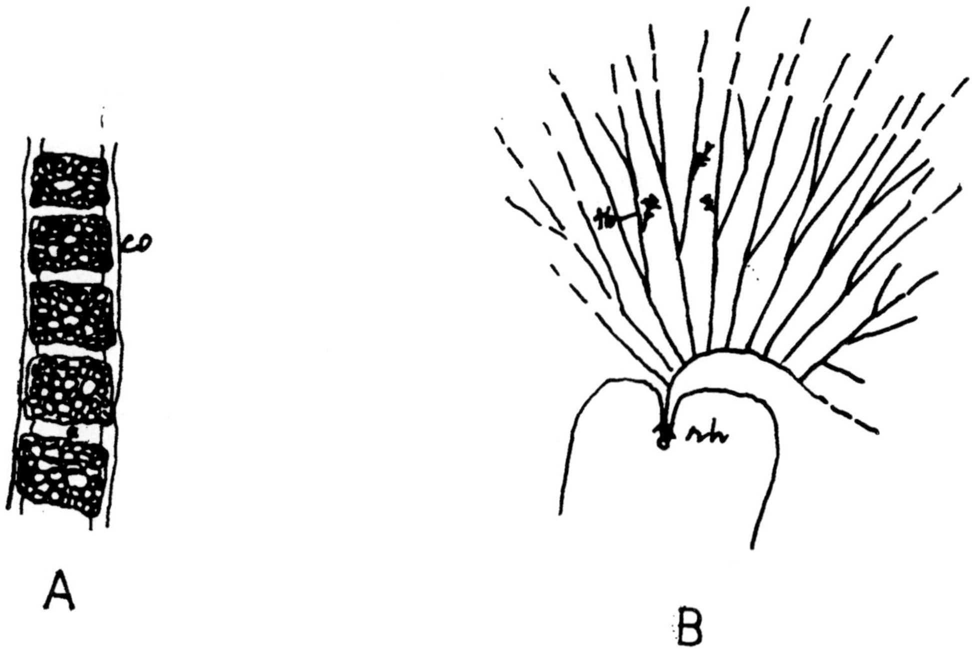

Fig. 112. A, B. Morphology.

(a, axial cell; co, cortex; fb, fertile branch; rh, rhizoid)

A. *Ceramium* : Structure of mature thallus. (after Fritsch, 1945).

B. *Radiathamnion* : Schematic diagram of thallus structure and branching pattern (after Mills and Kraft, 1981).

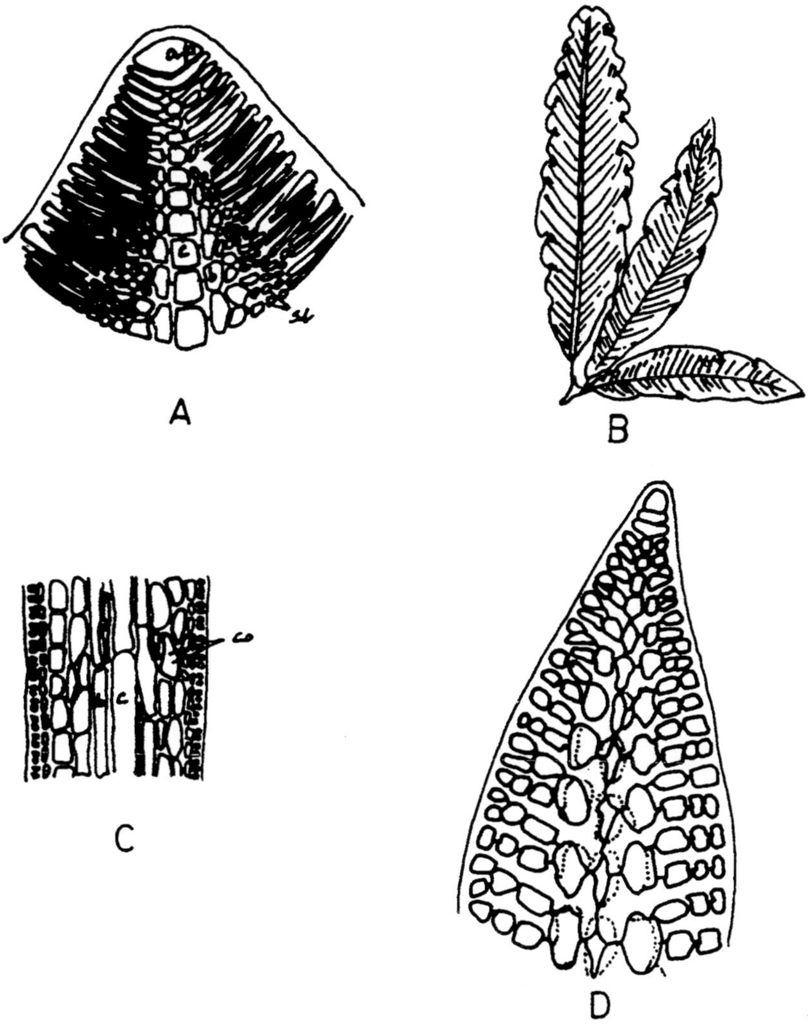

Fig. 113 A-D. Morphology

(ap. apical cell; b, basal cell of thallus; c, axial thread; co cortex; sl, secondary laterals)

A. *Hypoglossum* : Apex of thallus (after Fritsch, 1945)

B. *Delesseria sanguinea* : Habit of thallus. (after Fritsch, 1945)

C. *Delesseria sanguinea* : Longitudinal section of midrib. (b, elongate cells around axial thread; e, surface layer of midrib). (after Fritsch, 1945)

D. *Branchioglossum* : Thallus showing first order and initials of second order cell rows (after Yoshida and Mikami, 1992).

In *Callithamnion* sp. the apical cell of the main filament divides transversely to produce a chain of cells. The cells, a little below the apex of this chain, divide to produce laterals. The cells of laterals, in turn, divide by oblique divisions to produce laterals of indeterminate growth that occassionally overgrow the main axis resulting in pseudodichotomy. The apical cell grows by deposition of wall material at its tip but the intercalary cells growth occurs by elongation of the lateral wall in a confined band (Fig. 111 A-D; Vijayaraghavan *et al.*, 1995).

Radiothamnion speleotis is found rooted in the apical pits of *Laurencia* and can be differentiated from the host trichoblasts by its large size and reddish tint. The attachment is through a single prominent basal cell by wall mucilage. The thallus sometimes produces rhizoids which penetrate for short distances into the cortex. The thallus consists of an arched main axis and except a few lowermost and uppermost cells, all other cells of the axis bear on the convex side a single lateral branch (Fig. 112 B). Distal cells of main axis and almost all the branches eventually elongate, become colourless and develop into hair-like structures. Further growth occurs by formation, from each branch, of adaxial branchlets. Pseudodichotomy may result due to elongation of second and third order branches to the same length as that of first order. All cells are uninucleate and contain plastids (Mills and Kraft. 1981).

In most Ceramiales sympodial branching is a common occurence and is basically of two kinds (Norris *et al.*, 1984).

(1) Ramisympodial (Crouanoid)

(2) Cellulosympodial (Dasyaceous).

In Ramisympodial, development of a given axis ceases and further growth of the plant is from new branches that arise in intercalary positions. In Cellulosympodial, the apical cell initiates a determinate lateral branch and the lateral cell or branch formed by subapical cell carries on the development of the thallus axis.

In Delesseriaceae the thalli are often delicately foliose, develop by close juxtaposition of numerous branches in one plane, possess midrib with lateral veins and remain mostly single layered except in the region of midrib and veins. The cells of the axial row and its branches are linked through primary pit-connections whereas basal and adjacent cells of the laterals which help in formation of the midrib are longitudinally joined by a number of secondary pit-connections. Such connections are also found among cells of the adjacent rows. In most genera, the fronds have a conspicuous apical cell. In Nitophylleae, however, the apical cell is found only in the younger thallus. The cells of principle axis cuts off four pericentrals, the first two in the plane of branching and the other two perpendicular to this plane. The latter two form the several layered ribs. The mode of formation of pericentrals differs from that of Rhodomelaceae and Dasyaceae. Most of the genera of Delesseriaceae grow by means of apical cell but a few taxa like *Apoglossum, Delesseria, Phycodrys* and members of Nitophylleae also show intercalary growth.

In *Hypoglossum* the frond is a flat structure and the central cell cuts off two pericentrals that act as secondary apical cells; undergo oblique divisions and form the pinnately arranged laterals wihich in turn cut off abaxially second order laterals, whose apical cells elongate and reach the margin of the delicate but compact frond. Slightly behind the apex, the central cell and the first formed pericentrals elongate and produce the thick midrib. There is no thickening anywhere and thus no veins are formed. The midrib is, however, formed by cutting off pericentrals parallel to the surface (Fig. 113 A; Fritsch, 1945).

In *Apoglossum,* the blades have a conspicuous midrib. The transverse pericentrals surround the central cell (Wynne and Norris, 1991). In *Delesseria sanguinea,* known as "Dock leaved Delesseria", fronds with distinct side veins are produced from the primary laterals. Secondary veins, formed from cell rows of the second order, are usually invisible but can be seen under the microscope. The laterals of higher order always arise abaxially. Except for the cells of the axial row, intercalary division occurs in all the cells. The midrib develops identical to that of *Hypoglossum* except the basal cells of primary laterals which divide transversely and are reduced to half the length of the axial cell. The surface layer of the midrib remains meristematic and cuts off a relatively thick cortex bound by a superficial layer of very small cells (Fig. 113 B,C). The branches arise from the midrib. The wings gradually wear-off and only the midrib remains bearing many young fronds that remain anchored tightly to the midrib through the penetrating hyphae (Fritsch, 1945).

The Delesserieae differ from the Nitophylleae in possessing a distinct apical cell and in the absence of intercalary divisions of the axial cell. Althrough *Phycodrys* belongs to Nitophylleae, it shows considerable similarity to *Delesseria.* In *Phycodrys,* the branching may take place upto fifth order. The apical cell is not very conspicuous. Intercalary divisions occur both in the axial and other cell rows. The lower cell formed by such a division retains the primary pit-connection, whereas the upper cell gives rise to accessory cell rows. The lobing of fronds is due to outgrowth of the primary laterals, whereas the marginal pinnules on the lobe are the result of elongation of secondary laterals. Primary and secondary veins are visible to the naked eye. In most Nitophylleae the apical cell ceases to function and marginal growth takes over (Fritsch, 1945). In *Cottoniella amaniensis,* the thallus is monosiphonous. It is very narrow at the base, broad in the mid region and again narrow and pointed at the apex. Only the older parts of the thallus are corticated (Wynne and Norris, 1991).

In *Branchioglossum,* the thallus is ribbon-shaped. The frond arises from a basal disc and is compressed at the lower part and flattened above. The thallus bears alternating branches from the margin. The blade is monostromatic except in the midrib region. In the lower part of the thallus, cortical cells cut off small cells. Branches arise from the terminal cells of the second order cell rows. No lateral veins are observed in the blade. Growth of the thallus takes place by the activity of transversely dividing apical cells whose derivative constitute the cells of first order. There is no intercalary division in the first order cells. The second order cells reach the margin and occasionally reveal intercalary division forming branches of the second order. The marginal cells in the middle to lower part of the thallus form spinal processes, giving a serrulate appearance (Fig. 113 D; Yoshida and Mikami, 1992).

In *Yoshidaphycus,* the thallus is ribbon-shaped, thin and membranous. Branches are borne alternately from the margin. A small stem arises from the holdfast and expands into a blade after a very short distance. Additional attaching discs are produced from the lower margins of the thallus. The midrib is not distinct. Except for the midrib the blade is monostromatic. In the mature thallus, many small filamentous branchlets arise from the margin and bear the reproductive structures. The apical organization and branching pattern is similar to *Branchioglossum* except that the fourth order branchlets may be formed. Two kinds of branches occur : (i) The marginal cell can adventitiously produce a branch (ii) The cells of first order give rise to a branch initial which grows along cells of the second order and emerges through the margin. These second kind of branches can also emerge before reaching the margins. These branches are referred to as endogenous (Fig. 114 A; Mikami, 1992).

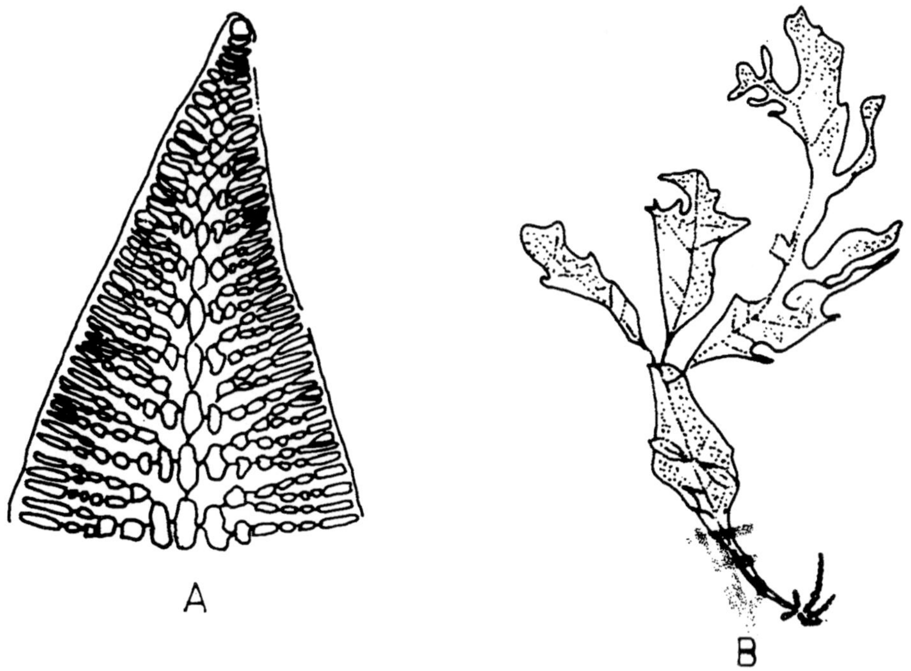

Fig. 114. A,B. Morphology.

A. *Yoshidaphycus* : Apical organization. (Yoshida and Mikani, 1992)

B. *Yamadaphycus* : Thallus of tetrasporangial plant. (after Mikami, 1973)

In *Yamadaphycus* the thallus is polystromatic. It consists of many leaf-like fronds that arise from the basal disc. Young fronds are found on the midrib. Each frond has a midrib and inconspicuous lateral veins. Cortex and medulla are well developed. The younger apices do not show intercalary division but mature fronds show such divisions among the cells of second order (Fig. 114 B; Mikami, 1973).

The family Rhodomelaceae has varied and complex morphology but is characterized by a basic polysiphonous structure and production of two kinds of laterals namely ordinary branches and trichoblasts. Unicellular hairs are absent. The thallus construction is of two types. Those which are radially constructed as in *Polysiphonia* and *Brongniartella* and those which have a bilateral or dorsiventral construction as in *Heterosiphonia.*

Polysiphonia is the most characteristic genus of the family. The daughter cells of the apical cell produce lateral branches before dividing longitudinally into central and pericentral cells. The pericentrals formation is identical to those of *Ceramium* but are of the same length as the axial cells. The laterals are of two kinds, ordinary branches and trichoblasts. The ordinary branches are polysiphonous and show unlimited growth. The trichoblasts are uniseriate, deciduous and bear the sex organs. The laterals are formed in a spiral succession. The pericentrals may be 4-20 in number and always develop in particular sequence. The pericentrals are connected to the axial cell by primary pit-connections and among themselves by secondary pit-connections.

Except the first few segments of the trichoblasts, the rest produce laterals which may often develop vigorously thus giving a pseudodichotomous appearance. The trichoblasts are arranged in a left handed spiral. The other laterals which are polysiphonous may either originate in place of the trichoblast or may arise from the basal cell of the trichoblast. These branches in the latter grow vigourously, displace the trichoblast and thus appear to originate from the main axis. The basal part of the branch is adpressed to the main axis and often lacks perientrals. In many *Polysiphonia* species, a few polysiphonous axes creep over the substratum and are firmly fixed by unicellular, thick walled and richly branched rhizoids which arise from the peripheral cells. The prostrate branches lack trichoblasts, are secondary in origin and arise endogenously from the central cell at the base of the primary erect axis. The branches of prostrate axes are also formed endogenously. These branches are monosiphonous and traverse between the pericentrals. These endogenous branches do not disturb the organization of central and pericentral cells. In many species the pericentrals undergo division to form the cortex made up of small cells (Fig. 115A). The overlying cortical cells are attached by means of secondary pit-connections to the underlying pericentrals.

In *Laurencia* the thallus has a polysiphonous axis which has an extensive cortication. The result of this extensive cortication is that pericentals are visible only at the apex. At the apices of branches there are obconical pits from which trichoblasts normally protrude. The thallus is attached to the substratum by means of a discoid holdfast. The blade is clavate in shape. The axial cell is much elongate. The extensive cortex is divisible into inner and outer cortex. The cells of inner cortex have dissected chloroplasts. The cortical cells are connected both longitudinally and laterally by pit-connections (Vandermeulen *et al.*, 1990).

A few taxa of Rhodomelaceae show specialization by aquiring bilateral and dorsiventral symmetry in addition to tier-like structure of *Polysiphonia. Odonthalia* and *Bostrychia* have a bilateral symmetry, but still can be allied to radially constructed type (Fig. 115B). But in *Herposiphonia* and *Pterosiphonia,* these tendencies become so strong that it becomes very difficult to include them in the radial pattern.

In *Herposiphonia* the thallus is dorsiventral and many of the species are epiphytic. The long axis is prostrate and attached by unicellular rhizoids. The apex is coiled and the convex face is towards the substratum. The branches arise one in each segment; and are formed in an acropetal succession and stand in four rows. The branches that occupy the sides are the laterals of unlimited growth and are prostrate. The dorsal oncs form the branches of limited growth and are erect. Three dwarf shoots are generally followed by a long shoot or an unlimited growth lateral. A few trichoblasts are borne at the tip of the dorsal branches which bear the reproductive organs (Fig. 115 C; Fritsch, 1945).

In the family Dasyaceae growth is sympodial. Persistent apical cell does not occur, but the successive segments produce laterally a new growing point that displaces the preceding apex. A lateral arising either from the basal or second or third segment continues the growth of the axis, whereas the upper parts of successive units of this sympodium are deflected to one side and simulate the laterals. These are actually pseudolaterals. In *Dasya*, the pseudolaterals are spirally arranged (Fig. 11.5 D, F) but *Dasyopsis* and *Heterosiphonia* reveal distichous arrangement and a pinnate habit. Main axis of the thallus and a few basal cells of pseudolaterals are polysiphonous. The number of pericentrals in *Dasya* and *Dasyopsis* is five and their pattern of formation is entirely different from that of Rhodomelaceae, whereas the number in *Heterosiphonia* is variable

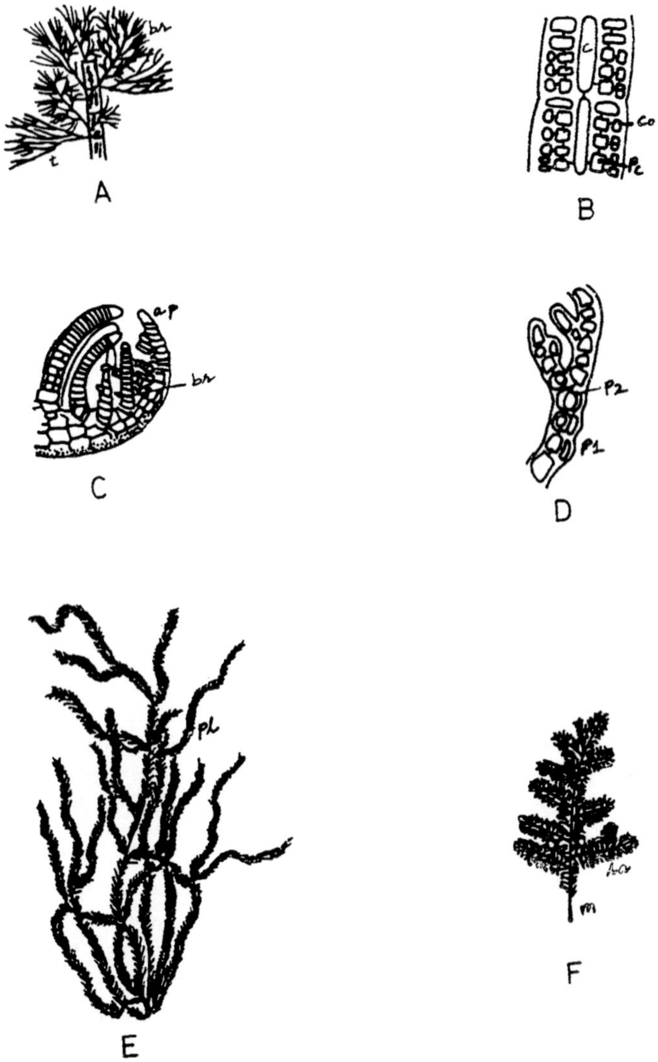

Fig. 115. A-F. Morphology

(after Fritsch, 1945)

(ap, apical cell; br, branch of unlimited growth; c, axial thread; co cortex; la, lateral axis; m, axis; pc, pericentrals; pl & p2, pericentrals; pl, pseudolaterals)

A. *Polysiphonia* : Part of the shoot system showing branches and trichoblasts.

B. *Bostrychia* : Diagram showing formation of cortical cells from pericentral cells.

C. *Herposiphonia* : Apex of thallus.

D. *Dasya* : Apice showing pattern of branching.

E. *Dasya* : Habit of plant.

F. *Thuretia* : Habit of plant.

and mode of formation is much like Rhodomelaceae. Branches of unlimited growth are produced by basal cells of trichoblasts. Adventitious lateral produced by upper ends of pericentral cells and sometimes from cortex are monosiphonous. The pericentral cell and the basal cell of the pseudolateral may cut off cells at these lower ends and these cells are connected to pericentrals and to each other through pit-connections. These cells form branched cortical threads which grow both upwards and downwards and form a thick cortex. Sometimes adventitious branches also participate in formation of cortex. The thallus of *Dictyurus* and *Thuretia* are reticulate (Fig. 115 F; Fritsch, 1945).

REPRODUCTION

SPERMATANGIA AND SPERMATIA

Spermatogenesis is highly conserved in Ceramiales. Spermatangia are generally initiated as subapical protrusions that are cut off by an oblique septa. A spermatangium normally differentiates to produce a spermatium. The wall of spermatangium remains intact after the release of spermatium and the new one may be formed within the old wall.

Ceramiaceae : Spermatangia may develop individually or in groups. In *Antithamnionella* (Whittick, 1980). 2 to 5 celled spermatangial branch develops either singly or in a group of 2 or 3 on the adaxial side of whorl branchlets. Each spermatangial branch bears a whorl of 2 or 3 spermatangia. In *Radiathamnion* (Mills and Craft, 1981) the spermatangial heads are borne mostly in pairs on small celled branchlets arising on primary or secondary laterals. They are elongate and composed of eight or nine fertile axial cells which directly cut off 2 to 4 spermatangia. In *Diplothamnion* sp., the spermatangia are borne in compact subspherical heads, that arise singly at the distal ends of lateral branch cell. Each head is composed of two axial cells that directly produce several tiers of pericentral cells or spermatangial mother cells. Spermatangia are spherical to ovoid and measure 4 or 5 μm in diameter (Fig. 116 D; Huisman, 1991). The male heads of *Wrangelia* have many branchlets borne on the globular terminal cell and enveloped by several one-celled threads (Fig 116 B; Fritsch, 1945).

Rhodomelaceae : The spermatangia are borne either in sori at blade tips, on either side of midrib, or on small protrusions. In *Laurencia,* the spermatangial branch ends are cup shaped and are broadly clavate to pyriform. Spermatangia are ovoid and contain a spherical nucleus each and are formed in dense clusters of repeatedly branched filaments in apical pits. Spermatangial trichoblasts terminate in inflated pyriform cells.

In *Polysiphonia* (Kylin, 1923) the male plant bears antheridial sori on a trichoblast consisting of 2-celled stalk surrounded by fertile region. The fertile region becomes polysiphonous and the pericentrals divide copiously to form a compact layer of mother cells, each of which gives rise to 2 or 3 spermatangia (Fig. 116 C).

Delesseriaceae : The spermatangia are found in sori. In *Yoshidaphycus* (Mikami, 1982), the spermatangial sori cover the surface of broad, lanceolate marginal branchlets except the midrib and margins.

In *Apoglossum,* the spermatangia occur in small islands separated by sterile cells and sori cover the entire blade from midrib to margins. In *Branchioglossum,* the sori are arranged in chevron-like patches diagonally on either side of the midrib. The midrib and margins lack sori (Fig. 116 A). In *Hypoglossum,* the sori are arranged in a pair and are separated by sterile midlines. The blade margins are sterile.

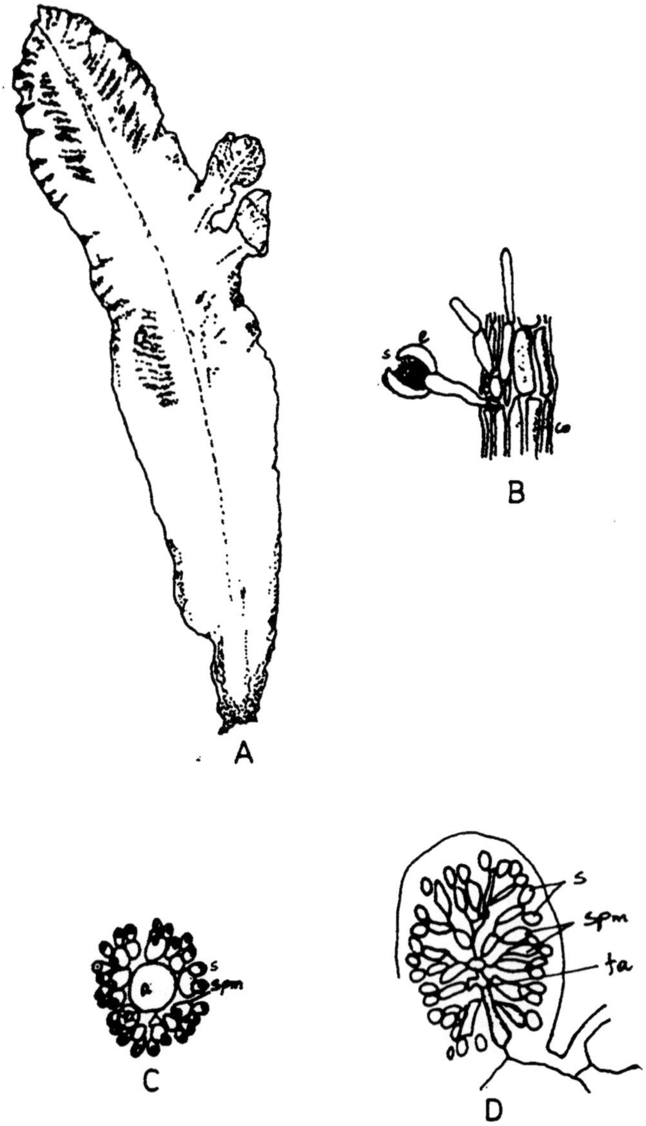

Fig. 116. A-D. Reproduction (Male plant).

(s, spermatium; a, axial cell; co, cortex; e, enveloping threads; fa, fertile axis; spm, spermatial mother cell)

A. *Branchioglossum* : Spermatangial blade (after Yoshida & Mikami, 1992)

B. *Wrangelia* : Spermatangial branch (after Fritsch, 1945).

C. *Polysiphonia* : Transverse section of spermatangial branch showing developing spermatia (after Fritsch, 1945).

D. *Diplothamnion* : Longitudinal section of spermatangial head (after Huisman, 1991).

Dasyaceae : The spermatangia are developed on pseudolaterals, although in some *Dasya* sp. they may be produced on adventitious branches. The sori development is identical to *Polysiphonia* (Fritsch, 1945).

CARPOGONIAL BRANCH AND CARPOGONIUM

The carpogonium is a transformed apical cell of a special lateral or a terminal filament containing a specific number of cells. In Ceramiales, the carpogonial branch is generally a 4-celled filament formed by a lateral initial cut off from periaxial cell.

The origin of carpogonial branch and auxiliary cell branch is controlled genetically and starts much before it is morphologically expressed. The time of transforming event may be when the carpogonial initial is cut from the supporting cell, or it may involve the apical cell of an already formed cortical filament, rhizoidal filament or a modified indeterminate axis. In the latter event the limit set for carpogonial branch are on the cytological level and the length of filament and position of supporting cell are poorly defined.

Ceramiaceae : In taxa with the simplest organization, as in *Antithamnion* and *Ballia,* it is difficult to differentiate between carpogonial branch and other laterals of limited growth. In other taxa the fertile lateral is represented merely by supporting cell and an overlying sterile cell, although the latter may divide. The sterile cell is cut off from the supporting cell prior to development of carpogonial branch. In *Crouania* and *Callithamnion,* the single fertile member of a whorl is always much reduced, and is represented only by the supporting cell. In *Callithamnion* procarp there is no sterile cell.

In *Antithamnion plumula* the carpogonial branch arises from the under side of the basal cell of the lateral and in the beginning the supporting cell does not differ from other sterile cells. In *Ceramium,* the procarp is formed just behind the apex. The supporting cell is the first formed pericentral of the segment and normally bears two carpogonial branches. Identical condition is seen in *Microcladia* and *Antithamnion* (Fig. 117 A).

In *Wrangelia penicillata* the procarps arise on special distichous laterals of limited growth, that are crowded together at apices of the main thallus branches. In these fertile laterals the pericentrals of upper 10 to 15 segments produce whorls of branchlets consisting of 2 or 3 cells. The slightly enlarged cell, of the oldest member of each whorl in the middle segment, functions as supporting cell and bears a curved carpogonial branch (Fig.117 B).

In *Griffithsia,* the fertile branchlets consist of a few multinucleate cells that arise from the apices of axial cells, but are soon pushed to one side by development of the laterals. The lowest cell of the branchlet later gives rise to envelope of cystocarp whereas the subterminal one produces the procarp. The latter cuts off three pericentrals, the basal of which bears the carpogonial branch whereas those on right and left develop into two celled branchlets.

In *Spermothamnion,* the procarp develops on short laterals borne on lower parts of side branches. The lateral has cells with dense contents. The subterminal cell of these lateral cuts off three pericentrals, the first abaxially and other two on left and right flanks. One of these, later forms the supporting cell and supports the carpogonial branch. The carpogonial branch curves towards the adaxial side of fertile lateral. The entire procarp is embedded in mucilage (Fig. 117 C, 118 C; Fritsch, 1945; Huisman, 1985). In *Rhipidothamnion,* the procarps are formed on the subterminal cells of upright axis, from which three pericentral cells are produced, the second forming the supporting cell (Fig. 118 B; Huisman, 1985).

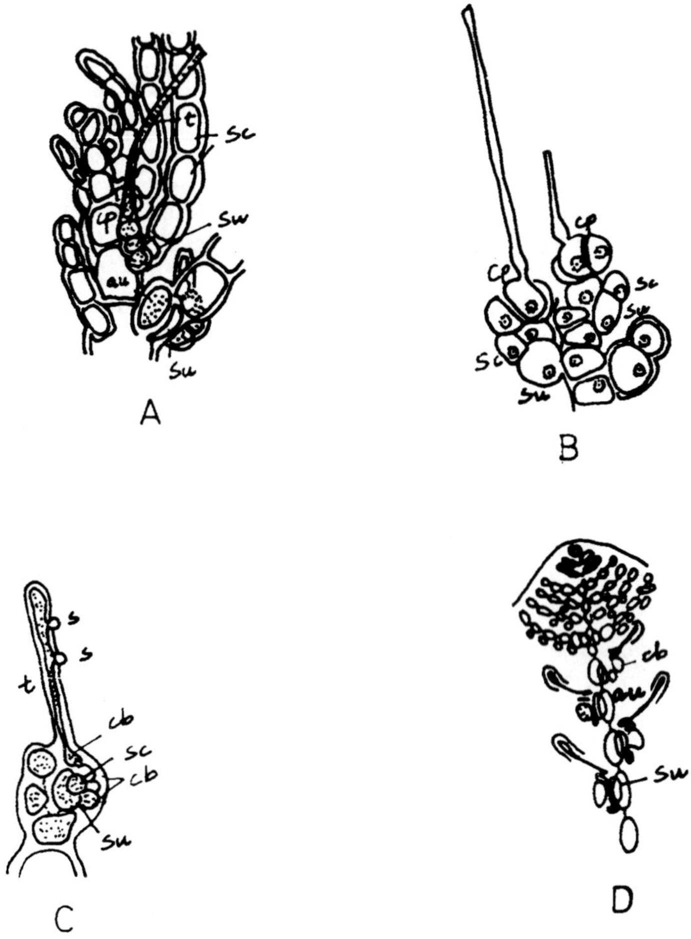

Fig. 117. A-D. Reproduction (Female plant)

(after Fritsch, 1945)

(au, auxiliary cell; cb carpogonial branch; cp, carpogonium; s, spermatium; sc, sterile cells; su, supporting cell {auxiliary mother cell}; t, trichogyne).

A. *Antithamnion* : A bıanch system with two carpogonial branches.

B. *Wrangelia* : Apex of fertile branch with two carpogonia.

C. *Spermothamnion* : Procarp at the time of fertilization.

D. *Delesseria* : Part of fertile frond showing carpogonial branch.

In *Callithamnion,* the procarp normally consists of a carpogonial branch and two auxiliary mother cells (pericentrals). Only one of these bears the curved carpogonial branch. The carpogonium lies midway between two mother cells and auxiliary cells are cut off from both mother cells. The procarps of *Ptilota plumosa* are borne on short laterals like that of *Spermothamnion* except that carpogonial branch is formed by first formed abaxial pericentral cell. In *Compsothamnion,* the second, out of three pericentral, forms the carpogonial branch. The procarp development in *Ptilothamnion* is identical to that of *Spermothamnion* except that the procarps are formed on the tip of laterals.

The carpogonial branch in *Callithamnion* sp. is 4-celled and arises from the lateral initials. It is arranged spirally and occurs on every fourth axial cell. The branch initiates on six days old axial cell and takes another four days to reach maturity (Fig. 119 A-D; Vijayaraghavan et al., 1995).

In *Spyridia,* the carpogonial branch is formed on adventitious laterals which consists of alternating sterile and fertile parts. The fertile parts bear branches alternatively on right and left and cut off 4-pericentrals per segment, out of which one forms the carpogonial branch.

In *Diplothamnion,* the procarps are formed subterminally on the indeterminate branches. The subapical cell of the axis produces two pericentral cells, of which one acts as the supporting cell and cuts off a lateral 4-celled carpogonial branch and a terminal, sterile cell (Fig. 118D; Huisman, 1991).

In *Radiathamnion* (Mills and Kraft, 1981), the procarp is formed on small celled laterals of secondary and tertiary branches. These are formed singly on the axis. The supporting cell is the basal cell of a short sterile branchlet which arises from the fertile axial cell and reaches 3 or 4 cells in length at the time of fertilization. The curved carpogonial branch initial is cut off from the supporting cell. The carpogonium has an elongate trichogyne. In *Antithamnionella* (Whittick, 1980), the first cell of a whorl branchlet becomes the supporting cell immediately after formation and divides laterally to produce a cell on its adaxial side. This cell divides to form the 4-celled carpogonial branch.

Delesseriaceae : The procarps in this family consists of two groups of sterile cells. One is cut off from the supporting cell before and the other after the formation of carpogonial branch. The first group is homologous to sterile cells of Ceramiaceae and the second is equivalent to the carpogonial branch itself. In *Polyneura* and *Hemineura* a second, although less developed, carpogonial branch is formed in place of the second sterile group.

In all members of this family the procarps arise from the axial cells of the fertile fronds. In *Delesseria sanguinea,* the fertile fronds consist of minute lanceolate leaflets. Each axial cell, of such a fertile leaflet, produces a pair of procarps except a few of the apical segment which develop only lateral pericentrals. The first step in their development is cutting off of tangential pericentrals in the plane of frond which function as supporting cell. Later the initial of carpogonial is cut off by an anticlinal wall. The carpogonial branches usually arise alternately to right and left in successive segments (Fig. 117 D). The trichogynes are inflated and project above the surface. The second cell in the carpogonial branch is enlarged. The carpogonia are borne in acropetal succession and as many as 50 may be found on one blade. Due to acropetal arrangement of carpogonia fertilization may be prolonged. Almost at the time of fertilization, a second sterile cell is cut off from the supporting cell and both the sterile cells form short threads. In *Membranoptera,*

Apoglossum and *Hypoglossum,* the procarp arrangement and structure are identical to those of *Delesseria,* except that in the former taxa the procarp arises from the midrib of ordinary vegetative fronds.

In *Grinnellia,* the procarps appear scattered but actually they arise from axial cells of fertile leaflets completely embedded in the parent thallus. The leaflets occasionally project above the surface and the procarps formed appear to be stalked. In *Claudea* and *Vanvoorstia,* the procarps arise from dorsal pericentrals. The procarps are irregularly distributed in *Phycodrys rubens* which appear as white dots near the margins or on special marginal proliferations of the older fronds. The fertile cells are large with dense contents and cut off tangential pericentrals that divide to form pairs of procarps. The trichogynes are perpendicular to the frond. The adjacent vegetative cells cut off tangential pericentrals. Both the sterile cells divide and the colourless elements thus formed cover the supporting cell and the carpogonial branch.

In *Polyneura,* the second group of sterile cells forms a reduced carpogonial branch. In *Cryptopleura,* the sterile group consists of a few cells. In *Cryptopleura, Nitophyllum, Acrosorium* and *Martensia,* the carpogonia are borne in terminal position.

Nitophyllum differs from the Delesseriaceae in that the fertile cell cuts off on either surface two pericentral cells. The first forms two or three cortical cells and the second functions as the supporting cell which bears the carpogonial branch along with a group of sterile cells. In *Martensia,* the margins of the longitudinal lamellae bear the procarps. The supporting cell appears to be cut off from the axial cell which is a part of single layered lamellae. The cells of carpogonial branch are multinucleate. In *Yamadaphycus* (Mikami, 1973), the procarps are borne in an acropetal order on central rows of proliferations produced by blade of female plant. The pericentral cell first cuts off mother cell of first group of sterile cells. Later it cuts off the mother cell of the carpogonial branch and the supporting cell. The mother cell of second group of sterile cell is cut off by the supporting cell between the first sterile cell and the basal cell of carpogonial branch. In *Branchioglossum* (Yoshida and Mikami, 1992) a few procarps are found in acropetal succession along the midrib of distal part of branches. The procarps are composed of a 4-celled carpogonial branch and two groups of sterile cells.

In *Yoshidaphycus* (Mikami, 1992), the procarps are borne in an acropetal order on central rows of proliferations produced by blade of female plants. The pericentral cell first divides to form mother cell of first group of sterile cells and later forms the mother cell of carpogonial branch and the supporting cell. The mother cell of second group of sterile cell is cut off by the supporting cell between the first sterile cell and the basal cell of carpogonial branch.

Rhodomelaceae : The procarps are usually formed on trichoblasts on suprabasal segment, which like the basal segment becomes polysiphonous. The fertile segment produces five pericentral cells. The first two cut off abaxially, the next two on the sides and the fifth on the adaxial side. The one on adaxial side gives rise to the procarp. First a lateral sterile cell is formed and later a 4-celled carpogonial branch develops followed by the formation of a second sterile cell. The first sterile cell is the vegetative cell of fertile lateral and second is homologous to the carpogonial branch. The cells of carpogonial branch are binucleate. These features are characteristics of *Polysiphonia.* Here the fertile trichoblast are borne just behind the apex.

The other members of Rhodomelaceae differ from *Polysiphonia* in the position of procarp. In many taxa the trichoblasts are branched as in *Polysiphonia, Pollexfenia, Herposiphonia* and

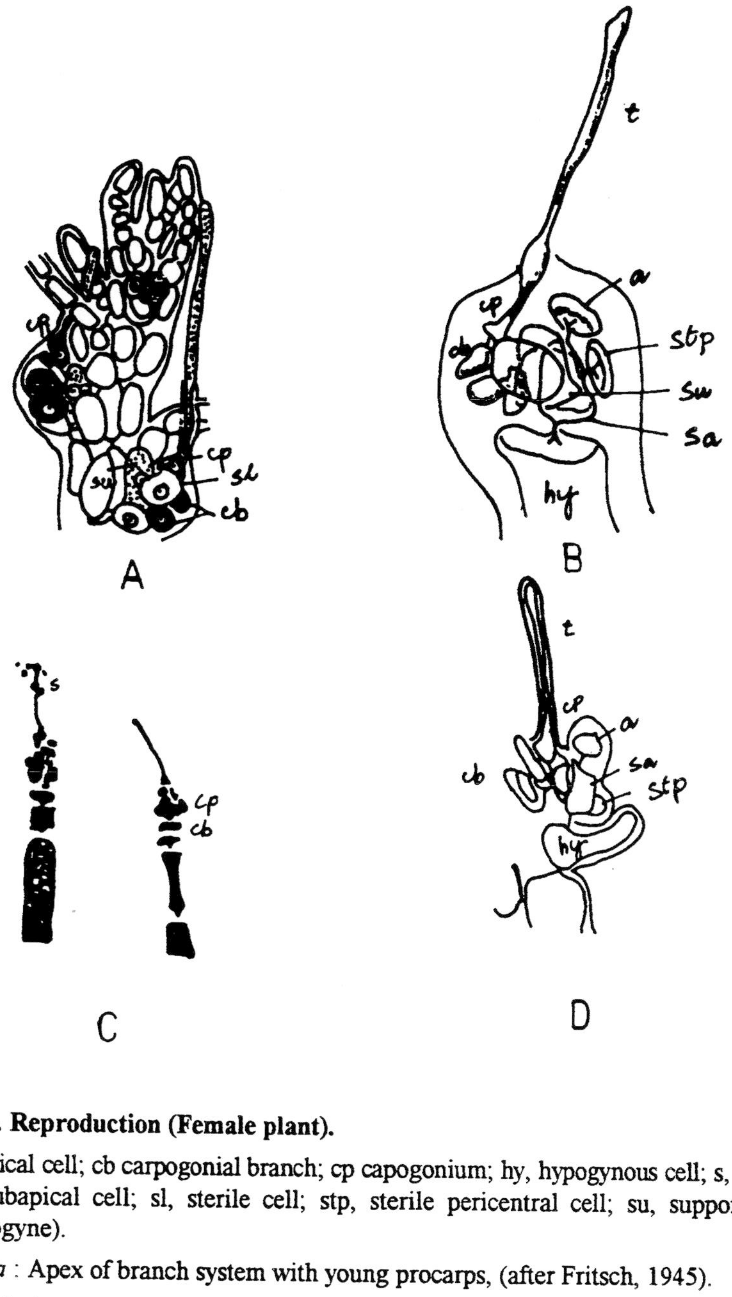

Fig. 118. A-D. Reproduction (Female plant).

(a, apical cell; cb carpogonial branch; cp capogonium; hy, hypogynous cell; s, spermatium; sa, subapical cell; sl, sterile cell; stp, sterile pericentral cell; su, supporting cell; t, trichogyne).

A. *Dasya* : Apex of branch system with young procarps, (after Fritsch, 1945).

B. *Rhipidothamnion* : A stage in development of procarp (after Huisman, 1985b).

C. *Spermothamion* : Mature procarp system with attached spermatia (after Huisman, 1985b).

D. *Diplothamnion* : Mature procarp (after Huisman, 1991).

Placophora. They may be simple as in *Rhodomela.* In *Laurencia* and *Chondria*, the trichoblasts consist of three segments, the second being larger than the other two. The procarps are found on trichoblasts still enclosed in apical depression. In *Laurencia,* only four pericentrals are formed from each fertile segment. In *Odonthalia,* the procarps are formed on adventitious branches, arising from the margins of the frond. Except for these differences, other characters are identical to the *Polysiphonia* type. In *Ptrosiphonia,* the procarps are formed on adventitious branches.

In *Cliftonaea* and *Leveillea,* the procarps are borne on laterals of 3 to 4 segments and arise from the lowest segment of the dwarf shoot. These laterals are borne endogenously (Fritsch, 1945).

Dasyaceae : The procarps are formed on sympodial axis in *Dasya* (Fig. 118 A) and *Dasyopsis* and on pseudolaterals in *Dasya arbuscula* and *Heterosiphonia.* Usually the third or sometimes the fourth pericentral bears the procarps. Only in *Heterosiphonia,* it is the fifth or the last formed pericentral which gives rise to the procarp. The development of procarp is similar to that of Rhodomelaceae.

The procarps of *Dasya* and *Dasyopsis* arise from successive segments of sympodial apex or more usually from adventitions branches. In *Heterosiphonia plumosa,* the procarps usually arise on fourth segment of a pseudolateral and have very elongate trichogynes. The sterile cells undergo considerable division before fertilization.

FERTILIZATION

Fertilization normally occurs by adhesion of spermatia to the surface layer of carpogonial wall. Generally spermatia are passively transported by water. But there are exceptions. In *Tiffaniella* (Ceramiaceae) the spermatia are released from a number of spermatangial heads into spermatangial strands which coalesce, expand, contract and rotate in water till they come in contact with the carpogonium (Fetter and Neushul, 1981).

The fusion of spermatium with carpogonium seems to require a binding substance which is excreted by the tip of the trichogyne (Broadwater and Scott, 1982). In *Callithamnion* (Ceramiaceae) the binding is not species-specific but occurs only in closely related species (Magrudur, 1984). Trichogynes normally persist and several spermatia may attach near the tip of trichogyne as in *Polysiphonia* (Rhodomelaceae) or along its sides.

The spermatia may initially attach itself to vegetative parts. In *Callithamnion,* the spermatia which have cone-shaped, fimbriate appendages on both ends, attach initially to vegetative hair cells and later to the nearby trichogyne. In *Hymenena* (Delesseriaceae), spermatia attach to the thallus surface and when the trichogyne breaks through the surface spermatia attach to it (Cole and Sheath, 1990). It might be possible that a chemotropic substance is the cause of trichogyne's directed movement.

Several nuclei from many spermatia enter the carpogonium but the neck is so narrow that it permits only one nucleus to pass through. In *Polysiphonia* and most other Ceramiales the vacuole shrinks immediately after one sperm nucleus has entered the carpogonium and thus brings it into contact with the egg nucleus. Along with this process the cytoplasm is pinched off, thus separating the trichogyne from the carpogonium and there by effectively hindering any supernumerary fertilization (Broadwater and Scott, 1982).

CARPOSPOROPHYTE

In Ceramiales the carposporophyte development is mediated by one or more auxiliary cells. The carpogonium may either produce a connecting cell or filament or may fuse directly with one or more auxiliary cells. The diploid nucleus is transferred to the auxiliary cell and gonimoblast filaments arise from it.

Drew (1954) and Papenfuss (1957) have distinguished two kinds of auxiliary cells–generative auxiliary cell and nutritive auxiliary cell. The former is responsible for production of gonimoblast filaments. But Cole and Sheath (1990) have restricted the term auxiliary cell only for generative auxiliary cell and the nutritive auxiliary cells have been termed either as nutritive cells, nutritive filaments or nutritive tissues.

The connecting cells are small, uninucleate, and cut off by the carpogonium with a minimal amount of cytoplasm. Ceramiales have very characteristic connecting cells. The nuclei are compact and surrounded by a clear hyaloplasm. The elongation of connecting cell occurs only when in contact with an auxiliary cell.

Procarps may be said to be present when the auxiliary cells are borne on the same branch system in close proximity to the carpogonium and absent when carpogonium and auxiliary cells are borne separately. It seems that procarpial condition has completely arisen anew together with evolution of connecting cells and auxiliary cells.

The carposporophyte may be divided structurally into 3 parts.

(i) outer photosynthetic tissue.

(ii) modified nonphotosynthetic inner gametophytic tissue

(iii) developing carposporophyte.

The outer compartment consists of assimilatory filaments or unmodified cortical tissues of vegetative system along with secondary assimilatory filaments or photosynthetic tissue generated before or after fertilization. In most of the members of Ceramiales a complex multilayered structure called a cuticle overlies the external tissues. The cuticle should be able to retain or delay the outward passage of metabolites that are needed for carposporophyte development. As most cystocarps are covered with cuticle and ostiole develops only after cystocarp maturation the preventive conduct of cuticle seems convincing.

The second compartment is composed of gametophytic tissues that are close to carposporangia or auxiliary cells and are so situated that they are readily contacted by developing carpospores. These cells are protein-rich. Some cells have increased number of nuclei, or their nuclei and nucleoli may be enlarged and contain amplified levels of DNA. These deeply staining cells are part of nutritive cells. The nutritive cells are of two kinds: one which lyses after each cell cycle and the other which persists. These persistent centres have been termed as nutrient processing centres (Cole and Sheath, 1990). These regions are characterized by cells which possess either numerous small nuclei, single large nucleus or moderate number of enlarged nuclei.

The innermost compartment consists of carposporophyte which may contain either simple gonimoblast filaments or complex structure including auxiliary cell and connecting filaments. In Ceramiales the latter is true.

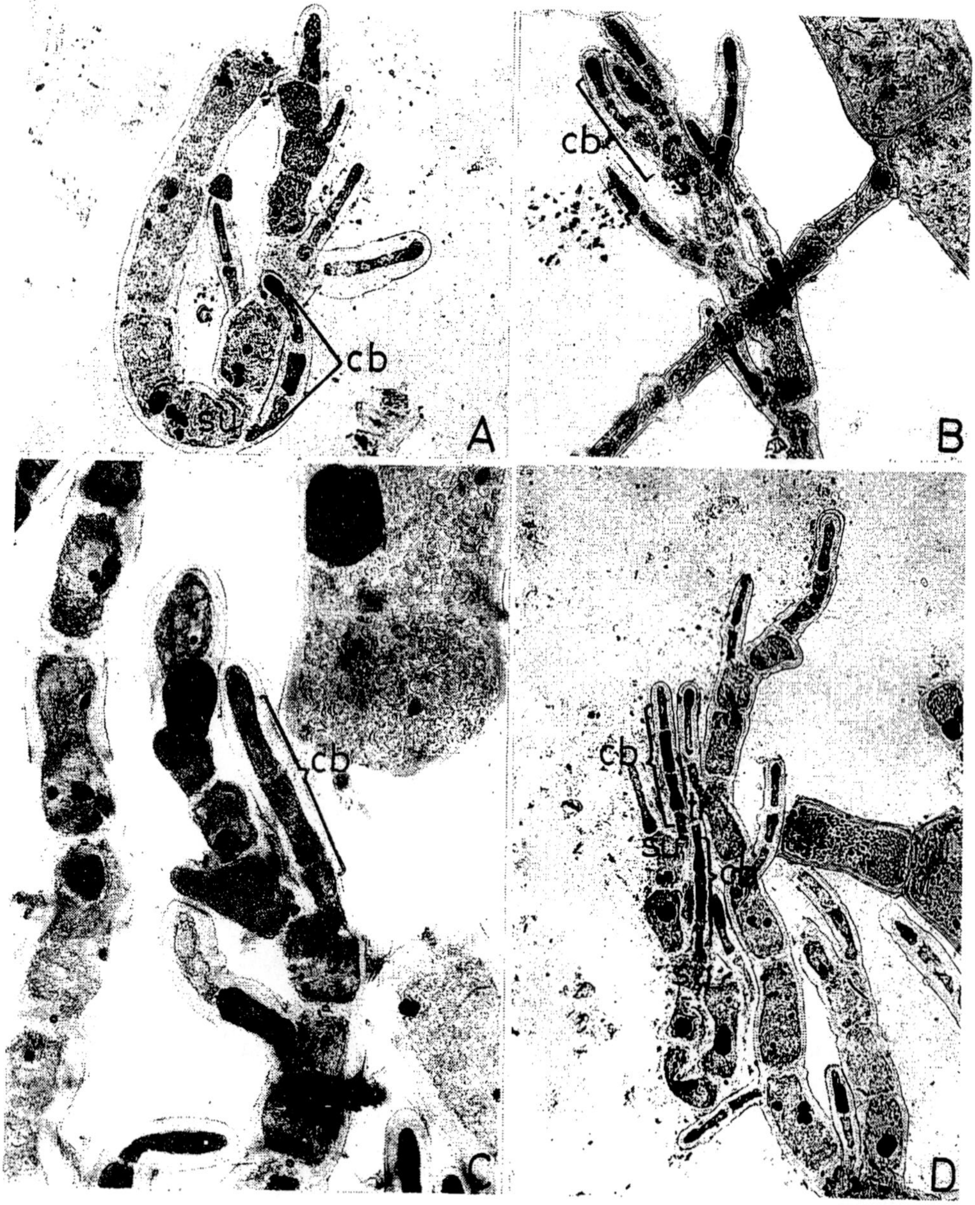

Fig. 119 A-D *Callithamnion* sp. in culture. Carpogonial branch. (after Vijayaraghavan *et al.*, 1995)

(*cb*, Carpogonial branch; *su*, supporting cell; *tr*, trichogyne)

A. Curved carpogonial branch arising from sixth axial cell below the apex.

B, C. Same, arising from fourth and fifth axial cell below the apex.

D. Carpogonia with well developed trichogynes.

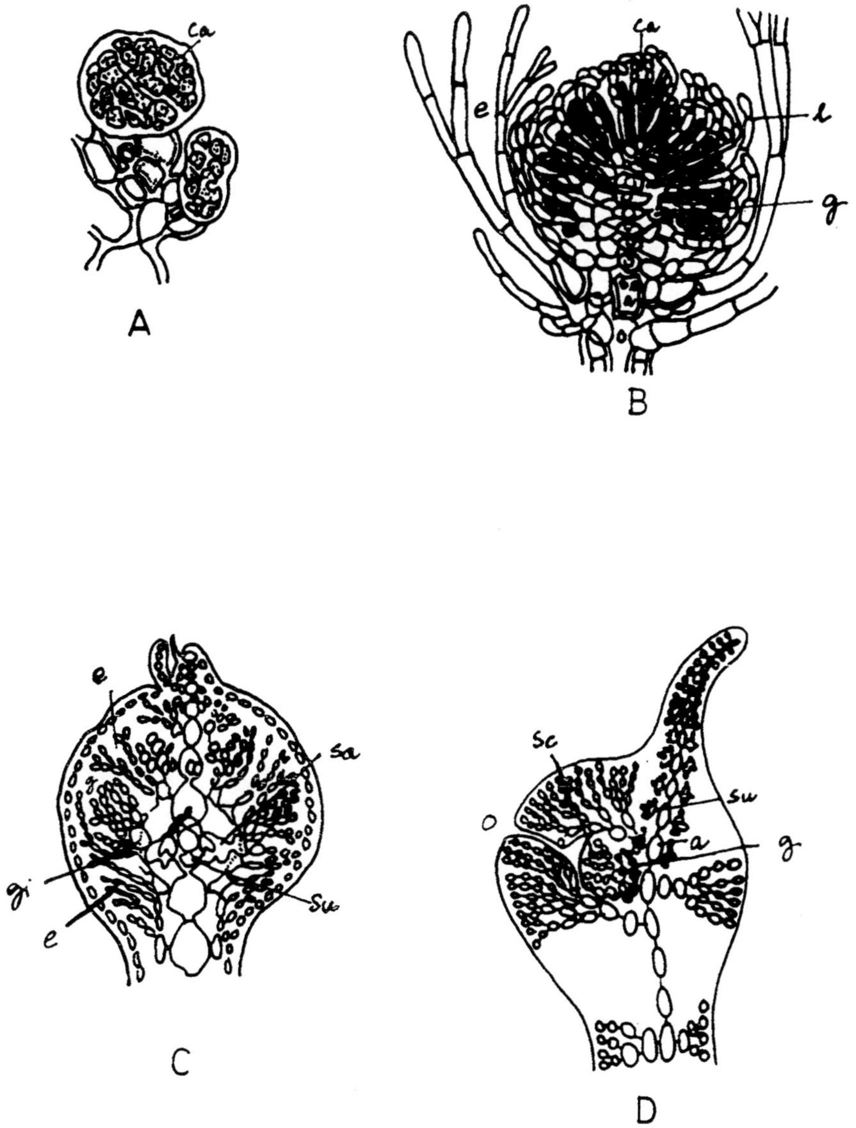

Fig. 120. A-D. Carposporophyte.

(after Fritsch, 1945)

(a, axial cell; ca, carposporangia; e, enveloping threads; g, gonimoblast; gi, gonimoblast initial; o, ostiole, sa, sterile axial cell; su, supporting cell).

A. *Anthithamnion* : Group of mature carpogonium

B. *Wrangelia* : A vertical section of mature cystocarp.

C. *Ptilota:* Longitudinal section of developing cystocarp.

D. *Delesseria sanguinea* : Longitudinal sections of developing cystocarp.

Certain modifications may occur in the relationship between the carposporophyte and the surrounding gametophyte tissue.

(i) enlargement and modification of structure in the existing pit-connections.

(ii) formation of central fusion cell by fusion between contiguous cells either through or around the pit-connections.

(iii) formation of new pit-connections or fusion of cells between gonimoblast filaments and noncontiguous gametophytic cells, occasionally with formation of a new and well defined placenta.

Ceramiaceae : Connection between fertilized carpogonium and auxiliary cell is usually through a special connecting cell. Gonimoblast lack compact envelope and most of the gonimoblast cells produce carpospores.

In *Antithamnion plumula,* the fertilized carpogonium gives rise to many small connecting cells which fuse with the auxiliary cell. The gonimoblast initial originates from the auxiliary cell and except the basal cell all the cells of gonimoblast filament bear carpospores. The carposporangia constitute several distinct groups called gonimolobes and are surrounded by loose envelope produced by surrounding sterile cells (Fig. 120 A). In *A. pacificum* and *A. spirographidis,* the gonimoblasts mature in succession. *Platythamnion* show similar features and the cystocarps in *Ballia* occurs in the axils of fertile laterals.

In *Wrangelia penicillata,* on each lateral usually only one out of many carpogonia is fertilized and concomitantly, all the branchlets of the fertile lateral develop and form a dense group overarched by surrounding sterile branches. The gonimoblast filaments grow along axial cells of fertile branch and form secondary pit-connections with the lower cells of the sterile filaments. The terminal carposporangia are formed on branches of creeping gonimoblast. The cystocarp formed on the fertile lateral, is traversed by an axial cell and is surrounded by an incomplete, almost compact envelope formed by sterile laterals (Fig. 120 B).

In *Griffithsia,* a special connecting cell is not produced and there is direct fusion between carpogonium and auxiliary cell. The fusion cells produce richly branched gonimoblasts of which almost all the cells form uninucleate carpospores in basipetal sequence. The pit-connections between cells of fertile branches enlarge. The supporting, the axial and the sterile cells fuse. Later the auxiliary cell fuses with the supporting cell. The enveloping threads are small and generally 2-celled. Identical cystocarp development is reported in *Halurus* and *Bornetia.*

In *Spermothamnion roseolum,* the last two pericentrals give rise to two auxiliary cells. One among these two pericentral cells is the supporting cell. The carpogonium, after fertilization, gives rise to connecting filaments which fuse with the auxiliary cells where the diploid nucleus is transfered. After receiving the diploid nucleus the auxiliary cell divides transversely into four parts, each gives rise to a gonimoblast filament with a carposporangium at the distal end.

The carpogonium in *Callithamnion,* after fertilization, divides longitudinally into two. Each chamber cuts off a small connecting cell which fuses with the protuberance from adjacent auxiliary cell. The nucleus of the auxiliary cell takes a basal position while the diploid nucleus moves toward the apex of the auxiliary cell. A septum later appears separating the upper diploid nucleus from the nucleus of the auxiliary cell. The upper compartment gives rise to gonimoblast threads. The carposporangial mass is naked and may be spherical or composed of successively maturing lobes.

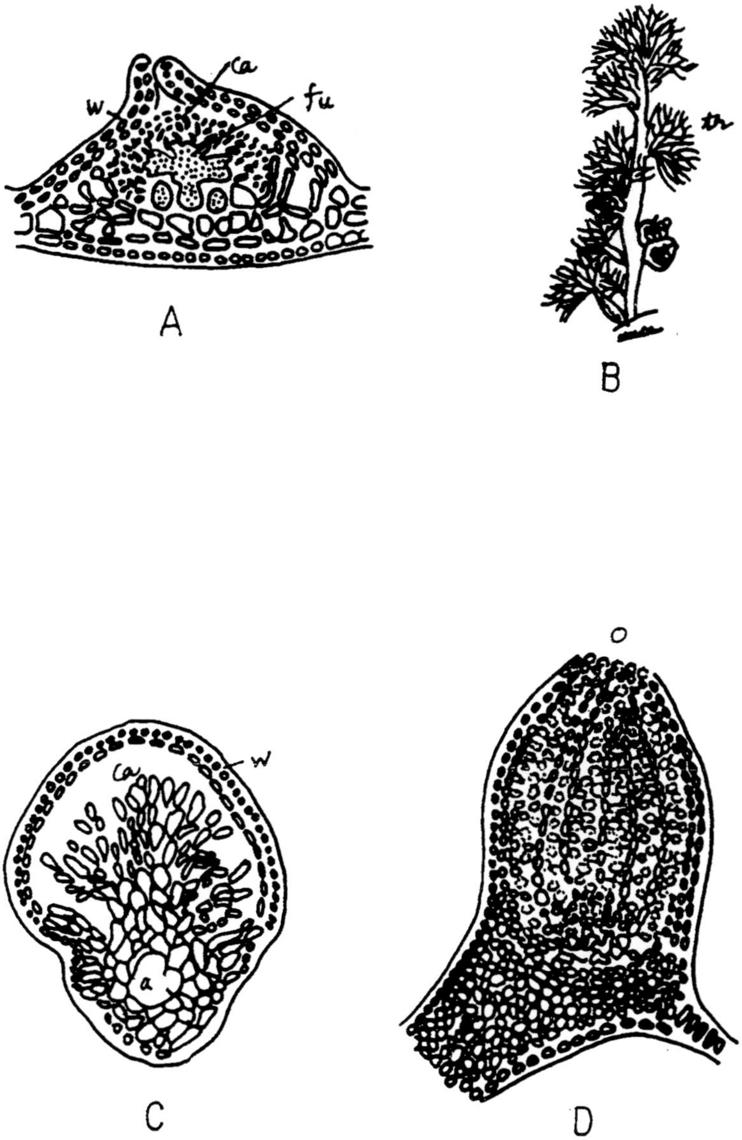

Fig. 121. A-D. Carposporophyte

(a, axial cell; Ca, carposporangia; fu, fusion cell; o, ostiole; tr, trichoblasts; w, wall of cystocarp).

A. *Hypoglossum* : A transverse section of cystocarp. (after Fritsch, 1945)

B. *Chondria* : A branch bearing cystocarps. (after Fritsch, 1945)

C. *Dasya* : A young cystocarp in section. (after Fritsch, 1945)

D. *Yoshidaphycus* : Section through a young cystocarp. (after Yoshida and Mikami, 1992).

In *Ptilota plumosa,* the fertilized carpogonium transfers the diploid nucleus by means of a connecting cell (Fig. 120 C). In *Spyridia,* the first of several pericentrals of each sterile segment gives rise to a filament , while the others, as well as the pericentrals of fertile segments whose procarp fails to develop further, grow into richly branched threads. The three remaining pericentrals of the fertile region cuts off cells that give rise to three gonimoblasts each. These cells put out processes which fuse with the large auxiliary cell. The gonimoblast cells produce carposporangia. The complete fertile branch gives rise to three lobed cystocarp and the carposporangia are surrounded on all sides by sterile threads (Fritsch, 1945).

In *Radiathamnion,* after fertilization, the first cell produced by the supporting cell fuses directly by a broad connection with the fertile axial cells and also gives rise to an elongate gonimoblast initial from its upper part. Eventually the supporting cell, fertile axial cell and auxiliary cell fuse to form the fusion cell. An apical gonimoblast initial is first and later two lateral initials develop. The carposporangia are found in chains. The branch axis in region of carposporophyte bears many additional laterals which envelope the carposporangial mass in a lax involucre (Mills and Kraft, 1981).

In *Antithamnionella,* after fertilization, a special connecting cell develops from the carpogonium which fuses with the auxiliary cell formed by supporting cell's adaxial side. The axial cell bearing the fertile branch, auxiliary cell and supporting cell fuse to form a fusion cell. The gonimoblast initial produces four or five cells, each of which form carpospore producing filaments that are aggregated in sequentially developing gonimolobes. Once fertilization has taken place the axial cell bearing the fertile whorl branchlets ceases to grow and growth of laterals is enhanced resulting in production of loose filamentous involucre of branchlets surrounding the carposporophyte (Whittick, 1980).

In *Diplothamnion,* after fertilization, the carpogonium fuses either directly or through a connecting filament with the auxiliary cell. Two or three gonimoblasts are produced from the auxiliary cell. Apical cell, sterile pericentral cell and sterile cell on supporting cell give rise to much branched, inner involucral filaments that surround the developing gonimoblasts. The basal cell of single lateral, which arises from hypogynous cell, forms simultaneously a secondary pit-connection with the cell lying just beneath it and forms an outward involucral branch. A fusion cell is finally formed that consists of subapical, supporting and auxiliary cells. The elongate branched filaments arise from the fusion cell and bear terminal clavate carposporangia.

In *Antithamnion,* the reported role of auxiliary cell in rejecting incompatiable fertilization is indicated during a cross between *Antithamnion sparsum* (male) X *A. defectum* (female). Normal gonimoblast development and release of carpospores occur. But this is not true for the reciprocal cross. The reason of failure of a reciprocal cross is attributed to the failure of the connecting cell to fuse with normally formed auxiliary cell.

Delesseriaceae : Only one procarp is generally fertilized on a leaflet and gives rise to only one cystocarp per leaflet. The lower cells of the fertile branch usually grow forming the pericarp. The carpogonium fuses directly with the auxiliary cell. Later the supporting and the axial cells also fuse forming a large fusion cell (Fig. 121 A). The auxiliary cell gives rise to branched gonimoblast filaments before formation of the fusion cell. The mature cystocarps are pedicellate and possess an ostiole (Fig. 120 D).

In *Claudea* and *Vanvoorstia,* the single cystocarp, on each fertile frond, bears tooth-like projections on its lower surface representing daughter blades that have failed to anastomose.

In *Phycodrys rubens,* after fertilization the auxiliary cell is formed. Nutritive materials accumulate in nearby axial cells and the two groups of sterile cells. At the same time the pericentrals form cystocarp wall. The gonimoblast initial divides vertically and each segment produces a branched thread where the upper cells form carpospores. A fusion cell is formed comprising axial, auxiliary and nearby cells.

In *Martensia*, the auxiliary cell connects itself to the second cell of carpogonial branch. No fusion cell is formed. The gonimoblasts are surrounded by nutritive filaments produced from the supporting cell and other cells. The mature cystocarp can be seen as large swelling on the margins of longitudinal lamellae (Fritsch, 1945).

In *Branchioglossum,* cystocarps are formed at the branching point of the thallus and almost occupy the whole width of the thallus. The fusion cell is formed and gives rise to gonimoblast filaments with three or four segments bearing the carposporangia in chains (Mikami and Yoshida, 1992).

In *Yoshidaphycus,* the cystocarp is pot-shaped with an ostiole. The pericarp has several rhizoidal cells. The gonimoblasts bear terminal carposporangia which are elliptical to long elliptical in shape (Fig. 121 D; Yoshida and Mikami 1992).

Rhodomelaceae : The auxiliary cell is cut off apically from the supporting cell and the gonimoblasts are formed in a sympodial manner. The cystocarp consists of an ostiole and the pericarp is formed by two lateral pericentrals of fertile trichoblast segments. The development of envelope starts before fertilization.

In *Polysiphonia,* the sterile cells divide transversely after fertilization and communication occurs between carpogonium and auxiliary cell. In *Polysiphonia decipiens* and *Laurencia pinnatifida,* a connecting cell is formed. The diploid nucleus passes to the auxiliary cell which fuses with the supporting cell. The fusion cell also includes the axial cell of fertile filament.

The members of Rhodomelaceae show a very characteristic method of gonimoblast development. The gonimoblast initial gives rise to a number of short threads - the terminal cells of these develop into a pear-shaped carposporangia. The subterminal cells grow out into two celled branches, the end cells of which produce carpospores. The fusion cell finally fuses with gonimoblast initial and with the sterile cells. The envelope starts developing before fertilization and is mainly produced by lateral pericentrals. The adaxial pericentral cell of basal segment also participates. The young envelope has two lateral valves and are closed at top with only the trichogyne projecting out. After fertilization, the two lateral valves unite and except the apical cells all others form two external pericentrals each. The cystocarps are stalked, the stalk formed of basal segment of a trichoblast, spherical or urn-shaped and the aperture is directed upwards. In *Brongniartella byssoides,* the cystocarps are identical to *Laurencia.* In *Ricardia, Amansia* and *Vidalia* the cystocarps are embedded in the thallus.

In *Rhodomela, Laurencia* and *Chondria* (Fig. 121 B), the mature envelope of cystocarp is made up of more than two layers due to tangential division of pericentrals. In *Bostrychia,* envelope develops only after fertilization. In *Laurencia,* the mature cystocarp is partially immersed in the thallus. The carposporangia are ovulate. As single fusion cell is not formed and the base of carposporophyte seems to consists of large and densely staining cells. A delicate network of dichotomously and trichotomously branched filaments radiate from this complex and lines the inner wall of the pericarp (Fritsch, 1945).

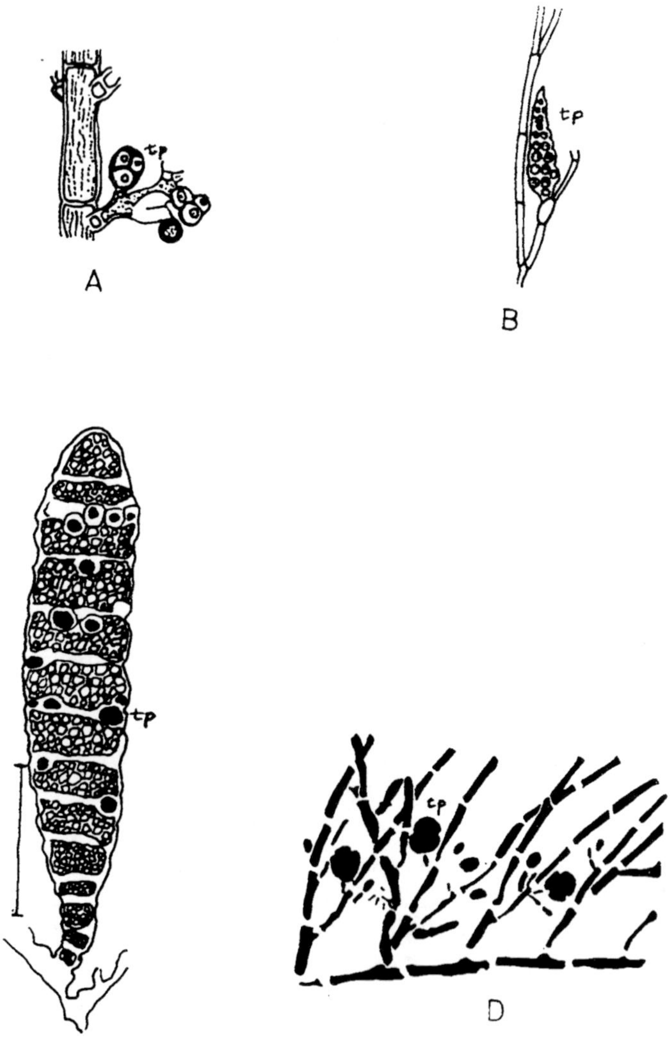

Fig. 122. A-D. Tetrasporophyte.

(tp, tetrasporangium)

A. *Antithamnion* : Tetraspores borne on tetrasporophyte (after Fritsch, 1945).

B. *Dasya* : Stichidia (after Fritsch, 1945).

C. *Ceramium* : Habit of mature tetrasporophyte (after Cormaci and Funari, 1991).

D. *Spermothamnion* : Tetrahedral sporangia borne terminally on single celled lateral branches (after Huisman, 1985b).

Dasyaceae : The cystocarp envelope appears after fertilization and gonimoblasts show monopodial branching. In *Dasya* (Fig. 121 C), *Dasyopsis* and *Heterosiphonia*, the general rule is one cystocarp per branch. One or sometimes two connecting cells are cut off from carpogonium which fuse with the auxiliary cell. The gonimoblast initials first form a row of large cells, which later cut off small cells that eventually form branched filaments. The carposporangia are formed in short chains and in basipetal manner. No fusion cell is formed and an ostiole is present. The envelope of cystocarp is produced by pericentral cells flanking the fertile pericentral and the adjacent pericentrals. The envelope is two layered except in *Dasyopsis* where it is three layered.

TETRASPOROPHYTE

These diploid plants resemble in morphology the gametophyte. The four families of Ceramiales differ markedly in development of tetrasporangia. The tetrasporangia form four uninucleate, equal sized spores through two or three cleavages. The spores are organized either in cruciate, zonate or tetrahedral manner.

Ceramiaceae : The tetrasporangia are found in diverse positions and are not grouped. In *Antithamnion* they occur in a series on the adaxial side of the laterals and may be sessile or borne on one or 2-celled stalk (Fig. 122 A). In *Spermothamnion* (Fig. 122 D) and *Compsothamnion*, they may be clustered on short branches. In *Crouania*, they are formed from the basal cells of laterals whereas in *Ptilota plumosa*, on threads produced by cortex of ultimate branches. In *Griffithsia*, the tetrasporangia bearing laterals are formed in a whorled manner at the apices of the branches. These special laterals are single celled and bear many tetrasporangia, which may be either protected by outgrowth from the lateral itself or from the parent segment. In *Wrangelia* identical protective branches are formed from the cell that bears the tetrasporangia.

In *Ceramium*, the tetrasporangia are formed by enlargement of cells of the cortex and are partly embedded (Fig. 122 C). In *Spyridia*, the tetrasporangia develops on filaments and in *Trailiella* is formed from the larger half on longitudinally divided cell of erect filaments.

In *Antithamnion, Crouania* and *Trailiella*, the spores are cruciate and in other taxa show tetrahedral arrangement.

Antithamnionella is different from *Antithamnion* in having tetrahedral spores. In *Ceramium* and *Seirospora* both cruciate and tetrahedral tetrads coexist. In *Callithamnion*, the dehiscence of tetrasporangia is through the detachment of the lid.

In *Diplothamnion*, one or two tetrasporangia arise in positions similar to that of spermatangia. The spores are arranged tetrahedrally (Huisman, 1991). In *Radiathamnion*, the tetrasporangia are borne on small celled branchlets of primary or secondary laterals. The spores are tetrahedrally arranged (Mills and Kraft, 1981).

In *Antithamnionella*, tetrasporangia are borne adaxially on whorled branchlets usually on a single cell pedicel. The mature tetrasporangia are divided cruciately but they mimic the tetrahedral form of division. However, their development is through the bisporangium and hence ontogenetically are cruciate (Whittick, 1980).

In Ceramiaceae, bisporangial plants have been reported in *Callithamnion, Crouania, Oohriella, Sierospora* and *Spermothamnion*. Not much is known about the life histories of such plants. In *Callithamnion byssoides*, bisporangia are normally found on those plants which lack tetrasporangia (Dixon and Price, 1981). Some times terminal sessile or stalked monosporangia are seen in *Callithamnion* sp. (Fig. 123 A-D; Vijayaraghavan et al., 1995; Mam, 1992)

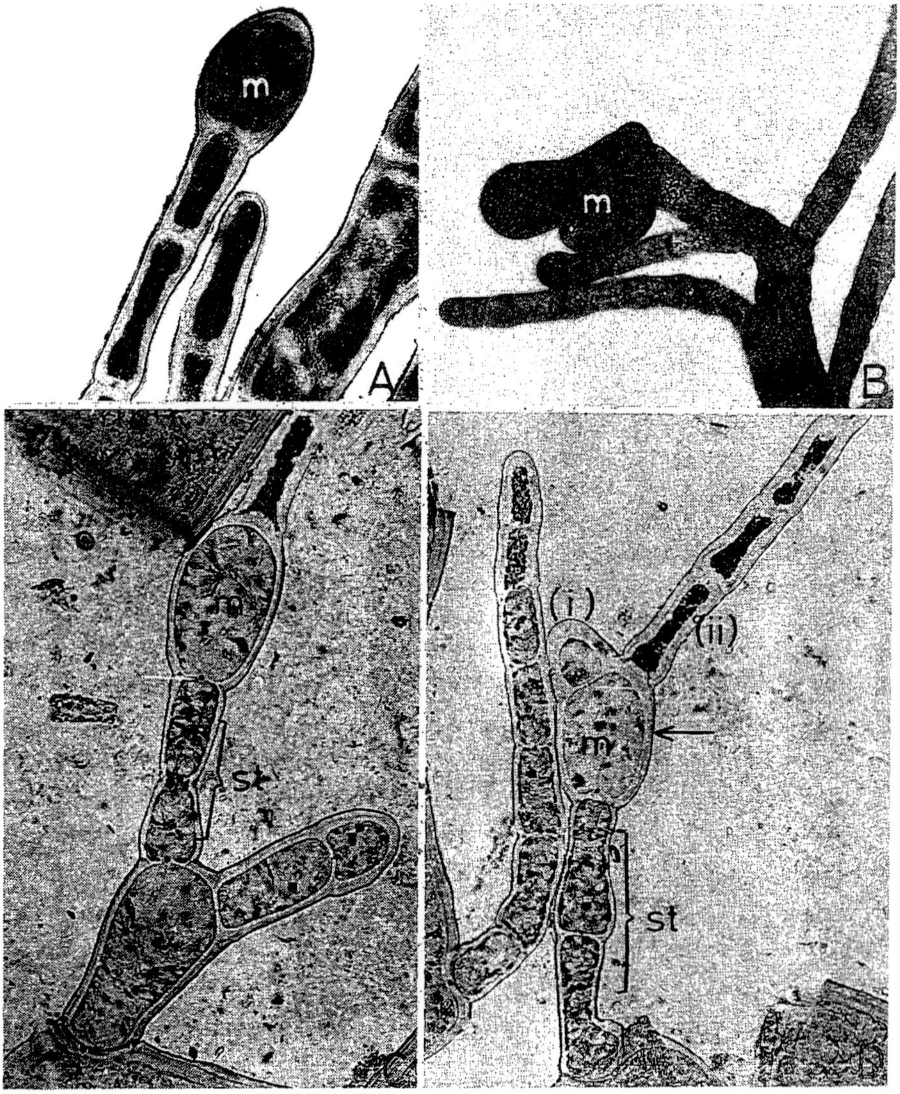

Fig. 123 A -D *Callithamnion* sp. in culture. Monosporangia

(after Vijayaraghavan *et al.*, 1995)

(*m,* monosporangia; *st,* stalk)

A. Terminally enlarged cell representing developing monosporangia.

B. Lateral branch showing sessile, germinating monosporangia.

C. Stalked germinating monosporangia.

D. Same, two (i and ii) filaments arising from the same monosporangia. (arrow)

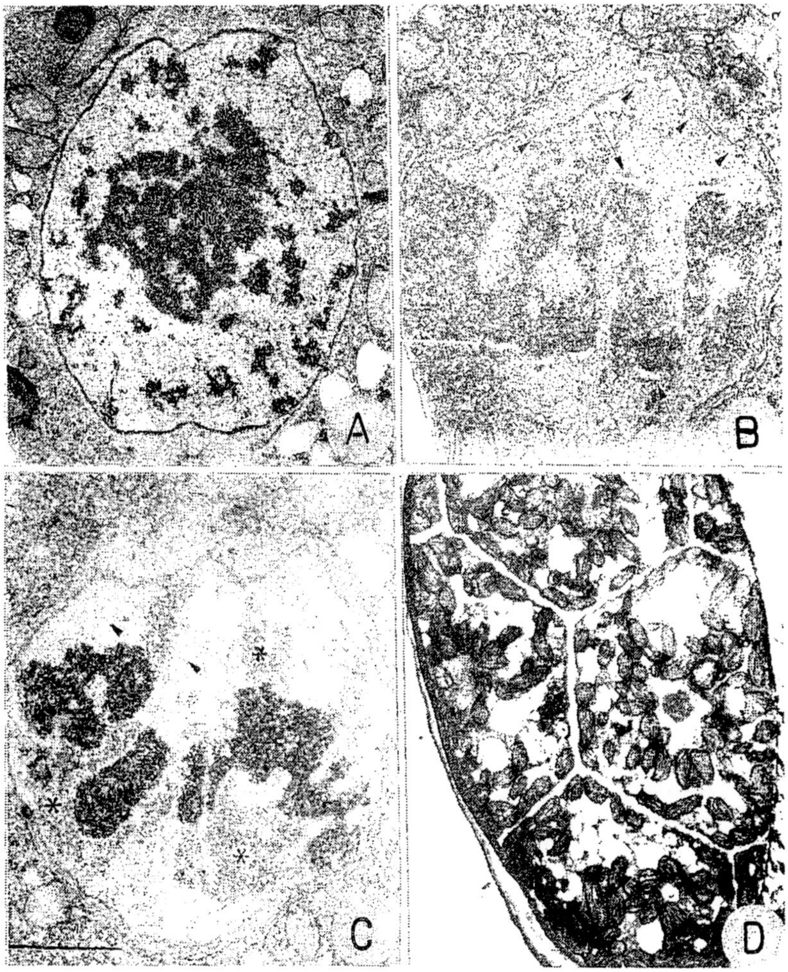

Fig. 124. A-D. Ultrastructure.

(after Cole and Sheath, 1990)

(Nu, nucleolus)

A. *Chondria* : Longitudinal section of late prophase nucleus showing condensing euchromatin and partially dispersed nucleolus.

B. *Polysiphonia* : Section of an early anaphase nucleus almost totally surrounded by perinuclear ER and microtubules attached to conspicuous layered kinetochores (arrowheads).

C. *Polysiphonia* : Section of a prometaphase nucleus showing nucleolar fragments (asterisks) associated both with the condensing chromosomes and intranuclear spindle microtubules (arrowheads).

D. *Pleonosporium vancouverianum* : Mature polysporangium showing regular cleavages.

Another kind of sporangia called polysporangia are found in Ceramiaceae. The term polysporangia includes sporangia with more than four spores formed in place of tetrasporangia (Drew, 1937). Polysporangia with spores in multiples of four have been reported in thirteen genera of Ceramiaceae. In *Tiffaniella snyderae* initials of polysporangia are multinucleate. Similarly the young initials of polysporangia in *Pleonosporium vancouverianum* (Sheath *et al.*, 1987) are multinucleate and are formed as laterals near the multinucleate apical cells. These lateral protrusions contain 2, 4 or 5 nuclei, each of which undergoes meiosis giving rise to polysporangia with 8, 16 or 20 uninucleate spores. Cleavage is simultaneous and centripetal (Fig. 124 D).

Still another kind of spore is found in Ceramiaceae. These are paraspores borne in parasporangia. According to Drew (1939) sporangia with more than four irregularly arranged spores that are able to reproduce the ploidy level of parent thallus and are formed at different places as that of tetrasporangia are parasporangia.

In *Ceramium strictum* a triploid strain reproduces by means of parasporangia (Rueness, 1973). *Callithamnion hookeri* also bears parasporangia (Dixon and Price, 1981). According to Guiry and Irvine (1989) alleles that control sexuality and tetrasporangial development can also lead to production of parasporangia if present in triploid state.

Delesseriaceae : The tetraspores are always found arranged in a tetrahedral manner. In *Delesseria sanguinea,* the tetrasporangia are formed in minute leaflets that lack midrib and are uniformly three layered. The surface layer cells from small branches which get covered by division of surrounding cells and the terminal cells of these branches bear the tetrasporangia. The surrounding cells later separate and form gaps through which the spores are shed.

The development of tetrasporangia is conserved throughout the family but their position varies. The tetrasporangia are formed in small proliferations in *Membranoptera;* appear as scattered outgrowths on the surface that bear them in *Grinnellia;* occur on both sides of midrib in *Hypoglossum, Apoglossum* and *Caloglossa* and are formed from pericentrals in the ultimate blades in *Claudea* and *Vanvoorstia.* In *Acrosorium, Cryptopleura, Phycodrys* and *Polyneura,* the sporangial initial arises from hypodermal cells whereas in the first three genera occasionally the central cells also form them.

The tetrasporangial sori in *Cryptopleura, Nitophyllum* and *Polyneura,* are irregularly scattered. In *Nitophyllum punctatum,* the division of intercalary cells result in formation of tetrasporangial initial and the pericentrals. These pericentrals further form tetrasporangial initials. In *Martensia,* the tetrasporangia occur in swollen areas on longitudinal lamellae of the thallus.

In *Apoglossum,* the sori occur as narrow acropetal patches of indefinite lengths beginning from blade apex. The tetrasporangia are formed by transverse division of lateral pericentrals. In *Branchioglossum pygmaeum,* the tetrasporangial sori are arranged in paired patches on either side of the midrib. In *Hypoglossum,* the sorus lies in form of a single broad patch near the apex of the blade and the pericentral forms the tetrasporangia (Wynne and Norris, 1991).

In *Branchioglossum,* the tetrasporangial sori are symmetrically arranged along the midrib on distal part of the branch. Tetrasporangia are arranged in one layer. Primordia of tetrasporangia may arise from second or third order cells including the lateral pericentrals but not from the cortical cells. The tetraspores may be arranged tetrahedrally or cruciately (Yoshida and Mikami, 1992).

In *Yamadaphycus,* tetrasporangia are borne on linear branches and arise from margins. The tetrasporangial initial are cut off from lateral pericentral cells and arranged symmetrically in two rows along the midrib. The tetrasporangia may divide tetrahedrally or cruciately (Mikami, 1973).

Rhodomelaceae : The tetraspores are mostly arranged in tetrahedral manner and are usually embedded in polysiphonous branches. But in *Lophocladia* and *Wrightiella* tetraspores are borne on trichoblasts. The fertile branches are often called stichidia. The tetrasporangial initial are generally formed by pericentrals and one tetrasporangium is produced in each segment. If the branch on which the tetrasporangia are borne is unbranched the sporangia are superposed but if the laterals are present then the arrangement of sporangia is generally spiral.

The fertile branchlet or stichidium generally possesses five pericentrals although the number may be different in vegetative branch. The fertile pericentral cuts off two superficial cover cells. After this the sporangial initial is produced on the apical side and a bearing cell is formed just below it. The bearing cell is linked by primary pit-connections with the axial and the cover cells. The mature spores are released through the passage formed by separation of the cover cells.

In *Rhodomela* and *Odonthalia,* the tetrasporangia originate from the two lateral of the six pericentrals. In *Bostrychia,* the sporangia is produced on ultimate branches and all the pericentrals of the segment normally become fertile. Sometimes the dorsal pericentrals remain sterile.

In *Chondria,* the tetrasporangial initials are formed in the apical-pit on the club-shaped branches. Three cover cells are produced. In *Laurencia* the initials are formed in apical-pits but are formed from young cortical cells. In *Ricardia* the young cortical cells produce tetrasporangial initials.

In *Polysiphonia* stichidia are present. In *Placophora* and *Pollexfenia,* the branches on which tetrasporangia are borne are in turn borne on the margin or surface of the flat thallus. In *Cliftonaea* and *Leveilla,* one sporangia is produced dorsally in each segment and is covered by many cover cells. The fertile branches of *Amansia* and *Vidalia* are profusely branched and two sporangia are produced per segment.

Dasyaceae : The tetrasporangia which are tetrahedrally arranged are borne on stichidia. In *Dasya,* the stichidia are formed on pseudolaterals or on adventitious branches and in *Heterosiphonia* on pseudolaterals. The number of pericentrals is 5 in *Dasya* and 6 or 7 in *Heterosiphonia.* Not all sporangia produce spores. One or two many remain sterile. The cover cells in *Dasya* are small and partially cover the sporangia. The stichidia in *Dasyopsis* are formed on adventitious branches and possess six pericentrals per segment and are repeatedly forked in *Dictyurus.*

LIFE CYCLE STUDIES

The life cycle is generally diplobiontic. In Ceramiales, it has been now proved cytologically that a distinct haploid sexual phase always alternates with a diploid asexual phase. A sharp segregation between sex organs and tetrasporangia occur but there are many exceptions. Although the widespread occurrence of an isomorphic alteration of haploid and diploid phases occurs, nevertheless, in certain genera, both the procarps and the tetrasporangia occur on the same individual. The exception is more common among members of Ceramiaceae. In *Callithamnion corymbosum*, the tetrasporangia are reported on male plants. In *Nitophyllum,* the fronds bear both cystocarps and tetrasporangia. The tetrasporangia are grouped around the rudimentary procarps which suggest a possible correlation between sporangial formation and procarp arrest.

The tetrasporangia, after initial multiplication, develop in a normal manner. Except one all other nuclei distintegrate. The single surviving nucleus does not divide and is liberated in the form of monospore. The monospore is haploid although the carpospore is diploid. In *Griffithsia, Polysiphonia*, and *Spermothamnion,* the tetrasporangia borne on sexual plants do not undergo reduction division and thus produce only haploid monospores. In *Spermothamnion* both the conditions are met with (i) the tetrasporangia are formed on the sexual plant itself, which also carries the carposporophyte (ii) the tetrasporophyte and sexual plants both occur on two different thalli. Schussing and Odle (1927) are of the view that no reduction division occurs in the first type, whereas Drew (1934) in *Spermothamnion turneri* reports that meiosis does occur in tetrasporangia found on plants bearing gonimoblasts and cystocarps. According to Drew such individuals are diploid and gonimoblasts of these plants are either tetraploid or triploid. The presence of such triploid carpospores has already been established.

The finding of two distinct kinds of life cycles has led to a theory which believes that, before the diplobiontic life cycle was established, both the sexual and the tetrasporophyte plants bore sporangia. Finally in many genera where both the stages in life cycle are separated, the tetrasporangia slowly disappeared from the sexual plant. In this case the tetrasporic and sexual individuals are often unequally represented and even their proportion in nature changes with season and place.

ULTRASTRUCTURAL AND HISTOCHEMICAL STUDIES

CELL WALL

The cell wall consists of two layers. The inner layer is firm, adjoins the cell cavity and consists mainly of cellulose. The outer layer is composed of pectic substances. In *Antithamnion* and *Griffithsia,* the cell wall increases to about ten times in thickness. The cuticle at the surface is probably a dense layer of pectin. In certain genera of Ceramiaceae the wall shows stratification which is normally evident where the wall growth is by means of apposition. In the apical cells of *Antithamnion, Bornetia, Callithamnion* and *Ceramium,* the tip of the protoplasts deposit successive layers. The outermost layers progressively burst and the wall thus reveals many funnel-shaped pieces.

The agarocolloids such as Funaran, Odonthalan, Porphyran and Sulphated galactans from *Ceramium, Laurencia* and *Polysiphonia* may be regarded on both chemical and biological grounds as analogues of carrageenan. As compared to polysaccharides, proteins are quantitatively minor components of red algal cell walls. The water soluble ones are found in mucilages whereas the other insoluble forms occur in the cuticle. The cells wall proteins are associated with cell adhesion (Pueschel, 1979), detoxification (Pederson *et al.*, 1979) and physical properties of thallus surface. A proteoglycan was extracted from *Laurencia* cell wall. In *Griffithsia,* a glycoprotein which is a wound-hormone known as rhodomorphin was extracted (Waaland and Cleland, 1974) which has an active protein part and a carbohydrate residue of a mannosyl and a glucosyl (Cole and Sheath, 1990).

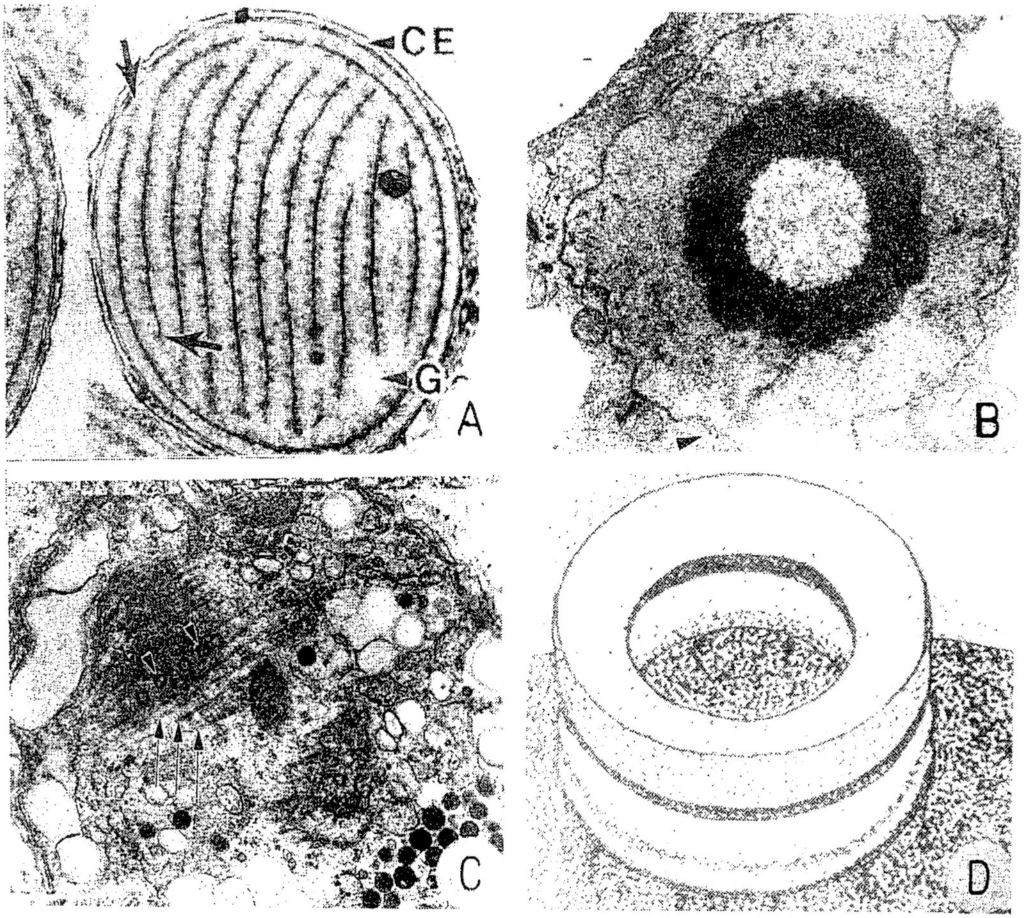

Fig. 125. A-D. Ultrastructure.

(after Cole and Sheath, 1990).

(E, chloroplast envelope: G, genophore; N, nuclear surface).

A. *Griffithsia* : Chloroplast with single encircling thylakoid.

B. *Dasya* : Diplotene nucleus with typical ''ring-shaped'' nucleolus. An NAO is also visible (arrowhead).

C. *Dasya* : Section of prophase nucleus showing an extranuclear ''cage'' of microtubules.

D. *Polysiphonia* : Diagrammatic representation of an NAO at the nuclear surface.

PIT-CONNECTION

Genetically related cells which are part of the same filament and derived from the same parent cell are connected by primary pit-connections. These pit-connections are single and circular. The pits may be narrow or occupy the entire septum. The main body of the pit consists of plug-core which mainly consists of proteins (Ramus, 1971; Pueschel, 1980). The plug-core is surrounded by pit-membrane on either side and thus is not in direct communication with the cytoplasm. The cap/pit-membrane is identical to a plasmamembrane. Many taxa of Ceramiales lack pit-membranes in gonimoblasts. In *Ceramium,* a strand of cytoplasm extends from pit across the central vacuole; and the cytoplasm where the threads fork is Y-shaped (Fritsch, 1945).

In the specialised taxa of Ceramiales (except the family Ceramiaceae) secondary pit-connections develop between unrelated cells. This may be in response to the special physiological need for translocation of food material. In *Polysiphonia,* secondary pit-connections are formed between overlying pericentral cells. In *Rhodomela subfusca,* several secondary pit-connections are seen between adjacent cells (Fritsch, 1945).

CHROMATOPHORES

The characteristic colour of red algal thallus is due to chromatophores. The pigments are water soluble phycobilins which are found in electron dense granules, the phycobilisomes. The phycobilisomes are arranged on stromal surface of photosynthetic membranes. Another ultrastructural feature of red algal plastid is lack of formation of grana. The thylakoids are unstacked and lack periplastid ER. The morphology of chromatophore varies from discoid to stellate. In *Laurencia,* the chromatophores appear to be dissected (Vandermeulen *et al.,* 1990). The shape of chromatophores often depends upon that of the cell and may vary appreciably in different cells of the same individual. In *Ceramium,* the cortical cell chromatophores are irregularly shaped or almost isodiametric, whereas in the axial cell they appear as elongate and ribbon-shaped. In Ceramiales, the chromatophore consists apparently of individual, flat thylakoids surrounded by a single peripheral thylakoid. This peripheral encircling thylakoid is referred to as girdle-lamella or girdle-band (Fig. 125 A).

Koslowsky and Waaland (1987) described the degeneration of chromatophores due to cytoplasmic incompatibility reaction induced by the fusion of somatic cells from two different strains of *Griffithsia pacifica* (see Cole and Sheath, 1990).

CARPOSPOROPHYTE

In *Polysiphonia novae-angliae*, the developing carpospores that are cut off from the large fusion cell have irregular and highly lobed nuclear profiles. The lobing increases the surface area. In the nucleus of carposporangia restricted portions of envelope are replaced by ER (Fig. 124 B). This intervening ER consists of a stack of 2 to 6 or more lamellae. The outermost portion of the stack lacks annuli and is continuous with the nuclear envelope. The others are located in nucleoplasm (Wetherbee and Wynne, 1973).

Reserves are often observed within or adjacent to the nucleus and frequently in proximity to the discontinuous region of nuclear envelope, substituted by ER lamellae. Mitochondria are found in abundance and in close association with golgi bodies. Four kinds of reserves are also found in cytoplasm.

(i) Striated vesicle : Cytoplasm of maturing carpospores typically contain scattered and elongate membrane bound cylindrical structures. Periodicity of striations is 320 A°. Similar striations are found in small nuclear protrusions thereby suggesting their production site.

(ii) Osmiophilic droplets : Spherical structures of electron-opaque materials are present in the cytoplasm of maturing carpospores. The unit membrane is difficult to distinguish. The electron density of inclusions strongly suggests a lipoidal composition. Such deposits are observed with nuclear evagination, which often appear constricted at the base thus suggesting a possible transport from nucleus.

(iii) Floridean starch : This is abundant in cytoplasm, not bound by any membrane and are most frequently located in proximity to stacked cisternae within the nucleus. They are more related to golgi bodies than proplastids.

(iv) Large fibrillar bodies : These cytoplasmic inclusions of irregular shape are composed of unevenly fibrillar material and are a conspicuous feature of the maturing carpospores. These inclusions are bound by a unit membrane.

Occurrence of irregularly shaped, highly convoluted nuclei has been reported in the carpospores of different species of *Polysiphonia,* as well as tetraspores of *Ceramium, Levringiella* and *Ptilota* (Whetherbee and Wynee, 1973).

CELL DIVISION

Mitosis

Ultrastructure of mitosis has been worked in *Griffthsia flosculosa* (Peyriere, 1971); *Pleonosporium vancoverianum* (Sheath *et al.,* 1987); *Apoglossum* (Dave *et al.,* 1982); *Membranoptera* (McDonald, 1972); *Dasya* (Philips and Scott, 1981) and *Polysiphonia* (Scott *et al.,* 1980). The dividing nucleus appears to have typical eukaryotic spindle composed of two opposing sets of interdigitating microtubules. The mature spindles are enclosed in the nuclear envelope except at the polar regions which have fenestrated gaps. The nucleolus disperses by the metaphase stage. The chromatin condenses and is arranged in the equatorial region. The chromosomes are connected to the spindle by means of kinetochores (Figs. 124 A-D). Each division pole contains ring-shaped "nucleus-associated organelle" (NAO) (Fig. 125 D). Normal events of mitosis from prophase to telophase occurs and an interzonal spindle midpiece persists until two separate daughter nuclei are formed.

A heterochromatin network is present in interphase-prophase nuclei. Prometaphase-anaphase chromosomes are relatively easy to recognize. At anaphase chromosomes move to poles, attach to spindle microtubules and at telophase can still be seen adpressed to the inner membrane of the nuclear envelope. The kinetochores in *Apoglossum, Dasya, Membranoptera* and *Polysiphonia* display a typical three layered structure. The number of microtubules attached to kinetochores is usually more than one and upto seven has been reported in *Polysiphonia* (Scott *et al.,* 1980).

The structures associated with the poles of spindle during nuclear division is NAO. Many taxa show homogeneity in the structure of NAO which consists of a pair of short, hollow, cylindrical structures called as "polar rings" (Fig. 125 D). The first alga for which a micrograph of NAO published is *Griffithsia.* The NAO at early prophase appears as a double-ringed structure with one doughnut-shaped portion (distal) directly above and attached to a second similarly shaped portion (proximal). The proximal portion appears to be attached by fine struts to the nuclear envelope. In *Polysiphonia,* during preprophase there are two NAOs attached to the nuclear envelope approximately 0.5 mm from each other (Scott *et al.,* 1980). As prophase progresses one migrates 180 degrees along the surface of the nucleus to form the opposing pole. At prometaphase the two halves of each NAO appear to "snap apart" such that the distal portion lies above and somewhat lateral to the proximal portion, which is still connected to the nuclear envelope. As evidence of tension, prior to the detachment of the proximal portion from the nucleus, the nuclear envelope subjacent to NAO deforms into an outward bulge called the nuclear envelope protrusion. The two halves of NAO remain separated throughout the rest of mitosis, although by late telophase the distal portion moves closer to the nuclear envelope. Later it lies close to the proximal portion, a position similar to the original prophase NAO configuration. This observation led to the conjecture that the two NAOs serve in a semiconservative fashion as templates for NAO biogenesis at either early prophase or late telophase (see Cole and Sheath, 1990).

Although spindle microtubules seem to arise from the NAO, actually they are not directly inserted upon NAOs. Instead the spindle microtubules may abut moderately dense "NAO associated material". The NAO are persistent features of postmitotic cells in *Polysiphonia* and also in cells that may be related to entirely different function than nuclear division.

The nuclear envelope throughout remains intact except at the spindle poles. By late prophase each pole has NAO surrounded by a ribosome free "zone of exclusion" that grow gradually in size during the time of NAO migration and apparently contain spindle precursors (Scott *et al.,* 1980). The NAOs in addition to being surrounded by "zone of exclusion" are embedded in moderately electron-dense material referred as NAO-associated material. During metaphase the nuclear envelope in the polar area becomes fragmented, leaving many small fenestrations in the nuclear envelope. The fenestrations develop probably due to dissolution of nuclear pores. The area subjacent to the NAO, however, remains intact.

Phillips and Scott (1981) did note a perinuclear "cage" of microtubules around prophase nuclei of *Dasya* that disappeared at metaphase (Fig. 125 C). They reported an identical structure in the mature, nondividing cells of *Polysiphonia.* McDonald (1972) documented at metaphase perinuclear microtubules after production of spindle in *Membranoptera.* Charleston (1984) noted microtubules radiating from nucleus in cells of *Antithamnion* and suggested their possible involvement in orientation of the division plane and in movement of daughter nuclei following anaphase. In Ceramiales, the spindle is completely within the nuclear envelope. The microtubules are not sharply focussed upon each NAO region and instead end abruptly at the cytoplasm/nucleoplasm interface. A few taxa have a system of membranes that totally or partially enclose the nucleus during mitosis. These membranes are known as PER (perinuclear ER) and help to separate the milieu of the dividing nucleus from the rest of the cell (Cole and Sheath, 1990).

Except in multinucleate cells, where cytokinesis may not be directly associated with karyokinesis, in all other red algal cells cytokinesis and karyokinesis appear to be temporally close, with incipient cleavage furrows present as early as prophase (Scott *et al.,* 1980).

Meiosis

Only in *Dasya* meiosis has been studied (Broadwater *et al.,* 1986). Most of the ultrastructural features including spindle organization are identical to that of mitosis and the main difference between meiosis and mitosis lies in prophase I, which is much more prolonged and complex as compared to mitotic prophase. Almost all the differences can be attributed to two basic processes of meiosis namely (i) chromosomal interactions (synaptonemal complex) and (ii) change from a diploid to haploid state. During diplotene, most components of cell increase greatly in size or number. These changes are associated with chromatin decondensation and apparent resumption of RNA synthesis. Conoomitantly electron-dense packets of material appear and the cytoplasm becomes more opaque. These are believed to be ''nuage'' which are cytoplasmic accumulations of ribosomal material (Broadwater *et al.*, 1986). The formation of nuage seems to be involved in changing the cell from sporophyte to gametophyte.

In postmeiotic cells at the prophase I numerous small vesicles are found at poles and to a lesser extent near nuclei. In *Dasya*, massive accumulations of smooth ER are found at the poles of meiotic cells during metaphase-anaphase I. In addition three extraordinary features were noticed. Oné of these, is the presence of ''ring nucleoli'' during diplotene. Nucleoli are characterized by unusually large central ''vacuole'' (Fig. 125 B). Polycomplexes (polymerized and units of synaptonemal complex) are present. Polycomplexes are normally found in postmeiotic cells, where they sometimes attain extremely odd morphologies. The delayed appearance of polycomplexes after dissolution of the synaptonemal complex implies that they may be more than a breakdown product. But the most puzziling was the behaviour of post division nuclei. At telophase II the four flat haploid nuclei are located far apart from one another. These nuclei round up and attain a central position in the nascent spores. Concomitantly the four nuclei move centripetally through massive deposits of starch to a central position in the tetrasporangium. The nuclei become very irregular in shape and are associated with annulate lamellae and large number of mitochondria. The four nuclei are seen to be closely adpressed to one another. Although enigmatic, this post division movement does explain ''uninuclear meiosis'' in red algae, meaning both divisions occur within the same nuclear envelope with no intervening interkinesis.

SELECTED REFERENCES

ABBOTT, I. A. 1976. *Dotyophycus pacificum* gen. et. sp. nov with a discussion of some families of Nemaliales (Rhodophyta). Phycologia 15 (2) : 125-132.

ABBOTT, I. A. 1984. Two new species of *Liagora* and notes on *Liagora farinosa* Lamouroux. Am. J. Bot 71 (8) : 1015-1022.

ABBOTT, I.A. and **YOSHIZAKI, M.** 1981. A second species of *Dotyophycus* that emphasizes the distinctness of a diffused gonimoblast. Phycologia 20 (3) : 222-227.

ABBOTT, I. A. and **YOSHIZAKI, M.** 1982. *Liagora valida* from sand Key Florida USA. Jpn. J. Phycol. 30 (1) : 9-14.

ABBOTT, I. A., ZHANG, Z. J. and **XIA B. M.** 1991. *Gracilaria mixta,* new spp. and other west pacific spp. of the genus. Pac. Sci. 45 : 12-27.

ABDEL-RAHMAN, M.H. 1981. The life history of *Acrochaetium unifilum.* Cryptogam. Algol. 2 (4) : 241-252.

ABDEL-RAHMAN, M. H. 1982. Photoperiodicity in *Acrochaetium asparagopsis.* 1. Response to a photoperiod of short day in the course of formation of tetrasporocyst. Physio Veg 20 (2) : 155-164.

ABDEL-RAHMAN, M. H. 1982. The involvement of an endogenous circadian rhythm in photoperiodic timing in *Acrochaetium asparagopsis.* Br. J. Phycol. 17 (4) : 389-400.

ABDEL-RAHMAN, M. H. 1985. The life history of *Acrochaetium gynandrum.* Cryptogam. Algol. 6 (1) : 1-12.

ABDEL-RAHMAN, M. H. and **BIDOUX C.** 1989. Life cycle of *Acrochaetium secundatum.* Cryptogam. Algol. 10. (3): 235-245.

ABDEL-RAHMAN, M. H. and **MAGNE, F.** 1983. A new type of life history in Rhodophyta. C. R. Seances Acad Sci Ser III Sci Vie 296 (14) : 641-644.

ABDEL-RAHMAN, M. H. and **MAGNE, F.** 1984. The life history of *Audouinella boryana Cals* Biol. Mar. 25 (1) : 1-14.

ABDEL-RAHMAN, M. H. and **MAGNE, F.** 1990. Carposporophyte variation in Acrochaetiales. Cryptogam. Algol. 11 (1) : 23-30.

ABDEL-RAHMAN, M. H.; MAGNE, F. and **BIDOUX, C.** 1990. Life cycle of *Acrochaetium corymbiferum.* Cryptogam. Algol. 11 (1) : 13-22.

AFAQ-HUSAIN, S. and **SHAMEEL. M.** 1990. The structure and reproduction of a new species *Helminthocladia nizamuddinii* (Nemaliales - Rhodophyta) from the coast of Pakistan. Bot. Mar. 34 (2) : 81-89.

AFONOSO-CARRILLO, J. 1989. Morphology, anatomy and vegetative reproduction of *Fosliella paschalis* (Corallinaceae. Rhodophyta). Phycologia 28 : 331-441.

AGHAJANIAN, J. G. 1978. The fine structure of the pit-connections of *Batrachospermum sirodotii* Skuja, Protoplasma 96 : 247-265.

AKATSUKA, I. 1979. A note on the male reproductive organs of *Gelidium japonicum* Okam and *Beckerella rubcostata* (Okam). Kylin (Gelidiaceae, Rhodophyta) Rev. Algol, N. S. (1) 17-20.

ALGARRA, P. THOMAS, J. AND **MOUSSEAU. A.** 1990. Phycobilisome heterogeneity in the red alga *Porphyra umbilicalis*. Plant Physiol. 92 : 570-576.

ANDERSON, N. S. DOLAN, T. C. S. and **REES. D. A.** 1965. Evidence for a common structural pattern in the polysaccharide sulphates of the Rhodophyceae. Nature. 205 : 1060-1062.

ANDERSON, R. J. and **STEGENGA, H.** 1985. A crustose tetrasporophyte in the life history of *Nothogenia erinacea* (Galaxauraceae). Phycologia 24 (1) : 111-118.

APT, K.E. and **GIBOR, A.** 1991. The ultrastructure of galls on the red alga *Gracilaria ephihippisora*. J. Phycology 21 : 409-413.

BALAKRISHNAN, M. S. and **CHAUGULE, B. B.** 1980. Cytology and life history of *Batrachospermum mahabaleshwarensis*. Balakrishnan et Chaugule. Crytogam. Algol. 1 (2) : 83-97.

BALLANTINE, D. L. 1985. *Botryocladia wynnei* and *B. spinulifera* (Rhodymeniales, Rhodophyta) Taylor and Abbott from Puerto Rico. Phycologia. 24 : 199-204.

BALLANTINE, D.L. and **WYNNE, M.J.,** 1988. The life history and development of *Hypoglossum rhizophorum* (Delesseriaceae, Rhodophyta) Taylor and Abbott from Puerto Rico. Phycologia. 24 : 199-204

BALLANTINE, D. L. and **WYNNE M. J.** 1988. The life history and development of *Hypoglossum rhizophorum* (Delesseriaceae, Rhodophyta) in culture, a new deepwater species from the Caribbean. J. Phycol. 24 : 8-12.

BALLESTEROS, E. 1988. Species composition and structure of infralittoral *Corallina elongata*. Community on the costa Beava (Northwestern Mediterranean). Invest Pesq 52 : 135-151.

BHATIA, BELA and **VIJAYARAGHAVAN, M. R.** 1990a. Extracellular secretions in *Nizymenia australis*, a red alga. Phykos 29 : 17-21.

BHATIA, BELA and **VIJAYARAGHAVAN, M. R.** 1990b. Histochemical analysis of vegetative and reproductive structures in *Nizymenia australis*. Seaweed Res. Utiln. 13 : 119-127.

BHATIA, BELA and **VIJAYARAGHAVAN, M. R.** and **KRISHNAMURTHY, V.** 1990. Structure of attaching organ of *Nizymenia australis* Sonder (Nizymeniaceae, Gigartinales, Rhodophyta). Seaweed Res. Utiln. 13 : 11-14.

BHATIA, BELA. 1993. Structure, development and reproduction in a red alga *Scinaia forcellata* Bivona-Bernardi—histochemical and ultrastructural studies-Thesis Ph. D. University of Delhi, Delhi.

BHATIA, BELA and **VIJAYARAGHAVAN, M.R.** 1995a, Developmental and histochemical studies on spermatangium and spermatium release in *Scinaia forcellata* Bivona-Bernardi (Rhodophyta) Seaweed Res. Utiln. 17 : 177-182.

BHATIA, BELA and **VIJAYARAGHAVAN, M. R.** 1995 b. Carposporangial development in the red alga *Scinaia pseudocrispa* (*S. forcellata*) from India. Aquat. Bot. 52 : 193-203

BHATIA, BELA and **VIJAYARAGHAVAN, M. R.** 1996. Histochemical studies on monosporogenesis and monospore release in *Scinaia pseudocrispa* (= *S. forcellata*) Galaxauraceae, Nemaliales, Rhodophyta from Port Okha, India (Communicated).

BIRD, C. J. and **DE OLIVEIRA, E. C. DE.** 1986. *Gracilaria tenuifrons* sp. nov. (Gigartinales, Rhodophyta), a species from the tropical West Atlantic with superficial spermatangia. Phycologia 25 : 313-320.

BOILLOT, A. 1968. Sur l' existence d'um tetrasporophyte dans le cycle de *Scinaia furcellata* Bivona. Nemalionales. C. R. Acad. Sc. Paris. 266 (D) : 1831-1832.

BOILLOT, A. 1969. Sur l' development des tetraspore et l'dedification due gametophyte chez *Scinaia furcellata* (Turner) Bivona Rhodophyces (Nemalionales) C. R. Acad. Sc. Paris. 268 (D) : 273-275.

BOLD, H. C. and **WYNNE, M. J.** 1978. Introduction to the Algae. Structure and Reproduction. Prentice-Hall, New Jersey.

BONIN, D.R. and **HAWKES, M. W.** 1987. Systematics and life histories of New Zealand Bonnemaisoniaceae (Bonnemaisoniales, Rhodophyta) : I the genus *Asparagopsis.* N. Z. J. BOT. 25 : 577-590.

BOROWITZKA, M. A. and **VESK, M.** 1978. Ultrastructure of the corallinaceae, I. The vegetative cells of *Corallina officinalis* and *C. cuvierii.* Mar. Biol. 46 : 295-304.

BRAMWELL, M. D. and **WOELKERLING. Wm. J.** 1984. Studies on the distribution of *Fheonphyllum-fosliella* plants (Corallinaceae, Rhodophyta) on leaves of the seagrass *Amphibolis antarctica* (Cymodoceaeae) Aust. J. Bot.. 32 : 131-137.

BREEMAN, A. M. and **GUIRY, M. D.** 1980. The influence of tetrasporogenesis in *Bonnemaisonia hamifera* Mar. Biol. (Berl) 102 (1) : 5-14.

BROADWATER, S. and **SCOTT, J.** 1986. Three-dimensional reconstruction of the chondriome of the unicellular red alga *Rhodella reticulata.* J. Cell Sci. 84 : 213-219.

BRODIE J. and **GUIRY. M.D.** 1988a. Life history and reproduction of *Botryocladia ardreana* sp. nov. (Rhodymeniales, Rhodophyta) from Portugal. Phycologia. 27 : 109-130.

BRODIE, J. and **GUIRY, M. D.** 1988b. Life history and reproduction of *Cordylecladia erecta* (Rhodymeniaceae, Rhodophyta). Br. Phycol J. 23 : 347-364.

BRODIE, J., GUIRY, M. D. and **MUSUDA, M.** 1991. Life history and morphology of *Chondrus nipponicus* (Gigartinales) from Japan, Br. Phycol. J. 26 : 33-50.

BROWN, D. L. and **WEIER, T. E.** 1968. Chloroplast development and ultrastructure in the freshwater red alga *Batrachospermum.* J. Phycol. 4 : 199-206.

BULA-MEYER, G. 1985. *Champiocolax sarae* gen. et sp. nov. and adelphohemiparasite of the Champiaceae (Rhodymeniales, Rhodophyta). Phycologia 24 : 429-435.

CEREZO, A. S. 1971. The fine structure of *Chaetangium fastigiatum* xylan : studies of the sequence and configuration of the (1 → 3) - linkages. Carbohyd. Res. 19 : 209-211.

CEREZO, A.S. LEZEROVICH, A., LABRIOLA. R. and **REES, D. A** 1971. A xylan from the Red Seaweed *Chaetangium fastigiatum* Carbohyd. Res. 19 : 289-296.

CHAMBERLAIN, Y. M. 1984. Spore size and germination in *Fosliella, Pneophyllum* and *Melobesia* (Rhodophyta, Corallinaceae), Phycologia 23 : 433-442.

CHAMBERLAIN, Y.M. 1985. Trichocyte occurence and Phenology in four species of *Pneophyllum* (Rhodophyta, Corallinaceae) from the British Isles. Br. Phycol. J. 20 : 375-379.

CHAMBERLAIN, Y. M. 1987 Conceptacle production and life history in four species of *Pneophyllum* (Rhodophyta, Corallinaceae) from the British Isles. Br. Phycol. J. 22 : 43-48.

CHAMBERLAIN, Y. M. 1988. Observation on an autogenous fertile outgrowth of *Lithophyllum crouanii* (Corallinales, Rhodophyta) from Southern England. Phycologia 27 : 378-386.

CHAMBERLAIN, Y. M. 1990 The genus *Leptophyllum* (Rhodophyta, Corallinaceae) in the British Isles with descriptions of *Leptophytum bornetii, L. elatum* sp. nov. and *L. laeve*. Br. Phycol. J. 25 : 179-199.

CHEN, L. C-M., EDELSTEIN, T. and MACLACHALAN, J. 1969. *Bonnemaisonia hamifera* Hariot in nature and in culture J. Phycol 5 : 211-220.

CHIANG, Y. M. 1971. Observations on the development of spermatangia in some genera of Helminthocladiaceae. Phycologia 10 (213) : 163-168.

CHIHARA. M. 1962. Life cycle of Bonnemaisoniaceous algae in Japan. Sci. Rep. Tokyo Kyoiku Daigaku 11 B : 27-55.

CHRISTIAN, D., STADLER, T. and VERDUS. C. M. 1987. Structures and functions of the polysaccharides from the cell wall of *Gracilaria verrucosa* (Rhodophyceae, Gigartinales). Hydrobiol. 151/152 : 139-146.

COHEN, Z., VONSHAK, A. and RICHMOND, A. 1988. Effect of environmental conditions on fatty acid composition of the red alga *Porphyridium cruentum* : Correlation to growth rate. J. Phycol. 24 : 328-332.

COLE, K. M. 1990. Chromosomes, in Cole K. M. and Sheath R. G. (Eds). Biology of Red Algae. Cambridge University Press, Cambridge.

COLE, K.M. and HYMES, B. J. 1983. Karyotypes and reproductive seasonality of the genus *Bangia* (Rhodophyta) in British Columbia, Canada. J. Phycol. 19 : 136-145.

COLE, K.M., PARK, C. M., REID, P. E. and SHEATH, R. G. 1985. Comparative studies on the cell walls of sexual and asexual *Bangia atropurpurea* (Rhodophyta). I. Histochemistry of Polysaccharides. J. Phycol. 21 : 585-592.

COLE, K. M. and SHEATH, R. G. 1980. Ultrastructural changes in major organelles during spermatial differentiation in *Bangia* (Rhodophyta). Protoplasma. 102 : 253-279.

COLE, K.M. and **SHEATH. R. G.** 1990. (Eds.) Biology of the Red algae. Cambridge University Press, UK.

COLL, J. and OLIVEIRA FILHO, E. C. De. 1977. The nuclear state of 'reproductive' cells of *Porphyra leucosticta* Thuret in Le Jolis (Rhodophyta, Bangiales). Phycologia. 16 (3) : 227-229.

CORMACI, M. and FURNARI, G., 1991. The distinction of *Ceramium giacconei* sp. nov. (Ceramiales, Rhodophyta) in the Mediterranean sea from *Ceramium cingulatum*. Cryptogam. Algol., 12(1) : 43-53.

CORREA, J.A. and **McLACHLAN**, J. 1977. Variation in the fine structure of the cuticle of *Chondrus crispus* (Rhodophyta). Br. Phycol. J. 12 : 285-286.

CORREA, J. A. and **McLACHLAN, J**. 1988. Variation in the fine structure of the cuticles of *Chondrus crispus* (Rhodophyta). Br. Phycol. J. 23 : 285.

CORTEL-BREEMAN, A. M. and **HOEK, VAN. DEN.** 1970. Life-history studies on Rhodophyceae I. *Acrosymphyton purpuriferum* (J. AG.) Kyl. Acta Bot. Neerl. 19 : 265-284.

CORTEL-BREEMAN, A. M. 1975. The life-history of *Acrosymphyton purpuriferum* (J. Ag.) Sjost. (Rhodophyceae, Cryptonemiales). Isolation of tetrasporophytes. With some remarks on the tetrasporophyte of *Bonnemaisonia asparagoides* (Woodw.) C. Ag. (Nemalionales). Acta Bot. Neerl. 24 : 111-127.

CUNNINGHAM, E. M.and **GUIRY, M. D.** 1989. A circadian rhythm in the long day photoperiodic induction of erect axis development in the marine red algae *Nemalion helminthoides*. J. Phycol. 25 (4) : 705-712.

DAI, JIXON, BAO, Z. M. 1988. Developmental studies on protoplasts of *Porphyra haitanensis*. Acta Genet. Sin 15(4) : 299-302.

DAWES, C. J., SCOTT, F.M. and **BOWLER, E**. 1961. A light and electron-microscopic survey of algal cell walls. 1. Phaeophyta and Rhodophyta. Amer. J. Bot. 48 (10) : 925-934.

DEASON. T. R., BUTLER, G. L. and **RHYNE,** C. 1983. *Rhodella reticulata* a new coccoid rhodophytan alga (Porphyridiales). J. Phycol. 19 : 104-111.

DECREW, C. T. and **WEST, J.A.** 1981. Life histories in the Phyllophoraceae (Rhodophyta : Gigartinales) from the Pacific Coast of North America. J. Phycol. 17 : 240-250.

DECEW, T. C. and **WEST, J. A**. 1981. Investigation on the life histories of three *Farlowia* species (Rhodophyta : Cryptonemiales, Dumontiaceae) from pacific north America. Phycologia 20 : 342-351.

DELEPINE, R. and **LAMBART, C**. 1981. The female apparatus of the Rhodophycean *Chaetangium fastigiatum* and the existence of a gonophore in the species C. R. Seances Acad. Sci Ser. III Sci Vie 292 (1) : 127-131.

DELIVOPOULOS, S. G.and **KUGRENS, P.** 1984. Ultrastructure of carposporogenesis in the parasitic red alga *Faucheocolax attenuata* Setch. (Rhodymeniales, Rhodymeniaceae). Amer. J. Bot. 71(9) : 1245-1259.

DELIVOPOULOS, S. G. and **KUGRENS P.** 1985. Ultrastructure of carposporophyte development in the red alga *Gloiosiphonia verticillaris* (Cryptonemiales, Gloiosiphoniaceae). Amer. J. Bot. 72 : 1926-1938.

DELIVOPOULOUS, S. G. and **TSEKOS, I.** 1985. Ultrastructure of the fusion cell in *Gracilaria verrucosa*.New Phytol. 101 : 605-612.

DIXON, P. S. 1959. The structure and development of the reproductive organs and carposporophyte in two British species of *Gelidium*. Ann. Bot. 23 (91) : 397 - 407.

DIXON, P. S. 1963. The taxonomic implications of the 'pit-connexions' reported in the Bangiophycidae. Taxon 12 108-110.

DIXON, P. S. 1982a. Life histories in the Florideophyceae with particular reference to the Nemaliales sensu lato. Bot. Marina 25 : 611 - 621.

DIXON. P. S. 1982b. Rhodophycota. In Synopsis and Classification of Living Organisms, (Ed.)

Sybil, P. McGraw Hill Book Co., New York.

DUCKWORTH, M. and **YAPHE, W**. 1971. The structure of agar I. fractionation of complex mixture of polysaccharides. Carbohyd. Res. 16 : 189-197.

DUCKWORTH, M., HONG. C. K. and **YAPHE, W.** 1971. The agar polysaccharides of *Gracilaria* species. Carbohyd. Res. 19 : 1-9.

DURAIRATNAM, M. 1987. Studies of the yield of agar, gel strength and quality of agar of *Gracilaria edulis* (Gmel.) Silva from Brazil. Hydrobiol. 151/152 : 509-512.

EDELSTEIN. T. 1970. The life history of *Gloisiphonia capillaris* (Hudson) Carmichael. Phycologia. 9 : 55-59.

EDELSTEIN., T. CHEM. L. 1971. Further observation on *Gloiosiphonia capillaris* (Hudson) Carmichael in culture. Phycologia. 10 : 215-219.

EDYVEAN, R. G. J. and **FORD. H.** 1987. Growth rates of *Lithophyllum incrustans* (Corallinales, Rhodophyta) from south west Wales. Br. Phycol. J. 22 : 139-146.

EISEMAN, N. J. and **MOE. R. L.** 1981. *Maripelta atlantica* sp. nov. (Rhodophyta, Rhodymeniales) a new deep water alga from Florida. J. Phycol. 17 : 299-308.

EVANS, L. V., CALLOW, M. E., PERCIVAL, E. and **FAREED**, V. 1974. Studies on the synthesis and composition of extracellular mucilage in the unicellular red alga *Rhodella*. J. Cell. Sci. 16 : 1-21.

FREDERICQ, **SUZANNE** and **HOMMERSAND, M**. 1990. Taxonomy of *Melanthalia obtusata* var. *abscissa* and its placement in Gracilariales (Rhodophyta). Cryptogam. Algol. 2 : 4-11.

FRESHWATER, WILSON, D. and **KAPRAUN, D. F.** 1989. Spheroplast isolation and propagation of *Porphyra carolinensis* Coll et Cox for mariculture. Aquaculture. 81 (2) : 101-110.

FRITSCH, F. E. 1945. The Structure and Reproduction of Algae. Vol. II. Cambridge Univ. Press, London.

FUJIMORI, YASUSHI, KATSUHITO NAKAMURA and **GORO TAMURA** 1991. Cysteine synthase from a red alga *Porphyra yezoensis* : Purification and properties. Japn. J. Phycol. 39 (1) : 7-14.

FUJITA, D. 1988. *Fosliella zostericola* (Roslie) Seagawa (Rhodophyta, Corollinales) in culture. The life cycle of an epiphytic crustose coralline alga. Japn. J. Phycol. 36 : 48-51.

FUJITA, Y. and **MIGITA, S**. 1987. Fusion of protoplasts from thalli of two different colour types in *Porphyra yezonsis* Ueda and development of fusion products. Japn. J. Phycol. 35 : 201-208.

FUJITA, Y. and **SAITO, M**. 1990. Protoplast isolation and fusion in *Porphyra* (Bangiales, Rhodophyta).Hydrobiologia. 204/205 : 161-166.

GABRIELSON, W.P. 1983 Vegetative and reproductive morphology of *Eucheuma isiforme* (Solieriaceae, Gigartinales, Rhodophyta). J. Phycol. 19 : 45-52.

GABRIELSON, P. W. and **GARBARY, D. J**. 1987. A cladistic analysis of Rhodophyta : Florideophycean orders. Br. Phycol. J. 22 : 125-138.

GANTT, E. and **CONTI, S. F.** 1965. The ultrastructure of *Porphyridium cruentum*. J. Cell Biol. 26 : 365-375.

GANTT, E. and **CONTI, S. F.** 1966. Granules associated with the chloroplast lamellae of *Porphyridium cruentium.* J. Cell. Biol. 29 : 423-434.

GANTT, E., SCOTT, J. and **LIPSCHULTZ, C.** 1986. Phycobiliprotein compostion and chloroplast structure in the freshwater red alga *Compsopogon coeruleus* (Rhodophyta). J. Phycol. 22 : 480-484.

GARBARY, D. J. and **GABRIELSON, P. W.** 1990. Taxonomy and Evolution in Cole, K. M. and Sheath, R. G. (Eds.) Biology of Red Algae, Cambridge University Press, London.

GARBARY, D. and **VELTKAMP. C. J.** 1980. Observation on *Mesophyllum lichenoides* (Rhodophyta, Corallinaceae) with the scanning electron microscope. Phycologia 19 : 49-53.

GARGUILO, G. M. : FLORA De. MASI and **TRIPODI, G.** 1985. A study on *Gracilaria dendroides* sp. nov. (Gigartinales, Rhodophyta) from the bay of Naples. Br. Phycol. J. 20 : 357-364.

GARGIULO, G. M. : FLORA De MASI and **TRIPODI**, G. 1986. Structure and reproduction of *Halymenia asymmetrica* sp. nov. Phycologia 25 : 144-151.

GARGIULO, G. M. : FLORA De MASI, and **TRIPODI, G.** 1991. Karyology of *Bangia atropurpurea* (Rhodophyta, Bangiales) from Mediterranean and northeastern Atlantic populations. J. Phycol. 27 : 306-309.

GERWIC, H. W. and **LANG, J. N.** 1977. Structural, chemical and ecological studies on iridescence in *Iridaea* (Rhodophyta). J. Phycol. 13 : 121-127.

GOFF, LYNDA, J. 1982. Reproductive development of the parasitic red alga *Gardneriella tuberifera* (Solieriaceae, Gigartinales). J. Phycol. 18 : 295-304.

GOFF, LYNDA J. and **COLE**, K. 1975a. The biology of *Harveyella mirabilis* (Cryptonemiales, Rhodophyceae) 1. Cytological investigation of *Harveyella mirabilis* and its host, *Odonthalia flocossa,* Phycologia 12 : 237-245.

GOFF, LYNDA J. and **COLE, K.** 1975b. The biology of *Harveyella mirabilis* (Cryptonemiales, Rhodophyceae) II. Carposporophyte development is related to taxonomic affiliation of the parasitic alga, *Harveyella mirabilis.* Phycologia 14 : 227-238.

GOFF, LYNDA J. and **COLE. K.** 1976. The biology of *Harveyella mirabilis* (Cryptonemiales, Rhodophceae) III. Spore germination and subsequent development. Can J. Bot. 54 : 268-280.

GOFF, LYNDA J. and **COLEMAN, A. W.** 1984. Transfer of nuclei from a parasite to its host. Proc. Natl. Acad. Sci. USA 81 : 5420-5424.

GOFF, LYNDA J. and **COLEMAN, A. W.** 1984. Elucidation of fertilization and development in a red alga by Quantitative DNA Microspectrofluorometry. Developmental Biology 102 : 173-194.

GONZALEZ, M. J. and GOFF, L. J., 1989. The red algal epiphytes *Microcladia coulteri* and *M. californica* (Rhodophyceae, Ceramiaceae). I. Taxonomy, life history and phenology, J. Phycol. 25 : 545-558.

GORDON-MILLS, E. M. and KRAFT. G. T., 1981. The morphology of *Radiothamnion speleotis* gen. et. sp. nov., representing a new tribe in Ceramiaceae (Rhodophyta) from southern Australia. Phycologia, 20(2) : 122-130.

GORDON-MILLS, E. and **NORRIS, R. E., 1986.** *Shepleya elixithamnia,* a new species of Ceramiaceae (Rhodophyta) from south Africa. Phycologia, 25(2) : 160 - 167.

GORI, P. 1980. Ultrastructure of carpospores in *Gastroclonium clavatum* (Roth) Ardissone (Rhodymeniales). Protoplasma. 103 : 263-271.

GRETZ, M. R. ARONSON, J. M. and SOMMERFELD, M. R. 1980. Cellulose in the cell walls of the Bangiophyceae (Rhodophyta). Science 207 : 779-781.

GRETZ, M. R., SOMMERFELD, M. R. and ARONSON, J. M. 1982. Cell wall compostion of the generic phase of *Bangia atropurpurea* (Rhodophyta). Bot. Mar. 25 : 529-535.

GRETZ, M. R., SOMMERFELD, M. R., ATHEY, P. V. and **ARONSON, J. M.** 1991. Chemical composition of the cell walls of the freshwater red alga *Lemanea annulata,* J. Phycol. 27 (2) : 232-240.

GUINSLEY, K. B. 1970. The relationship between methoxyl content and gelling temperature of agarose. Carbohyd. Res. 13 : 247-256.

GUIMARAES, SILVIA MARIA, P. B., CORDEIROMARINO, M., FUJII, M. and **YOKOYA, N**. 1990. Occurence of *Helminthocladia calvadosii* in Brazil. Rev. Bras Bot. 13 (2) : 69-74.

GUIRY, M. D. 1974. A preliminary consideration of the taxonomic position of *Palmaria palmata* (Linnaeus) Stackhouse = *Rhodymenia palamata* (Linnaeus) Greville. J. Mar. Biol. Ass. UK 54 : 509-528.

GUIRY, M. D. 1990a. Sporangia and Spores in Cole, K. M. and Sheath, R. G. (Eds.) Biology of Red Algae, Cambridge Universtiy Press, London.

GUIRY, M. D. 1990b. The life history of *Liagora harveyana* from S. E. Australia. Br. Phycol. J. 25 (4) : 353-362.

GUIRY, M. D. and **IRVINE, D. E. G**. 1978. A critical assessment of infraordinal classification in the Rhodymeniales. Proc. Int. Seaweed Symp. 8 :

GUIRY, M. D., KEE, W. R. and **GARBARY, D. J.,** 1987 (88) Morphology, temperature and photoperiodic response in *Audouinella botyocarpa* Woelkerling from Ireland. G. Bot. Ital. 121 (5/6) : 229-246.

HAINES, H. H., MATULEWICZ, M. C. and **CEREZO, A. S.** 1990 Sulfated galactans from the red seaweed *Nothogenia fastigiata.* Hydrobilogia 204/205 : 637-643.

HARLIN, M. M.; WOELKERLING, WM. J. and **WALKER, D. I**. 1985. Effects of a hypersalinity gradient on epiphytic Corallinaceae (Rhodophyta) in Shark Bay. Western Australia. Phycologia 24 : 389-402.

HAWKES, M. W. 1978. Sexual reproduction in *Porphyra gardneri,* (Smith & Hollenberg) Hawkes (Bangiales, Rhodophyta). Phycologia 17 : 326-350.

HAWKES, M. W., and **SCAGEL, R. F.** 1986a. The marine algae of British Coloumbia and Northern Washington : division Rhodophyta (red algae), class Rhodophyceae, order Palmariales. Can. J. Bot. 64 : 1148-1173.

HAWKES, M. W. and **SCAGEL, R. F.** 1986b. The marine algae of British Columbia and Northern Washington : division Rhodohyta (red algae), class Rhodophyceae, order Rhodymeniales. Can J. Bot. 64 : 1549-1580.

HOMMERSAND, M. H. and **FREDERICQ, S.** 1988. An investigation of cystocarp development in *Gelidium pteridifoluim* with a revised description of Gelidiales (Rhodophyta) Phycologia 27(2) : 254-272.

HUISMAN, J.M. 1985a. The *Scinaia* assemblage (Galaxauraceae, Rhodophyta). A reappraisal. Phycologia **24** (4) : 403-418.

HUISMAN, J.M., 1985b. *Rhipidothamnion secundum* gen. sp. nov. (Ceramiaceae, Rhodophyta) from eastern Australia. Phycologia, 24(1) : 55-66.

HUISMAN, J.M. 1986. The red algal genus *Scinaia* from Australia. Phycologia 25 (3): 271-269.

HUISMAN, J.M. 1987. The taxonomy and life history of *Gloiophloea.* Phycologia **26** (2) : 167-174.

HUISMAN, J.M., 1991. *Diplothamnion gordoniae* sp. nov. (Ceramiaceae, Rhodophyta) from Rottnest Island, Western Australia. Phycologia, **30(2)** :117-123.

HUISMAN, J.M. and **BOROWITZKA, M.A**. 1990. A revision of the Australian species of *Galaxaura* (Rhodophyta, Galaxauraceae), with a description of *Tricleocarpa* gen. nov. Phycologia **29** (2) : 150-172.

HUISMAN, J.M. and **WOMERSLEY, H.B.S.** 1992. Cystocarp development in the red alga *Nothogenia fastigiata* (Galaxauraceae, Nemaliales). Phycologia **31** (3/4) : 359-364.

HUGO, A.A. and **MATSUHIRO**, B. 1987. Polysaccharides from nuclear phases of *Iridaea ciliata* and *I. membranacia.* Hydrobiol. **151/152** : 531-534.

HUTH, K. 1981. Paraspores in the chantransia phase of *Lemanea fluviatilis.* Arch. Protistenkd - **124** (4) : 437-448.

HYMES, B.J. and **COLE, K.M.** 1983. The cytology of *Audouinella hermannii* (Rhodophyta, Florideophyceae) II. Monosporongenesis. Can. J. Bot. 61 : 3377-3385.

JHA, A. 1996. Investigations into life-cycle of the red alga *Asparagopsis taxiformis* (Delile) Trevisan, with a note on *Falkenbergia rufolanosa* (Harvey Schmitz). Thesis Ph. D. University of Delhi. Delhi.

JONES, P.L. and **WOELKERLING, Wm. J.** 1984. An analysis of trichocyte and spore germination attributes as taxonomic characters in the *Pneophyllum - Fosiella* complex (Corallinaceae, Rhodophyta) Phycologia **23** : 183-194.

KAJIMURA, M. 1988. Three new deep-water species of *Scinaia* (Galaxauraceae, Rhodophyta) from the sea of Japan. Bot. Mar. 31 : 175-185.

KAJIWARA, T., KASHIBE, M., **MATSUI, K**. and **HATANAKA**, A. 1990. Volatile compounds and long-chain aldehydes formation in conchocelis filaments of a red alga, *Porphyra tenera.* Phytochemistry. **29** (7), 2193-2195.

KAPRAUN, D.F. and **FRESHWATER, D.W.** 1987. Karyological studies of five species of *Porphyra* (Bangiales, Rhodophyta) from the North Atlantic and Mediterranean. Phycologia. **26** (1) : 82-87.

KAPRAUN, D.F., HINSON, T.K. and **LEMUS, A.J.** 1991. Karyology and cytophotometric estimation of inter and intraspecific nuclear DNA variation in four species of *Porphyra* (Rhodophyta). Phycologia. **30** (5) : 458-466.

KASKA, D.D., POLNE-FULLER, M. and **GIBOR, A.** 1988. Developmental studies in *Porphyra* (Bangiophyceae). II Characterization of the major lectin binding glycoproteins in differentiated regions of the *Porphyra perforata* thallus. J. Phycol. **24** : 102-107.

KAWAGUCHI, S. 1991. Taxonomic notes on the Halymeniaceae (Rhodophyta) from Japan. I. *Halymenia acuminata* (Holmes) J. Agardh. Japn. J. Phycol. (Sorui) **39** : 329-336.

KING, R.J., PUTTOCK, C.F. and **VICKERY,** R.S., 1988. The taxonomic study on the *Bostrychia tenella* complex (Rhodomelaceae, Rhodophyta). Phycologia, **27(1)** : 10-19.

KURGRENS. P. and **DELIVOPOULOS, S.G.** 1985. Ultrastructure of auxiliary and gonimoblast cells during carposporophyte development in *Faucheocolax attenuata* (Rhodophyta). J. Phycol. 21 : 240-249.

KUGRENS, P. and **DELIVOPOULOUS, G.S.** 1986. Ultrastructure of the carposporophyte and carposporogenesis in the parasitic red alga *Plocamiocolax pulvinata.* J. Phycol.. **22** : 8-21.

KUGRENS, P. and **WEST, J.A.** 1973. The ultrastructure of an alloparasitic red alga *Choreocolax polysiphoniae.* Phycologia **12** : 175-185.

KRAFT, G. T. 1981. Rhodophyta : morphology and classification. In. Biology of Seaweeds, (Eds.) C. Lobban and M. J. Wynne Blackwell Scientific Publications. USA.

KRISHNAMURTHY, V. 1984. Chromosome numbers in *Porphyra* C. Agardh. Phykos. **23** (1 & 2) : 185-190.

KRISHNAMURTHY, V. and **SUNDARARAJAN, M.** 1985. Four new taxa of Nemaliales from India. Seaweed Res. Utiln. **9** (1 & 2) : 43-46.

KRISHNAMURTHY, V. and **SUNDARARAJAN, M.** 1986. Studies on some species of *Galaxaura* Lamouroux from India. Seaweed Res. Utiln. **9** (1 & 2) : 31-44.

KRISHNAMURTHY, V. and **SUNDARARAJAN, M.** 1987. Studies on some Indian Nemaliales. II The genus *Scinaia* Bivona-Bernardi. Seaweed Res. Until **10** (2) : 109-125.

KROEN, WILLIAM, K. and **RAMUS, J.** 1990. Viscous polysaccharides and starch synthesis in *Rhodella reticulata* (Porphyridiales, Rhodophyta). J. Phycol. **26** (2) : 266-271.

KUIPER, J. 1983. The life history of *Chromastrum alariae* (Acrochaetiaceae). Acta. Bot. Neerl **32** (3) : 129-152.

KUMANO, S. 1982. Development of carpogonium and taxonomy of six species of the genus *Sirodotia* from Japan and W. Malaysia. Bot. Mag. Tokyo **95** (1038) : 125-138.

KUMANO, S. and **ORLANDO, N.** 1990. *Batrachospermum macrosporum* Montage from South America. Japn. J. Phycol **38** (2) : 119-124.

KUMANO, S and **RATNASABAPATHY, M.** 1982. Studies on freshwater red algae of Malaysia : 3 Development of carposporophytes of *Batrachospermum cayennense* Montagne, *B. beraense* Kumano, and *B. hypogynum* Kumano et Ratnasabapathy. Bot. Mag. Tokyo **95** (1039) : 219-228.

KUMAR, H.D. 1990. Introductory Phycology. Affiliated East-West Ltd., New Delhi.

KYLIN, H. 1956. Die Gatterngen der Rhodophyceen. Gleerups, Lund.

LAHAYA, M., ROCHAS, C. and YAPHE, W. 1985. A new procedure for determining the heterogeneity of agar polymers in the cell walls of *Gracilaria* spp. (Gracilariaceae, Rhodophyta). Can. J. Bot. **64** : 579-585.

LANGE, W., WILHELM, C., WEHRMEYER, W. and MORSCHEL, E. 1990. The supramolecular structure of Photosystem II - Phycobilisome complex of *Porphyridium cruetum.* Bot. Acta. **103** : 250-257.

LEBEDNIK, P.A. 1977. Post-fertilization development in *Clathromorphum, Melobesia* and *Mesophyllum* with comments on the evolution of the Corallinaceae and the Cryptonemiales (Rhodophyta) Phycologia **16** : 379-406.

LEE, Y-P. 1983. Sexual reproduction in *Audouinella alariae* from the N. Atlantic Ocean. Bull. Korean Fish Soc. 16 (3) : 265-272.

LEE, Y-P. 1985. Notes on reproduction in *Rhodochorton purpureum* with special reference to Hokkaido plants. Korean. J. Bot. **28** (1) : 45-46.

LEE, Y-P. and **KUROGI, M.** 1983. The life history of *Audouinella alariae* in nature and culture. J. Fac Sci Hokkaido Univ Serv Bot. **13** (1) : 57-76.

LEE, I.K., WEST, J.A. and **HOMMERSAND, M.H.** 1988. *Binghamiopsis caespitosa* gen. et sp. nov. (Lomentariaceae, Rhodophyceae) from the Eastern Pacific. Korean J. Phycol. **3**(1) : 1-13.

L'Hardy-Halos, M-Th. and **MAGGS. C.A.,** 1991. The novel life history in *Aglaothamnion diaphanum* sp. nov. (Ceramiales, Rhodophyta) from Brittany and the British Isles. Phycologia, **30(5)** : 467-479.

LINDSTROM, S. C. and **SCAGEL, R. F.** 1987. The marine algae of British Columbia, northern Washington, and southeast Alaska : division Rhodophyta (red algae), class Rhodophyceae, order Gigartinales, family Dumontiaceae, with an introduction to the order Gigartinales. Can. J. Bot. 65 : 2202-2232.

LINDSTROM, S.C. 1988. The Dumontiaceae, a resurrected tribe of red algae (Dumontiaceae, Rhodophyta). Phycologia **27** : 89-102.

LITTLER, M.M.; LITTLER, S.D. and **TAYLOR, R.P.** 1987, Functional similarity among isomorphic life-history phases of *Polycavernosa debilis* (Rhodophyta, Gracilariaceae). J. Phycol. **23** : 501-505.

LOBBAN, C.S. and **WYNNE, M.J.** 1981(Eds) The Biology of Seaweeds, Blackwell Scientific Publications. U.S.A.

MAGGS, C.A. and **GUIRY, M.D.,** 1982. The taxonomy morphology and distribution of species of *Scinaia* Biv. Bern (Nemaliales, Rhodophyta) in N.W. Europe. Nord. J. Bot. 2 : 517-523.

MAGGS, C.A. and **GUIRY, M.D.** 1985. Life history and reproduction of *Schmitzia hiscockiana* sp. nov. (Rhodophyta, Gigartinales) from the British Isles. Phycologia **24** : 297-310.

MAGNE, F. 1961. Sur la Caryologie de deus Rhodophyces. consider eas fusquicicamme *a* cycle cytologique entierment haplophasique. C.R. Acad. Sci. Paris 252 : 4023-4024.

MAGRUDER, W.H. 1984. Reproduction and life history of red alga *Galaxaura oblongata* (Nemaliales). J. Phycol **20** (3) : 402-409.

MANOJ KUMAR 1996. Structure, ultrastructure and reproduction in *Gracilaria corticata* J. Agardh. Thesis Ph. D. University of Delhi, Delhi.

MAM, BABITA 1992. In vitro studies of *Callithamnion* sp. (Ceramiaceae, Ceramiales) : Growth and differentiation. Thesis M. Phil. University of Delhi, Delhi.

MATTY, P.J. and **JOHANSEN, H.W.** 1981. A histochemical study of *Corallina officinalis* (Rhodophyta, Corallinaceae) Phycologia **20** : 46-55.

MAUSUDA, MICHIO and **HORIUCHI, K.** 1988. Additional notes on the life history of *Nemalion vermiculare*. Japn. J. Phycol **36** (3) : 231-236.

McBRIDE, D. L. and **COLE, K.** 1971. Electron microscopic observations on the differentiation and release of monospores in the marine red alga *Smithora naiadum*. Phycologia **10** : 49-61.

MENG, J. and **SRIVASTAVA, L.M.** 1991. Partial purification and characterization of Floridoside phosphate synthase from *Porphyra perforata*. Phytochemistry. **30** (6) : 1763-1766.

MICHAEL, R., McKAY, L. and **GIBBS,** S.P. 1990. Phycoerythrin is absent from the pyrenoid of *Porphyridium cruentum* : Photosynthetic implications. Planta. **180** : 249-256.

MIKAMI, H., 1973. *Yamadaphycus*, a new genus of the Delesseriaceae (Rhodophyta). Phycologia, **12(3/4)** : 139-143.

MILLAR, A.J.K., 1986. *Baldockia verticilata* (Griffithsieae, Ceramiales), a new red algal genus and species from eastern Australia. Phycologia, **25(1)** : 87-97.

MILLAR, A.J. K. and **KRAFT, G.T**. 1984. The red algal genus *Acrosymphyton* (Dumontiaceae, Cryptonemiales) in Australia. Phycologia **23** : 135-145.

MITMAN, G.G. and **PHINNEY, H.K.** 1985. The development and reproductive morphology of *Halosaccion americanum* (Rhodophyta, Palmariales) J. Phycol. **21** : 578-584.

MOE, R.L. 1979. *Minium parvum* gen. et sp. nov., a crustose member of the Rhodymeniales (Rhodophyta). Phycologia. **18**(1) : 38-46.

MOE, R.L. 1986. *Hymenocladiopsis crustigena* (Rhodymeniaceae), A new genus and species of marine Rhodophyceae from the Antarctic Peninsula. Phycologia. **25**(1): 1-9.

MUKAI, L.S., CRAIGIE, J.S. and **BROWN, R.G.** 1981. Chemical composition and structure of the cell walls of the conchocelis and thallus phases of *Porphyra tenera* (Rhodophyceae). J. Phycol. **17** : 192-198.

MUMFORD, T.F. and **COLE, K.** 1977. Chromosome numbers for fifteen species in the genus *Porphyra* (Bangiales, Rhodophyta) from the west coast of North America. Phycologia. **16** (4) : 373-377.

NASHIER, INDU 1996. A detailed study from gametophytes to tetrasporophyte in the red alga *Solieria robusta* (Greville) Kylin. Thesis Ph. D. University of Delhi, Delhi.

NICOLSON, D.H. and **NORRIS, J.N.** 1983. Ordinal orthography : Nemalionales Vs Nemaliales. Taxon **32** (92) : 288-292.

NONOMURA, A.M., 1979. Development of *Janczewskia morimotoi* (Ceramiales) on its host *Laurencia nipponica* (Ceramiales, Rhodophceae). J. Phycol., **15** : 154-162.

NONOMURA, A.M. and **WEST, J.A.**, 1981. Host specificity of *Janczewskia* (Ceramiales, Rhodophyta). Phycologia, **20(3)** : 251-258.

NORRIS, R.E. 1987. Reproduction in *Sarcodia dentata* comb. nov. (Gigartinales, Rhodophyceae) with comments on the Sarcodiaceae. Br. Phycol. J. **22** : 147-155.

NORRIS, R.E., WOLLASTON, E.M. and **PARSONS, M.J.**, 1984. New terminology for sympodial growth in the Ceramiales (Rhodophyta). Phycologia **23(2)** : 233-237.

NUVN, J. R., PAROLIS, H. and **RUSSELL, I.** 1973. Polysaccharide of the red alga *Chaetanguim erinaceum* Part I isolation and characterization of the water soluble xylan. Carbohyd. Res. 26 : 169-180.

OAKLEY, B.R. and **DODGE, J.D.** 1974. The ultrastructure and cytochemistry of microbodies in *Porphyridium.* Protoplasma. **80** : 233-244.

O'KELLY, C.J. and **BACA, B.J.,** 1984. The time course of carpogonial branch and carposporophyte development in *Callithamnion cordatum* (Rhodophyta, Ceramiales). Phycologia, **23(4)** : 407-417.

PASRICHA, SEEMA 1993. Ultrastructural and histochemical studies on reproduction in a potential carrageenophyte *Laurencia pedicularioides* Boergs. with emphasis on carposporophyte development. Thesis Ph. D. University of Delhi, Delhi.

PARKINSON, P.G. 1983. The typification and status of the name *Chaetangium* algae. Taxon **32** : 605-610.

PATRONE, L.M. BROADWATER, S.T. and **SCOTT, J.L.** 1991. Ultrastructure of vegetative and dividing cells of the unicellular red algae *Rhodella violacea* and *Rhodella maculata.* J. Phycol. **27** : 742-753.

PEDERSON, M., ROOMANS, G.M. and **HOFSTEN, A.V.,** 1980. Cell inclusions containing bromine in *Rhodomela confervoides* (Huds.) Lamour. Ceramiales). Phycologia, **19(2)** : 153-158.

PENROSE, D. 1991. *Spongites fruticulosus* (Rhodophyta, Corallinaceae), the type species, in southern Australia. Phycologia **30** : 438-448.

PENROSE, D. and **WOELKERLING, Wm. J.** 1991. *Pneophyllum fragile* in southern Australia implications for generic concepts in the Mastophoroideae (Corallinaceae, Rhodophyta) Phycologia **30** : 495-506.

PENROSE, D and **WOELKERLING, Wm. J.** 1992a. A reappraisal of *Hydrolithon* and its relationship to *Spongites* (Rhodophyta, Corallinaceae) Phycologia **31** : 81-88.

PENROSE, D. and **WOELKERLING, Wm. J.** 1992b. *Hydrolithon cymodoceae* (Fosile) comb. nov. (Rhodophyta, Corallinaceae) in southern Australia and its relationship to *Fosliella.* Phycologia **31** : 89-100.

PLANTINO, M.E. and **OLIVEIRA, D.C.E.** 1988. Sterility barriers among species of *Gracilaria* (Rhodophyta, Gigartinales) from the Sao Paulo Littoral, Brazil. Br. Phycol. J. **23** : 267-271.

POLNE - FULLER, M. and **GIBOR, A.** 1984. Developmental studies in *Porphyra* I. Blade differentiation in *Porphyra perforata* as expressed by morphology, enzymatic digestion and protoplast regeneration. J. Phycol. **20** : 609-616.

POLNE - FULLER, MIRIAM, and **GIBOR, A.** 1990. Developmental studies in *Porphyra* (Rhodophyceae) : III Effect of culture conditions on wall regeneration and differentiation of protoplasts. J. Phycol. **26** (4) : 674-682.

PRICE, I.R. and **KRAFT, G.T.** 1991. Reproductive development and classification of the red algal genus *Ceratodictyon* (Rhodymeniales, Rhodophyta). Phycologia. **30** (1) : 106-116.

PUESCHEL, C.M. 1977. A Freeze - etch study of the ultrastructure of red algal pit-plugs. Protoplasma **91** : 15-30.

PUESCHEL. C.M. 1979. Ultrastructure of tetrasporogenesis in *Palmaria palmata* (Rhodophyta) J. Phycol. **15** : 409-424.

PUESCHEL, C.M. 1980. A reappraisal of the cytochemical properties of rhodophycean pit-plugs. Phycologia **19** (3) : 210-217.

PUESCHEL, C.M. 1988a. Cell sloughing and chloroplast inclusions in *Hildenbrandia rubra* (Rhodophyta, Hildenbrandiales) Br. Phycol. J. **23** : 17-23.

PUESCHEL, C.M. 1988b. Secondary pit-connections in *Hildenbrandia* (Rhodophyta, Hildenbrandiales) Br. Phycol. J. **23** : 25-32.

PUESCHEL, C.M. 1989. An expanded survey of the ultrastructure of red algal pit-plugs. J. Phycol 25 : 625-636.

PUESCHEL, C.M. 1990. Cell structure in Cole, K.M. and Sheath, R.G. (Eds.) Biology of Red Algae, Cambridge University Press, London.

PUESCHEL, C.M. and **COLE, K.M.** 1982. Rhodophycean pit-plugs : An ultrastructural survey with taxonomic implications. Amer. J. Bot. **69** (5) : 703-720.

PUESCHEL, C.M. and **COLE, K.M.** 1985. Ultrastructure of germinating carpospores of *Porphyra varietata (*Kjellm). Hus (Bangiales, Rhodophyta). J. Phycol. **21** : 146-154.

PUESCHEL, C.M. and **MAGNE, F.** 1987. Pit plugs and other ultrastructural features of systematic value in *Rhodochaete parvula* (Rhodophyta, Rhodochaetales). Cryptogam. Algol. **8** : 201-9.

PUESCHEL, C.M. and **PARTHASARATHY, M.V.** 1984. X-ray microanalysis of phytoferritin in *Constantinea* (Cryptonemiales, Rhodophyta) Phycologia **23** : 465-469.

RAISCO, N., MARIANI, P. VECCHIA, F.D. and **TREVISAN, R.** 1991. The vegetative thallus of *Pterocladia capillacea* (Gelidiales, Rhodophyta) II Pit connections. Bot. Mar. **34** : 187-194.

RAMM- ANDERSON, S.M. and **WETHERBEE, R.** 1982. Structure and development of *Nemalion helminthoides* (Nemalionales, Rhodophyta). J. Phycol. **18** : 133-141.

RAMUS, J. 1969a. Pit connection formation in the red alga *Pseudogloiophloea.* J. Phycol, 5 : 57-63.

RAMUS, J. 1969b. Dimorphic pit connections in the red alga *Pseudogloiophloea*. J. Cell. Biol. **41** : 340-345.

RAMUS, J. 1972. The production of extracellular polysaccharide by the unicellular red alga *Porphyridium aerugineum.* J. Phycol. **8** : 97-111.

RAMUS, J. and **ROBINS, D.M.** 1975. The correlation of golgi activity and polysaccharide secretion in *Porphyridium* J. Phycol. **11** : 70-74.

RAO, P. S. 1970. Systematics of Indian Gelidiales. Phykos **9** : 63-78.

REED, R.H. 1985. Osmoacclimation in *Bangia atropurpurea* (Rhodophyta, Bangiales) : The osmotic role of Floridoside. Br. Phycol. J. **20** : 211-218.

REED, R.H., COLLINS, J.C. and **RUSSELL, G.** 1980. The effects of salinity upon galactosyl - glycerol content and concentration of the marine Red Alga *Porphyra purpurea* (Roth) C. Ag. J. Exp. Bot. **31** (125) : 1539-1554.

RICKER, R.W. and **KRAFT, G. T.**. 1979. Morphology of the subarctic red alga *Cenacrum subsutum* gen. et sp. nov. (Rhodymeniales) from Macquarie Island. J. Phycol. **15** : 434-44.

ROBINS, P. A. and **KRAFT, G. T.** 1985. Morphology of the type and Australian species of *Dudresnaya* (Dumontiaceae, Rhodophyta). Phycologia **24** : 1-34.

Rueness, J., 1973. Speciation in *Polysiphonia* (Rhodophyceae, Ceramiales) in view of hybridization experiments : *P. hemisphaerica* and *P. boldii*. Phycologia, **12(3/4)** : 107-109.

SAHOO, D. and **VIJAYARAGHAVAN, M. R.** 1986. Vegetative and reproductive structures of *Centroceros clavulatum* (C. Agardh) Montague. Phykos 25 : 24-28.

SAUNDERS, G. W., MAGGS, C. A.. and **McLACHLAN, J. L.**. 1989. Life-history variation in *Rhodophysema elegans* (Palmariales, Rhodophyta) from the North Atlantic and crustose *Rhodophysema* spp. from the North Pacific. Can. J. Bot. **67** : 2857-2872.

SCOTT, J. 1983. Mitosis in freshwater alga *Batrachospermum ectocarpum*. Protoplasma. **118(1)** : 356-370.

SCOTT, J. 1984. Electron microscopic contributions to red algal phylogeny. J. Phycol. **20** : Suppl. : 6.

SCOTT, J. BOSCO, C., SCHORNSTEIN, K. and **THOMAS, J.**, 1980. Ultrastructure of cell division and reproductive differentiation of male plants in the Florideophyceae (Rhodophyta) : Cell division in *Polysiphonia*. J. Phycol., **16** : 507-524.

SCOTT, J. L. and **DIXON, P. S.** 1973. Ultrastructure of spermatium liberation in the marine red alga *Ptilota densa*. J. Phycol., **9** : 85-91.

SCOTT, J., THOMAS, J. and **SAUNDERS, B.** 1988. Primary pit-connections in *Compsopogon coeruleus* (Balbis) Montagne (Compsopogonales, Rhodophyta). Phycologia. **27** : 327-333.

SCOTT, F. J. 1982. **WETHERBEE R.** and **KRAFT, G. T.** 1982. The morphology and development of some prominently stalked southern Australian Halymeniaceae (Rhodophyta; Corallinaceae) I. *Cryptonemia kallymenioides* (Harvey) Kraft comb. nov. and *C. undulata sonder* J. Phycol, **18** : 245-257.

SEARLES, R. B. and **LEWIS, S. M.** 1983. North carolina marine algae : A new species of *Helminthocladia* with a reappraisal of the generic limits of *Helminthocladia* and *Helminthora*. J. Phycol. **19(2)** : 164-172.

SHEATH, R. G., COLE, K. M. and **HYMES, B. J.**, 1987. Ultrastructure of polysporogenesis in *Pleonosporium vancouverianum* (Ceramiaceae, Rhodophyta). Phycologia **26(1)** : 1-8.

SHEATH, R. G., MORRISON, M. O., COLE, K. M., VANALSTYNE, K. L. 1986. A new species of freshwater Rhodophyta, *Batrachospermum carpocontortum*. Phycologia **25(3)** : 321-330.

SHEATH, R. G. VAN ALSTYNE, K. L. and **COLE, K. M.** 1985. Distribution. seasonality and reproductive phenology of *Bangia atropurpurea* (Rhodophyta) in Rhode Island, U.S.A. J. Phycol. **21** : 297-303.

SHIMIZU, T. and **MASUDA, M.** 1983. The life history of *Farlowia irregularis* Yamada (Cryptonemiales, Rhodophyta) Jap. J. Phycol. **31** : 202-207 (In Japaneses)

SILVA, P. C. and **JOHANSEN, H. W.** 1986. A reappraisal of the order corallinales (Rhodophyceae) Br. Phycol. J. **21** : 245-254.

SPARLING, S. R. 1961. A report on the culture of some species of *Halosaccion, Rhodymenia* and *Fauchea*. Amer Jour. Bot. **48** : 493-499.

SRINIVASAN, K. S. 1969. Phycologia Indica (Incones of Indian Marine Algae) Vol. I. Botanical Survey of India, Calcutta.

SOUTH, G. R. and **WHITTICK, A.** 1976. Aspects of the Life-history of *Rhodophysema elegans* (Rhodophyta, Peyssonneliaceae). Br. Phycol. J. **11** : 349-354.

SOUTH, G. R. and **WHITTICK, A.** 1987. Introduction to Phycology. Blackwell Scientific Publications, Oxford.

STEEL, R. L. 1986. Genetics of *Champia parvula* (Rhodymeniales, Rhodophyta): Mendelian inheritance of spontaneous mutants. J. Phycol. **22** : 538-542.

SYENECK, R. S. and **PAINE R. T.** 1986. Ecological and taxonomic studies of shallow-water encrusting Corallinaceae (Rhodophyta) of the boreal northern Pacific. Phycologia **25** : 221-240.

STENGENGA and **VROMAN, M.** 1986. *Pterocladiophila hemisphaerica* (Rhodophyta, Cryptonemiales) in the Caribbean, Acta Bot. Neerl. **35** : 1-4.

TALARICO, L. GUIDA, G., MURANO, E. and **PIAEQUADIO, A. M.** 1990. Ultrastructure of the cell wall of *Gracilaria cf. verrucosa* (Gracilariales, Rhodophyta). Effect of steam explosion. Hydrobiol. **204/205** : 597-601.

TAIT, M. I. **MILNE, A. M., GRANT, D., SOMERS, J. A., STAPLES, J., LONG, W. F., WILLIAMSON, F. B.** and **WILSON, S. B.** 1990. *Porphyra* cell cultures : Isolation, growth and polysaccharide production. J. Appl. Phycol. **2(1)** : 63-70.

THRIB, H. H. and **BENSON-EVANS, K.** 1985. The effect of temperature on the growth of *Lemanea* thalli and carpospore germination. Arch. Hydrobiol. **103(3)** : 341-346.

TRICK, H. N. and **PUESCHEL, C. M.** 1991. Cytochemical evidence for homology of the outer cap-layer of red legal pit-plugs. Phycologia 30 : 196-204.

TRIPODI, G., GARGUILO, G. M. and **FLORA DE MASI**. 1986. Electron microscopy of membranous bodies and genophore in chloroplasts of *Botryocladia botryoides* (Rhodymeniales, Rhodophyta). J. Phycol. **22** : 560-563.

TSEKOS, I. 1981. Growth and differentiation of golgi apparatus and wall formation during carposporogenesis in *Gigartina teedii* (Rhodophyceae). J. Cell Science, **52** : 72-84.

TSEKOS, I. and **BOSABALIDIS, A. M.** 1988. Proteinaceous inclusions in specialized thallus cells of *Chylocladia verticillata* (Rhodymeniales, Rhodophyta). Phycologia. **27(1)** : 20-24.

TURVEY, J. R., and **WILLIAMS E. L.** 1970. The structures of some Xylans from red algae. Phytochemistry. **9** : 2383-2388.

TSEKOS, I, and **REISS, H.D.** 1990. The supramolecular organization of photosynthetic membranes of the thallus stage of *Porphyra leucosticta.* Thuret. (Bangiales, Rhodophyta) visualized by Freeze-Fracture. Bot. Acta. **103** : 244-249.

USOV, A. I., ADAMYANTS, K. S. A. and **YAROTSKII, S. V.** 1975. Polysaccharides of algae XVIII. Acetolysis of sulfated mannans from *Nemalion vermiculare.* Sur Zhurnal Obshchei Khimii **45(6)** : 1377-1381.

USOV, A. I. ADAMYANTS, K. S., YARCTSKII, S. V. and **ANOSHINA, A. A.** 1975. Polysaccharides of algae : XVI. Study of the structure of sulfated mannan of the red alga *Nemalion vermiculare.* Sur. Zhurnal Obschei Khimii **45(4)** : 916-921.

USOV, A. I. ADAMYANTS, K. S., YAROTSHI, S. V., ANOSHINA, A. A. and **KOCHETKOV, N. K.** 1974. Polysaccharide of algae XIV. Isolation of sulfated mannan and neutral xylan from the red alga *Nemalion vermiculare.* Sur Zhurnal Obshchei Khimii **44(2)** : 416-420.

VAN DEN HOEK, C. and **CORTEL-BREEMAN A. M.** 1969. The life histories of *Acrosymphyton purpuriferum* (J. Ag) Kyl. and *Halymenia floresia* (Clem.) Ag. in unialgal cultures. Br. Phycol. J. 4 : 213.

VAN DEN HOEK, C. and **CORTEL-BREEMAN, A. M.** 1970 Life-history studies on Rhodophyceae III. *Scinaia complanata* (Collins) Cotton. Acta. Bot. Neerl. 19 : 457-467

VAN DER MEER, J. P. 1981. The life history of *Halosaccion ramentaceum.* Can. J. Bot. **59** : 433-436.

VAN DER MEER, J. P. and **CHEN, L. C. M.** 1979. Evidence for sexual reproduction in the red algae *Palmaria palmata* and *Halosaccion ramentaceum.* Can. J. Bot. **57** : 2452-2459.

VAN DER MEER, J. P and **TODD**, E. R. 1980. The life history of *Palmaria palmata* in culture. A new type for the Rhodophyta. Can. J. Bot. **58** : 1250-1256.

VANDERMEULEN, H., GARBARY, D. J. and **GUIRY, M. D.**, 1990. *Laurencia minuta* sp. nov (Ceramiales, Rhodomelaceae), a dimunitive red alga from the Gulf of Aqaba (Red Sea). Br. Phycol. J., **25** : 237-244.

VERDUS, C. M., CHRISTIAEN, D. MORVAN H. and **STADLER T.** 1986. Etude ultrastructurale et. cytochimique de la parol cellulaire chez *Gracilaria verrucosa* (Rhodophyceae). Can. J. Bot. **64** : 96-101.

VIJAYARAGHAVAN, M. R. and **BHATIA, BELA.** 1991. Developmental and histochemical studies on *Nizymenia australis* Sonder (Nizymeniaceae, Gigartinales, Rhodophyta). Proc. Indian Natn. Sci. Acad, **B57** : 69-76.

VIJAYARAGHAVAN, M. R. and **BHATIA, BELA.** 1995a. Structure, Ultrastructure and differentiation of utricle in *Scinaia pseudocrispa (= S. forcellata)* (Nemaliales, Rhodophyta) from Port Okha, India. Phytomorphology. 45 : 207-212.

VIJAYARAGHAVAN, M. R. and **BHATIA, BELA** 1995b. Reproductive biology of Nemaliales with special reference to *Scinaia pseudocrispa* (Clemente) Wynne (*Scienaia forcellata* Bivona-Bernardi) -- An overview. J. Indian bot Soc. (Platinum Jublee Volume) 74A : 5-13.

VIJAYARAGHAVAN, M. R. and **BHATIA, BELA** 1996. Developmental and histochemical studies on the spermatangium and carposporophyte of *Scinaia forcellata* (Nemaliales, Rhodophyta) from Port Okha, India, Nova Hedwigia, **112** : 135-146.

VIJAYARAGHAVAN, M. R., NASHIER INDU, and **MAM BABITA** 1995. In vitro studies of *Callithamnion sp.* (Ceramiaceae, Ceraminales) : Growth and Differentiation. Seaweed Res. Utlin. **17.** 13-22.

WAALAND, J. R., DICKSON, L. G. and **CARRIER, J. E.** 1987. Conchocelis growth and Photopheriodic control of conchospore release in *Porphyra torta* (Rhodophyta). J. Phycol. **23** : 399-406.

WAALAND, J. R., DICKSON, L. G. and **WATSON, B. A.** 1990. Protoplast isolation and regeneration in the marine red alga *Porphyra nereocystis. Plants.* ***181*** : 522-528.

WALKER, R. and **MOSS. B**. 1984. Mode of attachment of six epilithic crustose Corallinaceae (Rhodophyta) Phycologia **23** : 231-329.

WEST, J. A. and **CALUMPONG, H. P.** 1988. *Dawsoniocolax bostrychiae* (Choreocolacaceae, Gigartinales), an alloparasitic red alga new to Australia. Phycologia **27** : 463-468.

WETHERBEE, R. and WYNNE M.J., 1973. The fine structure of the nucleus and nuclear associations of developing carposporangia in *Polysiphonia novae-angliae* (Rhodophyta). J. Phycol., **9** : 402-407.

WHITTICK, A., 1980. *Antithamnionella floccosa* (O. F. Mull.) nov. comb. : a taxonomic re-appraisal of *Antithamnion floccosum* (O. F. Mull.) Kleen (Rhodophyta : Ceramiaceae). Phycologia, **19(1)** : 74-79.

WILCE, R. T. and **DAVIS, A. N.** 1984. Development of *Dumontia contorta* (Dumontiaceae, Cryptonemiales) compared with that of other higher red algae. J. Phycol. **20** : 336-351.

WILKS, K. M. and **WOELKERLING, Wm. J.** 1991. Southern Australian species of *Melobesia* (Rhodophyta, Corallinaceae) Phycologia **30** : 507-533.

WOELKERLING, Wm. J. 1971. Morphology and taxonomy of Audouinella complex (Rhodophyta) in southern Australia. Austral. J. Bot., Suppl. No. 1. 91.

WOELKERLING, Wm. J. and **IRVINE, L. M.** 1986. The neotypification and status of *Mesophyllum* (Rhodophyta, Corallinaceae) Phycologia **25** : 379-396.

WOELKERLING, Wm. J. 1988. The Coralline Red Algae : An Analysis of the genera and subfamilies of nongeniculate Corallinaceae (Rhodophyta). British Museum (Natural History) and Oxford University Press, London.

WOELKERLING Wm. J. and **Foster, M. D**. 1989. A systematics and ecographic account of *Synarthrophyton schielianum* sp. nov. (Rhodophyta, Corallinaceae) from the Chatham Islands. Phycologia **28** : 39-60.

WOLLASTON, E. M., 1984. Species of Ceramiaceae (Rhodophyta) recorded from the International Indian Ocean Expedition, 1962. Phycologia, **23(3)** : 281-299.

WYNNE, M. J., 1985. Evidence of the transfer of *Cyclospora curtissiae*. J. Agardh to *Carpoblepharis*_(Ceramiaceae, Rhodophyta). Phycologia, **24(1)** : 49-54.

WYNNE, M. J. 1985. Two new species of *Tayloriella* (Rhodomelaceae, Rhodophyta) from the north eastern North Pacific. J. Phycol., **21** : 107-114.

WYNNE, M. J. and **BALLANTINE, D. L.,** 1991. *Laurencia iridescens* sp. nov. (Rhodomelaceae, Ceramiales) from the Caribbean Sea). Phycologia, **30(5)** : 394-401.

WYNNE, M. J. and NORRIS, R. E., 1991. *Branchioglossum pygmaeum* sp. nov. and new records of other Delesseriacious algae (Rhodophyta) and Natal, South Africa. Phycologia, **30(3)** : 262-271.

XUE-WU, L. and **GORDON, M. E.** 1987. Tissue and cell culture of New Zealand *Pterocladia* and *Porphyra* species. Hydrobiologia. **151/152** : 147-154.

YAMAMOTO, H. 1990. In vitro life history and spermatangial pattern of *Gracilaria arcuata* Zanardini from Philippenes. Bull. Fac. Fish. Hokkaido Univ. **41** : 145-148.

YAMAMOTO, H. 1991. Life history of *Gracilaria salicornia* (C. Ag.) Dawson (Gracilariaceae, Rhodophyta) *in vitro*. Jpn. J. Phycol: (Sorui) **39** : 55-66.

YOSHIDA, T. 1959. Life cycle of a species of *Batrachospermum* found in Northern Kyushu Japan J. Jour. Bot. **17** (1) : 29-42.

YOSHIDA, T. and **MIKAMI, H.,** 1992. *Branchioglossum spiniferum* sp. nov. (Delesseriaceae, Rhodophyta) from north Kyushu, Japan. Jpn. J. Phycol, (Sorui) **40** : 379-384.

YOSHIZAKI, M. 1986. The morphology and reproduction of *Thorea okadai*. Phycologia **25(4)** : 476-481.

YOUNG, D. N. 1978. Ultrastructural evidence for a secretory function in the ''gland cells' of the marine red alga *Botryocladia pseudodichotoma* (Rhodymeniaceae). *Protoplasma*. 94 : 109-126.

ZABLAKIS, E. 1987. The red alga *Scinaia furcata* sp. nov. (Galaxauraceae, Nemaliales) from Hawaii. Phycologia **26(1)** : 53-58.

ZOHANG, W. CHAPMAN, D. J. PHINNEY, O. B. SPRAY, C. R., **YAMANE, H**. and **TAKAHASHI,** N. 1991. Identification of cytokinins in *Sargassum muticum* (Phaeophyta) and *Porphyra perforata* (Rhodophyta). J. Phycol. **27** : 87-91.

ZUPAN, J. R. and **WEST, J. A.,** 1988. Geographic variation in the life history of *Mastrocarpus papillatus* (Rhodophyta). J. Phycol., **24** : 223-229.

INDEX

H

R